DATE DUE

*The chemistry
of the liquid
alkali metals*

The chemistry
of the liquid
alkali metals

C. C. Addison

Emeritus Professor of Inorganic Chemistry
University of Nottingham, England

A Wiley–Interscience Publication

JOHN WILEY & SONS
Chichester · New York · Brisbane · Toronto · Singapore

Library of Congress Cataloging in Publication Data:

Addison, C. C., 1913–
  The chemistry of the liquid alkali metals.

  1. Alkali metals. 2. Liquid metals. I. Title.
QD172.A4A34   1984   546'.38   84-7496
ISBN 0 471 90508 9

The British Library Cataloguing in Publication Data:

Addison, C. C.
  The chemistry of the liquid alkali metals.
  1. Alkali metals 2. Liquid metals
I. Title
  546'.38   QD172.A4

  ISBN 0 471 90508 9

Typeset by Spire Print Services Ltd, Salisbury, Wilts.
Printed by St. Edmunsbury Press Ltd., Suffolk.

Contents

Preface

This book has been written with three main aspects of the subject in mind. Firstly, the alkali metals are easily converted to the liquid state; the liquids act as solvents, and reactions can occur between dissolved species. This medium therefore provides an important extension to the chemistry of non-aqueous solvents, especially since the reactions are now influenced by the electronic, as opposed to the more usual molecular, environment. I hope that chapters dealing with the nature and reactivity of dissolved species, solvation, solubilities and electrical conductivity of solutions and so forth will encourage comparison with molecular solvents. Because of the unique chemical and physical nature of these liquid metals, the laboratory techniques required are necessarily different from those normally available in the chemical laboratory. I have therefore included chapters on such aspects as manipulation of the liquids, analysis and purification, which should be of assistance to a research worker new to the field.

Secondly, the liquid alkali metals, particularly lithium and sodium, are finding ever-increasing large-scale applications, from storage batteries to sodium-cooled reactors and the possible fusion reactors of the future. Many of the technological problems encountered (e.g. corrosion) are chemical in origin, and my colleagues and I have no doubt that our research has benefitted from the close collaboration between laboratory and technical scale experimental work, which has now been maintained for over 25 years. By providing this broad coverage of the chemistry of the liquid metals, with reference wherever appropriate to technical as well as academic aspects, it is hoped that the book will help towards continued collaboration.

Thirdly, I have been assured by many teachers that the attention given to liquid metals in university and college courses is restricted by the lack of a suitable text on the subject. With the liquid alkali metals in particular, there is a wide gap between the limited treatment given in most standard college texts, and the advanced level at which the subject is discussed at many current international conferences. It is the aim of the present volume to bridge this gap. I have therefore tried to keep the text as readable as possible, with cross-referencing between the various chapters. For the same reason, though quantities are normally expressed in SI units, I have not attempted to adhere rigidly to this system, since correlation between this text and the original literature is sometimes easier in terms of the more classical units.

In writing the book, I am deeply indebted to my colleagues Dr. R. J. Pulham, Dr. M. G. Barker and Dr. Peter Hubberstey of the Chemistry Department, University of Nottingham, for their assistance. Not only have they helped with discussions related to the manuscript itself, but the development of this area of research owes much to their own originality, initiative and experimental skill. I am also grateful to Mrs. Margaret Pulham for her enthusiasum for the project, and for the care with which she prepared the typescript.

A large part of the book was written during the tenure of a Leverhulme Emeritus Fellowship, and it is a pleasure to express my gratitude to the Leverhulme Trust for this award.

C. C. ADDISON

University of Nottingham
February 1984

Chapter 1

Some basic physical and chemical properties

1.1 Introduction

Extensive work on the physics of the liquid metallic state has been carried out for many years in most of the developed countries of the world. Much of this work is concerned with pure metals, or mixtures of pure metals, for the understandable reason that in many cases the presence of impurities, particularly non-metals, would invalidate the measurements. The texts listed under References 1 and 2 provide a fairly comprehensive survey of the types of physical measurement which have been carried out. In contrast, the chemistry of (and in) liquid metals is centred essentially on the behaviour of these so-called impurities, i.e. of elements and compounds which can exist in a metallic environment, often in very dilute solution. Publications on chemical themes tend to be scattered over journals of chemistry, metallurgy and materials science, but the collection of over 480 references which is presented at the end of this book will enable the reader to gain ready access to the relevant literature.

This chapter is intended only to provide some selected factual information and some general comments, for easy reference and as a background for more specific aspects which are discussed in later chapters.

1.2 Atomic properties

Figure 1.1 shows the relationship between the alkali metals and other metals of the periodic table. The alkali metals appear at the extreme left-hand side of the table, and are shown in heavy type in Figure 1.1. Some atomic properties are collected in Table 1.1. These metals have a single s electron outside a noble gas core, and thus represent an extreme in physical properties and chemical behaviour. They have low ionization energies, caesium having the lowest ionization energy of any element in the periodic table. The resulting M^+ ions are spherical and of low polarizability, so that the simple chemistry of their reactions with other elements is essentially that of the M^+ ions. Some

1

Main groups		1	2											3	4	5	6	7	8
Period	Electron shells																		
1	1s	H																H	He
2	2s,2p	Li	Be											B	C	N	O	F	Ne
3	3s,3p	Na	Mg				Transition metals							Al	Si	P	S	Cl	Ar
4	4s,3d,4p	K	Ca	Sc	Ti	V	Cr	Mn	Fe	Co	Ni	Cu	Zn	Ga	Ge	As	Se	Br	Kr
5	5s,4d,5p	Rb	Sr	Y	Zr	Nb	Mo	Tc	Ru	Rh	Pd	Ag	Cd	In	Sn	Sb	Te	I	Xe
6	6s,4f,5d,6p	Cs	Ba	La	Hf	Ta	W	Re	Os	Ir	Pt	Au	Hg	Tl	Pb	Bi	Po	At	Rn
7	7s,5f,6d	Fr	Ra	Ac															

Lanthanides(4f)	Ce	Pr	Nd	Pm	Sm	Eu	Gd	Tb	Dy	Ho	Er	Tm	Yb	Lu
Actinides (5f)	Th	Pa	U	Np	Pu	Am	Cm	Bk	Cf	Es	Fm	Md	No	Lr

Figure 1.1   Periodic table of the elements

covalency does occur, and there is evidence for the presence of dimers (e.g. Li_2, Na_2) and higher aggregates in alkali metal vapours.

The ionization energies are known with some accuracy (Table 1.1) and are positive in sign; by normal convention the quoted values represent the energy required to achieve the process $M \rightarrow M^+ + e$. In contrast, electron affinities are much less precise, and are often obtained by extrapolation. Values obtained by different workers vary quite widely.[3] The values given in the table are those quoted by Zollweg;[4] at least they indicate the order of magnitude of electron affinities, and the variation down the group. They are given a positive sign in the table because of the different convention used for electron affinities, but represent the energy released in the process $M + e \rightarrow M^-$. Electron affinities are small compared with ionization energies, but it is of interest that even for alkali metal atoms the addition of an electron to a metal atom is exothermic. However, this state can only be realized where the alkali metal atom is isolated in its environment, e.g. in the

Table 1.1   Atomic properties of the alkali metals

	Li	Na	K	Rb	Cs
Electronic configuration	$[He]2s^1$	$[Ne]3s^1$	$[Ar]4s^1$	$[Kr]5s^1$	$[Xe]6s^1$
Atomic number	3	11	19	37	55
Atomic weight	6.94	22.99	39.10	85.47	132.9
Atomic radius (metallic) (Å)	1.225	1.572	2.025	2.16	2.35
Ionisation energy (first electron)					
(eV)	5.39	5.14	4.34	4.18	3.89
(kJ g atom^{-1})	520	496	419	403	376
Electron affinity					
(kJ g ion^{-1})	60	52	45	41	38
Electronic work function (eV)	2.42	2.28	2.24	2.09	1.81

metal vapour. Compounds of Na^-, for instance, are never found because the overall energy balance in a Born–Haber cycle (say in the production of sodium chloride from its elements) includes the much larger electron affinity of chlorine, and the formation of an Na^- compound is not energetically favourable.

There is also considerable variation in published values for the work function, i.e. the amount of energy required to remove an electron from the bulk metal.[5] Values vary with the method used, but variations arise also because of the extreme difficulty in obtaining a perfectly pure alkali metal surface. The work function is again a periodic function of the atomic number, and the alkali metals have the lowest values of any of the elements.

1.3 Physical properties of the liquid metals

Some important properties are collected in Table 1.2. Where appropriate, and unless otherwise stated, values are quoted for 200°C, at which temperature all the alkali metals are liquid. Water and mercury are probably the two liquids with which the reader will be most familiar, and corresponding values for these two liquids are included in the table for comparison.

More than any other group in the periodic table, the alkali metals show clearly the influence of increasing atomic mass and size on the physical properties of the bulk element. Thus, melting and boiling points, viscosity, surface tension, heats of fusion and vapourization and specific heat, all decrease along the series Li to Cs. Vapour pressures, and the volume change on fusion, increase steadily along the series. There are some properties, however—for example density and thermal and electrical conductivity—for which a discontinuity occurs. These properties have sometimes been regarded as passing through a maximum, or minimum, at sodium, but this is not a satisfactory approach. Reference to Table 1.2 shows that each of these properties changes regularly from Cs to Na, and it is lithium which is to be regarded as anomalous. This is due largely to the relatively small size of the Li atom and the Li^+ ion; the s^1 electron of lithium is based on a core of two rather than eight electrons, so that the Li^+ ion has an exceptionally high charge/radius ratio. This explains many of the anomalous properties of lithium salts, but we may also think of the Li^+ ion as the core which remains in the liquid metal when the s electron is released into the conduction band, so that the small size of Li^+ can also be related to the anomalous properties of liquid lithium. In some cases there is an actual reversal, at lithium, of the trend shown by the heavier metals between Cs and Na (e.g. thermal and electrical conductivities). In others the trend is maintained but the values are anomalous; thus, the melting and boiling points of lithium are exceptionally high, and the density, though it is the lowest, is high relative to the atomic weight.

Two other aspects deserve brief reference. Firstly, the melting points of the metals fall rapidly with atomic weight, and at caesium the melting point is below blood temperature. Caesium contained in a glass vessel can therefore

Table 1.2 Physical properties of the liquid alkali metals

	Li	Na	K	Rb	Cs	H$_2$O	Hg
Melting point (°C)	180.5	97.8	63.2	39.0	28.5	0	−39
Boiling point (°C)	1317	883	754	688	671	100	357
Liquid range (°C)	1138	785	697	649	677	100	496
Density (kg/m^3 at 200°C)	507	904	797	1390	1740	990 (50°C)	13100
Viscosity (cp) at 200°C	0.565	0.450	0.300	0.346 (39°C)	0.350 (105°C)	0.55 (50°C)	1.01
Surface tension at 200°C (Nm^{-1} × 10^{-3})	398	195	103	76 (39°C)	74 (105°C)	68 (50°C)	436
Vapour pressure (torr) at 200°C	10^{-9}	1.5 × 10^{-4}	0.006	0.04	0.08	92 (50°C)	17
Heat of fusion (kJ mol^{-1})	2.93	2.64	2.39	2.20	2.09	6.01	2.3
Heat of vapourization (kJ mol^{-1})	147	99.2	79.1	75.7	66.5	41.1	59.1
Specific heat at 200°C (J mol^{-1} deg^{-1})	3.51	1.34	0.786	0.463	0.447	4.18 (50°C)	0.134
Thermal conductivity at 200°C (J cm^{-1} s^{-1} deg^{-1})	0.42	0.84	0.56	0.31	0.21	0.0067 (50°C)	0.12
Electrical resistivity at 200°C (µΩ cm)	29.1	13.5	20.6	35.8	56.6	>10^6	114
Volume increase on melting (%)	1.5	2.17	2.41	2.54	2.66	−0.014	3.6

be melted in the hand. Secondly, the density and viscosity of liquid sodium are not greatly different from the values for water, so that operations such as pouring, pumping, stirring and general manipulation of the fluid in the laboratory on a litre scale are much more like those for water than for mercury, which is the only other liquid metal with which most chemists will have had previous experience.

1.4 Chemical properties of the liquid metals

The vigorous chemical reactivity of the alkali metals with most other elements is already well known, and with most reagents increases with electropositive nature from lithium to caesium. Thus, lithium reacts gently with water, sodium reacts vigorously, potassium ignites, and rubidium and caesium explode. On the other hand liquid lithium reacts readily with (and dissolves) nitrogen, whereas the other alkali metals do not react with nitrogen.

In principle the reactivities of the liquid metals resemble those of the solid metals, but the rates are usually higher because of the higher temperatures involved. Also, the products of reaction (e.g. hydrides, oxides) can often dissolve in the liquid metal, so that a clean surface is maintained and reaction is not inhibited by film formation. It is because of this high reactivity that special apparatus is required for the analysis and laboratory handling of the metals, which is discussed in later chapters. Indeed, a large part of the book is concerned with the behaviour of other elements in a liquid alkali metal environment, so that the theme of chemical reactivity need not be developed further at this stage.

Special methods are also required for the isolation of the metals from their ores, ranging from electrolysis of molten salts (Li and Na) to distillation from mixtures of their molten chlorides with less volatile metals such as calcium or sodium. These processes will not be discussed in this volume, as they are well documented in many standard inorganic chemistry texts.

1.5 The colours of the alkali metals

When free from surface contamination, the metals Li, Na, K and Rb are silver-bright and lustrous. Caesium, however, is pale gold in colour in both solid and liquid states; coupled with its low density, and the low viscosity of the liquid, it must surely be the most attractive element in the periodic table. Earlier workers attributed the colour to the presence of trace amounts of oxide impurity, and certainly the gold colour does darken on addition of oxygen. However, intensive purification of the metal by all available methods in the writer's laboratory over several years has not succeeded in removing the colour, and we have no doubt that caesium metal is indeed pale golden in colour.

Any substance which appears coloured does so because of the absorption of radiation in the visible region (400–700 nm), and this absorption arises because of the promotion of electrons from a lower to a higher energy state. The colour of caesium arises from the same sort of reasons as are used[6] to explain the better-known colours of metallic copper and gold, which will therefore be outlined briefly. The free electrons in a metal exist in energy (conduction) bands, and electrons in the s, p or d atomic orbitals of individual atoms give rise to energy bands with s, p or d character in the bulk metal. Depending on the position of the metal in the periodic table, and thus the energy bands available, and depending also on the energy of the applied radiation, promotion of electrons may or may not be possible. Promotion of electrons is also determined by the extent to which the s, p and d bands overlap. It is the uppermost electron energy levels which are significant, and these are said to constitute the Fermi surface. In the case of copper, X-ray data indicate that s, p and d bands overlap, and electrons in the rather narrow d band are about 2.3 eV below the Fermi surface. These electrons can be excited to levels above the Fermi surface by green, blue or violet light, but not by light at the red end of the visible spectrum. The latter is therefore reflected, which accounts for the red colour of copper metal. In the case of silver, the d band lies 4 eV below the Fermi surface. Ultraviolet radiation is necessary for electron excitation; light in the visible range is almost totally reflected, and hence silver metal shows no colour. Brass represents an inter-

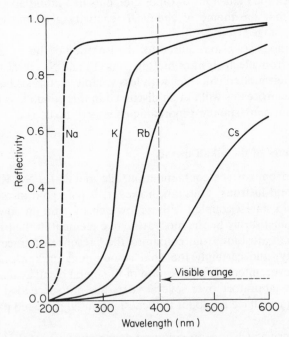

Figure 1.2 Reflectivity of alkali metals for normal incidence

mediate stage. The addition of zinc to copper increases the number of electrons per atom and alters the Fermi surface level; radiation of lower wavelength than for copper is required for electron excitation, and the colour changes from the orange–red of copper to the yellow of brass. A similar situation is responsible for the colour of metallic gold.

With sodium, the profile of the energy band is largely s in character, and there is little overlap of the 3s band with 3p and 3d bands. Electron excitation is difficult, and would require radiation of wavelengths far below the visible region. Visible light is therefore almost totally reflected, and the lighter alkali metals are silver in appearance. As the atomic number of the metal increases, overlap with d orbitals makes a significant contribution to the band structure. The various energy levels come closer together, and the band structure becomes more complicated. The wavelength required for electron excitation moves closer to the visible, and although the particular electron transition which takes place in caesium has not been determined, it is clear that it can be achieved, with caesium, by radiation at the blue end of the visible spectrum. This is illustrated by reflectance measurements on the alkali metals[7] using normal incidence (Figure 1.2). With the lighter metals, reflectivity falls off sharply as wavelength decreases, but is almost complete over the visible range. With caesium, however, reflectivity is very small in the blue region of the visible spectrum, and then rises steeply towards the red end of the spectrum, which is consistent with its characteristic golden colour.

Chapter 2

Manipulation of the liquids

2.1 Introduction

Because of the high chemical reactivity of the liquid alkali metals, special techniques are necessary for their manipulation. This chapter deals with some of the basic features which are common to all liquid metal work; the methods described will be those applicable to laboratory-scale work, but the successful handling of the liquid metals on a large scale is based on much the same principles.

To a large extent, the chemistry of the liquid alkali metals involves the use of these liquids as solvents, and they are therefore required to be of high purity. As available commercially, the metals contain impurities which may be metallic or non-metallic in nature. The removal of these impurities will be the subject of the next chapter; however, once obtained in a pure state the metals will only remain so if they are kept in a closed vessel, isolated from the atmosphere. Even so, the purified liquid metal may then take up impurities from the containing vessel or from the gas used to fill the enclosed space (the cover gas), so that the choice of container materials and the nature of the cover gas are of first importance.

2.2 The cover gas

All the alkali metals react with hydrogen and oxygen, but only lithium reacts with nitrogen. Therefore nitrogen is acceptable as the cover gas for the metals Na, K, Rb and Cs, but not Li. However, the nature of the reaction to be carried out must also be kept in mind; for example, reactions of solutions of calcium in liquid sodium cannot be studied under nitrogen as cover gas, since the dissolved calcium would readily take up nitrogen from the cover gas. Calcium is the main impurity in liquid sodium, and when the latter is used on a large scale (e.g. as a coolant), reaction between dissolved calcium and nitrogen cover gas could lead to precipitation of calcium nitride, and consequent blocking of a flowing sodium circuit. With this proviso, however, nitrogen is an excellent cover gas for all the metals except lithium, and because of its cheapness and ready availability is coming into increasing use.

Argon is the ideal cover gas, and because of its chemical inertness it is universally applicable. As with nitrogen, however, scrupulous purification is necessary; a measure of this is given by the fact that argon which is available commercially with greater than 99.99 per cent purity will cause a film to form at the clean surface of a liquid alkali metal almost immediately, and so requires further careful purification. An indication of the purities of the best argon and nitrogen available in quantity are given in Table 2.1.

A number of methods for the purification of argon are listed below. None of these methods taken individually will deal with all impurities in the cover gas, and it is necessary to put together a train consisting of two or more purification steps, designed to suit the particular experiment concerned.

(1) Water is the most serious impurity, not only because of its rapid reaction with the alkali metals, but also because it appears to enhance the rate at which the metals can react with other impurities present in the cover gas. Conventional drying agents such as phosphorus pentoxide or magnesium perchlorate are not able to reduce the water content of the high-purity gases referred to in Table 2.1.[10]

(2) Molecular sieves have been used extensively as drying agents, and have proved to be highly efficient at removing water (and also carbon dioxide and hydrocarbons) from the gas stream. In a typical apparatus the gas is passed through a column of the sieve measuring about 1m × 20mm, contained in a glass U-tube. The sieve (Linde molecular sieve, Grade 4A, Union Carbide Ltd.) is previously activated by heating for 24 h at 300°C and 10^{-5} torr.

(3) Oxygen and nitrogen can be removed by passing the gas through a column of calcium turnings or titanium granules. The reaction becomes more efficient at high temperatures, and the metals are therefore contained in a steel tube heated to 700°C. The use of chips of uranium metal for this purpose, which works well, is discouraged nowadays on health and safety grounds.

(4) Metal oxides are sometimes used in purification, and a furnace of copper oxide at 700°C removes hydrogen, carbon monoxide and hydrocar-

Table 2.1 Impurities in cover gases

Impurity	Maximum impurity levels (ppm by volume)	
	99.998% argon[8]	99.997% nitrogen[9]
O_2	2	5
N_2	10	—
H_2O	2	2
Hydrocarbons	1	1
H_2	1	0.0
CO	1	1
CO_2	1	0.5

bons from the gas stream. Manganese II oxide provides another interesting example, and is used specifically for oxygen removal. The reagent is prepared by heating a slurry of fireclay (2 parts) and manganese II carbonate (1 part) at 200°C for 24 h. The resultant brown cake formed by the reaction

$$2MnCO_3 + \tfrac{1}{2}O_2 \quad \rightarrow \quad Mn_2O_3 + 2CO_2$$

is broken into lumps and transferred to the reaction column. It is then reduced at 400°C by a stream of hydrogen, when the reagent turns to the bright apple green colour of MnO:

$$Mn_2O_3 + H_2 \quad \rightarrow \quad 2MnO + H_2O$$

The column is maintained at 100°C during passage of the argon gas. As oxygen is taken up from the gas, the column reverts from bright green to brown, so that the colour change indicates when regeneration is necessary.

(5) Whatever purification train is used, it is desirable that the gas should be given a final treatment by submitting it to intimate contact with the liquid metal to be used in the experiment. For reactions involving liquid sodium, a convenient low-temperature glass bubbler is shown in Figure 2.1. Sodium–potassium alloy (70 wt % K, which has a melting point of $-12°C$) is added through tap T_1 and rests on the sintered glass disc S_1 under argon. Slight increase in argon pressure overcomes surface tension forces, and the liquid falls on to the sintered disc S_2. The apparatus is flushed with argon through the bypass, after which the bypass tap T_2 is closed. Argon entering at T_3 then passes through disc S_2, and through the alloy as a stream of fine bubbles; the disc S_1 serves as a splash trap.

If sodium–potassium alloy (termed NaK) is considered to be too hazardous, it can be replaced by liquid sodium alone, in which case the bubbler is immersed in an oil bath at 100–150°C; at these temperatures attack of liquid sodium on Pyrex glass is slow. The composition of the liquid metal in the bubbler can also be adjusted to suit specific needs. For example, nitrogen can be efficiently removed from the gas stream using a solution of about 5 per cent barium in sodium.

(6) The above methods refer particularly to cases in which pure argon is to flow through a piece of apparatus, or where the volume of cover gas required is small. Reactions are sometimes carried out, however, in a dry box, the internal volume of which may be up to a cubic metre, and in such cases extra precautions are necessary. The dry box should contain an open steel tray containing liquid sodium, and maintained at about 150–200°C. Any adventitious contamination of the pure argon supplied to the box can then be detected by clouding of the sodium surface. If the experiments involve liquid lithium, then a liquid lithium pool at 250°C should be exposed in the dry box.

(7) Much disappointment can be avoided if it is remembered that water vapour, carbon dioxide, oxygen and nitrogen can diffuse through plastic or rubber tubing and contaminate the gas stream. Connecting tubes and joints should always be of glass or metal, depending on temperature.

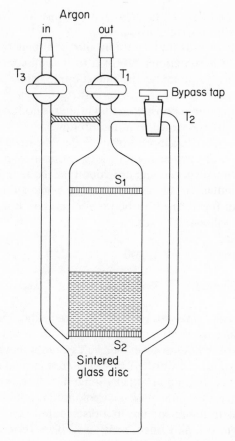

Figure 2.1   Sodium–potassium bubbler

(8) The methods listed above refer to the purification of argon. However, most of the methods also apply to nitrogen, provided that allowance is made for the fact that nitrogen reacts readily with lithium, and with the alkaline earth metals. Thus, methods discussed under (1), (2), (4), (5) and (7) are directly applicable to the purification of nitrogen. Method (3) is clearly not applicable, and under (6) only sodium would be used in a dry box charged with nitrogen.

2.3   Container materials

Glass

In the chemical laboratory, liquids are normally contained in glass vessels, and it is necessary to consider first to what extent the liquid alkali metals and glass are compatible. In fact, glass has very limited use. Laboratory experience will be discussed first, and the reasons for the reaction of the metals with glass will then be considered.

No glass is inert towards liquid lithium, but the 'penetration' of glass by liquid lithium which is sometimes referred to is a misconception. Chemical attack of glass by liquid lithium begins almost immediately on contact, so that the glass first darkens in colour and its structure then collapses. 'Lithium glass' contains some lithium phosphate and lithium fluoride, but this does not materially improve its resistance to liquid lithium.

Liquid sodium attacks glass more slowly. Soda glass need not be considered seriously because of the readiness with which it fractures, but Pyrex glass can be used for short-time experiments provided that the temperature is not too high. Attack is evident by discoloration of the glass through yellow and brown to black, but it is difficult to be precise on the rate of attack; a rough guide would be as follows:

Temperature (°C)	200	250	300
Discoloration by sodium visible in:	3–4 days	about 1 day	within several hours

Differential thermal analysis of liquid sodium–silica mixtures gives the reaction temperature as 370°C, at which temperature reaction is rapid and complete. Pure sodium does not wet clean glass, but impurities in the sodium (particularly the oxide) aid wetting; this enhances intimate contact of metal with glass, and hence the rate of attack. Some dissolved metals (e.g. barium) have a similar effect. Once the glass is discoloured it readily cracks, so that it is hazardous to retain liquid sodium in a discoloured vessel.

Various other glasses, and silica itself, have been tested, but pronounced changes in the $SiO_2:Al_2O_3:B_2O_3:Na_2O$ ratios seem to have little effect on the rate of attack by liquid sodium. 'Sodium resistant glass'[11] is prepared by flashing a coating of a high-borate glass on to the surface of soda glass; it has a somewhat greater resistance to sodium vapour, but offers no real advantage for containment of liquid sodium. There are a few special applications, however, in which such glasses find a use. For example, the sodium–sulphur cell (Chapter 18) requires a hermetically and ionically insulating seal between the sodium and sulphur electrode compartments, and this seal must possess an ultimate lifetime of over 50,000 h at 350–400°C. This has been accomplished using a suitable sealing glass.[12] None of the pure barium alumino-borate glasses tested (e.g. B_2O_3 55–65, Al_2O_3 15–25, BaO 20 mol %) were suitable because of their high expansion coefficients and rapid stabilization rates; but additions of silica (5–25 mol %) reduced the expansion coefficients and improved the working properties of barium alumino-silicate glasses without adversely affecting their resistance to sodium.

Attack on glass by liquid potassium, rubidium and caesium is perhaps rather slower than, but broadly similar to, that of sodium at comparable temperatures. However, because of their lower melting points, these liquids can be handled at lower temperatures, at which the rate of attack is much slower. For example, a sample of liquid sodium–potassium alloy kept in a Pyrex glass tube for over 3 years at room temperature showed no sign of attack, and a sample of caesium metal, which had been melted frequently with hand-heat during the same period, also gave no sign of attack on the Pyrex container. However, a word of caution should be added here. Oxygen is an important element in liquid sodium chemistry, and solutions of oxygen in sodium at concentrations as low as 5–10 ppm have important industrial significance. Liquid sodium may well extract small amounts of oxygen from glass before attack on the glass becomes evident, so that glass containers should not be used in experiments involving very dilute solutions of non-metals.

The alkali metal–glass interaction can be attributed to two processes; firstly, the penetration of the glass by alkali metal atoms, and secondly the chemical reduction of the components of glass by the metal. Penetration is probably the initial step with each metal, and the rate might be expected to vary with atomic size. Reduction can be interpreted in terms of two rections:

$$4M + SiO_2 \rightarrow 2M_2O + Si$$
$$4M + 3SiO_2 \rightarrow 2M_2SiO_3 + Si$$

Both the elemental silicon produced, and the penetrated metal atoms, can give colour to the glass. Laboratory experience shows that lithium is much more reactive than the other metals, and this is consistent with its small atomic size, and with the enthalpy values (ΔH°_{500}) for the first of the above reactions (Table 2.2). With liquid lithium, reduction to silicon is complete. Silica is wetted by lithium at its melting point, and the wetting, penetration and total reduction leads to rapid collapse of the glass. Table 2.2 suggests that, with the other alkali metals, complete reduction is thermodynamically less favourable, and the formation of ternary oxide now becomes significant. Values for the free energy of formation of the ternary oxides are not available with sufficient accuracy, but in the series Na→Cs, it is likely that ΔG of formation for the ternary oxides M_2SiO_3 will become greater than the M_2O value, so that the free energy change in the second reaction shown above becomes progressively larger than for the complete reduction. Other ternary

Table 2.2   Enthalpy changes for reaction $4M + SiO_2 \rightarrow 2M_2O + Si$

Alkali metal M	ΔH°_{500} kJ/mol
Li	−311
Na	+5
K	+114
Rb	+174
Cs	+200

oxides may also be produced; for instance, if the liquid sodium has a high oxygen content, Na_4SiO_4 is also formed.

The silicon produced on reduction is likely to be present in the glass as the finely divided element, but some may be present as the silicide of one of the metals present, and is not readily detected by chemical methods. When glass which has been darkened by immersion in an alkali metal is treated with water, an effervescence of hydrogen occurs. This may indicate the presence of silicon or low-valent silicon compounds, but could also arise, of course, from any alkali metal atoms which have penetrated into the glass. Evidence for the production of silicon has, however, been obtained by a somewhat indirect method. If powdered glass is immersed in liquid sodium contained in a nickel vessel, and the whole is heated to reaction temperature, some of the silicon produced dissolves in the sodium, and is then taken up by the nickel container to give a surface layer of nickel silicide, which has been identified by X-ray methods.[13]

The interpretations above have been based on the reactions of silica alone. Because the many types of glass available contain other added oxides in widely varying amounts, there is inevitably a lack of precision in discussing the glass reactions, and it is adequate in this context to consider glass as a mixture of oxides. Nevertheless, the same general principles apply. Fe_2O_3 is reduced by sodium to $Fe + Na_2O$ below 400°C, and Cr_2O_3 is reduced at about 400°C to $NaCrO_2 + Cr$. Al_2O_3 is stable to pure sodium, but if the sodium contains much dissolved oxygen $NaAlO_2$ is formed, which does not react further with sodium. The oxide B_2O_3 resembles SiO_2 in its behaviour; sodium boride is probably produced and the ternary oxide $Na_4B_2O_5$ has been identified as a product. The X-ray pattern of the sodium $-B_2O_3$ reaction is more complicated than in the case of SiO_2, and suggests that other ternary oxides are also formed.

Transition metals

In the laboratory and on an industrial scale, the liquid alkali metals are handled, almost entirely, in equipment constructed from transition metals, either singly or alloyed. Obviously, only those metals which show no appreciable solubility in the liquid alkali metals can be used as containers, and this solubility increases from left to right in the periodic table. Thus, the 'refractory' metals, i.e. the metals of the titanium group (Ti, Zr), the vanadium group (V, Nb, Ta) and the chromium group (Cr, Mo, W) are all useful as container materials. However, these metals are all expensive, and the physical properties of some of them (e.g. high melting point, malleability etc.) render them difficult to fabricate. As a result, their use tends to be restricted to laboratory work and to parts of industrial plant where special properties are required. The element which finds by far the widest use is iron; by alloying this with chromium and nickel, together with small amounts of one of the refractory metals, a wide range of steels can be produced which are resistant to attack by the alkali metals—up to and sometimes beyond 600°C—and from which equipment on a small or large scale can be fabricated. Nickel represents the limit in the transition block of the periodic table beyond which it is not possible to find a container metal; beyond this the other nickel group metals (Pd, Pt) and the heavier metals of copper group (Ag, Au) and the zinc group (Cd, Hg) show increasing solubility in the alkali metals.

As early as 1950, publications were appearing which made it clear that the early transition metals would provide satisfactory container materials for the liquid alkali metals.[14] Since that time there has been extensive research into the formulation of alloys, particularly steels, which will best withstand attack by the liquid metals under a wide range of temperature and experimental conditions, and this continues to be a vigorous area of research. It is not possible in this chapter to do justice to the considerable progress which has been made, but the following remarks will indicate some of the factors which determine the choice of a suitable metallic container. If the alkali metals could be obtained in absolute purity, it is unlikely that corrosion of the container metals mentioned above would ever occur to any extent. But it has to be accepted that, in practical terms, there is no such thing as a 'pure' liquid alkali metal, and this applies particularly to lithium. The dissolved impurities, essentially hydrogen, oxygen, nitrogen or carbon, which may be present in as little as a few parts per million, carry much of the responsibility for corrosion, so that a knowledge of the solubility of these non-metals in the liquid metals is of paramount importance.

Steel provides by far the most important container material, and the wide range of compositions available makes it possible to devise steels which are appropriate for the various experimental conditions encountered in handling the alkali metals. Once again, liquid lithium presents some unique problems, and it will be convenient to deal first with the elements sodium to caesium. Of these elements, experimental evidence relates almost entirely to liquid sodium, but similar principles can be regarded as applicable to potassium, rubidium and caesium. The compositions of two of the steels in common use are given in Table 2.3.

Where an experiment is of short duration, the temperature relatively low, and the quantity of liquid sodium is small, the choice of steel to be used is not of major importance, and most stainless steels will be suitable. Many laboratory experiments clearly fall into this category. However, this is not always the case, and we must then consider the type of problem which arises

Table 2.3 Compositions of two common steels

| AISI type | Weight % of elements added to iron | | | | | | |
	Cr	Ni	Mn	Si	C	Mo	Ti
316	16–18	11–14	2.0	0.75	0.08	2–3	—
321	17–20	10–15	2.0	0.75	0.08	—	0.4

most obviously in a fast nuclear reactor, where many tonnes of liquid sodium circulate in steel containers. Compatibility over periods of years has then to be considered, and problems which are not immediately obvious become quite apparent over these long periods.

The problems arise from the ability of flowing sodium to transfer alloying elements in the steel from one part of the circuit to another, by dissolving them from a region where their 'chemical potential' is high to one where the potential is low; the phenomenon, which is referred to again in Chapter 17, is termed *mass transfer*, and can result in changes in the physical properties of the chosen steels.

The latter fall broadly into one of two types, ferritic and austenitic. These terms have a structural connotation, and are related to the structures of pure iron. At temperatures up to about 900°C, iron has a body-centred cubic structure; this form is usually termed *ferrite*, and the term is often extended to describe the solid solutions formed by iron with alloying elements where atoms of the latter replace atoms of iron. In these 'substitutional alloys' the essential ferritic structure is retained. In the temperature range 900–1400°C, pure iron has a face-centred structure, and the term *austenite* is used to describe this form of iron, and also those solid solutions with alloying elements which retain the austenitic structure. The transition temperature (910°C in pure iron) is varied considerably by the presence of metallic or non-metallic alloying elements, so that at any given temperature a steel may have ferritic or austenitic structure.

Chromium and nickel are the chief metals involved in the formation of substitutional iron alloys, and non-metals (e.g. carbon, nitrogen, boron) usually form interstitial alloys, in which the foreign atoms occupy spaces between the metal atoms; in such alloys the metal structure is distorted, but remains essentially intact. The extent to which the non-metals can enter interstitially depends on the structure of the host lattice; as expected, carbon can enter into the face-centred cubic (austenitic) modification to a much greater extent than into the body-centred cubic (ferritic) form.

The above remarks give no more than a simplified account of an important area of metallurgy, but they suffice as a background to the behaviour of steels in prolonged contact with liquid sodium. Mass transfer occurs because the elements concerned have small, but significant, solubility in liquid sodium. Carbon and nickel have been studied extensively, partly because of the profound changes in the mechanical properties of a steel which can result when these elements are extracted by sodium. They will therefore be used as examples, but we should remember that mass transfer of the other additive elements can also occur.

Some general principles have been established (see Reference 464). Thus, the transfer of interstitial elements (e.g. carbon) is more rapid than for substitutional elements, which is to be expected since removal of an interstitial atom does not involve a fundamental change in the crystal structure of the metal. As a result, the possible depth of depletion of an

Table 2.4 Approximate solubilities of some metals at 650°C in liquid lithium and liquid sodium (wt ppm)[16,118]

Solute metal	In liquid lithium	In liquid sodium
Ni	1000	1.7
Cr	15	0.12
Fe	16	1.0

Solubilities in the alkali metals are treated in greater detail in Chapter 5.

interstitial element by sodium is much greater than for a substitutional element. In a system built entirely of a single steel, carbon will move from the high- to the low-temperature regions, which is consistent with the increasing solubility of carbon in liquid sodium with increasing temperature (see Chapter 5). In a circuit composed of a ferritic and an austenitic steel, carbon is likely to be transferred from the ferritic to the austenitic steel. These are experimental observations, and before it is possible to predict with any precision the direction or extent of carbon transfer, it will be necessary to consider in more detail the strength of the chemical bonding of carbon in the metal. A useful approach[15] is to regard the carbon as present in the form of carbides (e.g. Fe_3C, $Cr_{23}C_6$) in a ferritic or austenitic matrix phase, and to determine the relative stability of these carbides towards liquid sodium.

Because of their significant solubility (see Table 2.4) liquid sodium also has the ability to remove alloying elements from the surface of container steels, and to deposit them at other (usually lower temperature) regions of the circuit. Figure 2.2 illustrates this in the case of nickel. Removal of nickel may

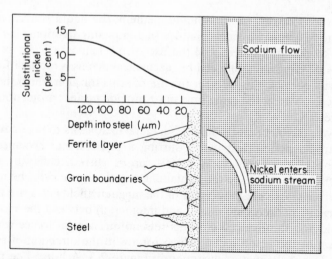

Figure 2.2 Removal of nickel from the surface of an austenitic steel by liquid sodium. (Reproduced by permission of the Central Electricity Generating Board)

result in transformation of the crystal structure of the surface regions to that of pure iron and ferritic steels. The rate at which this occurs depends on temperature and a number of purely mechanical factors, but is also influenced (and enhanced) by the presence of dissolved oxygen in the sodium.

The chemistry of the liquid lithium–steel interactions differs from that of the other alkali metals in two important respects. Firstly, nitrogen has a considerable solubility (1.7 mol % at 420°C) in liquid lithium, and replaces oxygen as the element mostly responsible for corrosion of steels by liquid lithium. Secondly, the solubility of iron and the alloying elements (especially nickel) is much higher than in liquid sodium. This is illustrated by the values in Table 2.4. The very high solubility of nickel means that nickel-containing steels are unsuitable as containers for liquid lithium over long periods, and even for short-time laboratory experiments at temperatures over 500°C it is desirable to contain liquid lithium in pure iron vessels. The refractory metals have low solubilities, and it is reported that vanadium alloys do not suffer significant losses of material at temperatures up to 700°C in pure lithium. Although corrosion by liquid lithium is therefore ruled by these dissolution processes, the extent of dissolution is greatly enhanced by non-metals dissolved in the liquid metal. Dissolved oxygen and carbon act in this way, but their effects can be minimized by a suitable selection of alloy.[16] On the other hand, the introduction of nitrogen into most transition metals and their alloys from liquid lithium occurs readily, and this nitriding has a highly deleterious effect. This applies even to refractory metals (e.g. titanium[17]), and this problem can only be dealt with by careful purification of the liquid lithium.

2.4 Stirring

In reactions carried out in a liquid medium, the liquid is normally stirred during the reaction. Molecular liquids (e.g. aqueous solutions) can often be exposed to the atmosphere and the use of mechanical stirrers presents no problems. Stirring a liquid metal by mechanical means is more difficult, however, partly because this calls for the use of seals through which the stirrer is introduced, which at the same time must provide perfect protection from the atmosphere.

However, the fact that the liquid is a metal enables us to take advantage of the electromagnetic effect, when stirring a liquid metal becomes an easy operation. All the liquid alkali metals can be stirred readily by rotating a permanent magnet outside the containing vessel, whether this be made from steel or from glass. The movement of the magnetic field gives rise to induced eddy currents in the liquid metal, and interaction between the moving magnetic field and the induced currents sets up mechanical forces which cause movement of the liquid metal, which rotates in the direction of spin of the magnet. This technique is illustrated in Figure 2.3. A horseshoe magnet A, approximately equal in outside diameter to the diameter of the vessel, is attached to an electric motor and rotated at a speed of about 1000–3000 revolutions per minute. Smooth rotation of the liquid metal is prevented by

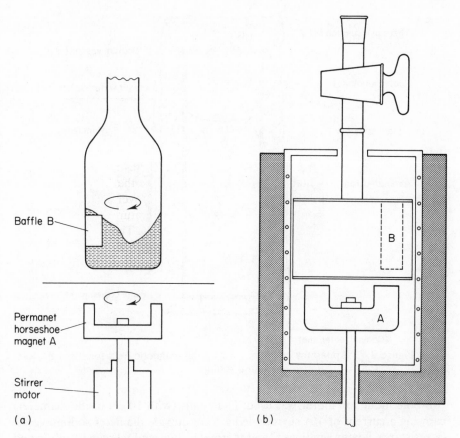

Figure 2.3   Magnetic stirring. (a) Glass vessel for short-term and low-temperature
experiments. (b) Steel assembly for high temperatures

insertion of a baffle B into the wall of the reaction vessel. The liquid pours
over this continuously, and this has the additional advantage of maintaining a
clean surface on the liquid metal. The degree of stirring can be varied by
altering the speed of rotation of the magnet, but depends also on the density
of the liquid metal. Thus, the method is not efficient for mercury, but vigorous
stirring of the alkali metals is readily achieved because of their low densities.
This method is cheap, and the equipment is easy to assemble.

The same principle has now been developed into equipment which has no
moving parts, and a greater degree of safety.[18,19] The method utilizes the
principle of the three-phase induction motor, and Figure 2.4 shows schematically an arrangement which has been used to stir liquid sodium. The induction
motor consists of two basic parts, the rotor and the stator. In this application,
the rotor is replaced by a vessel containing liquid sodium. When three-phase
current is applied to the windings, eddy currents are induced in the liquid
metal by the resultant rotating magnetic field, and metal movement is produced in the same direction as the field. In a typical arrangement,[18] a 1.5 h.p.

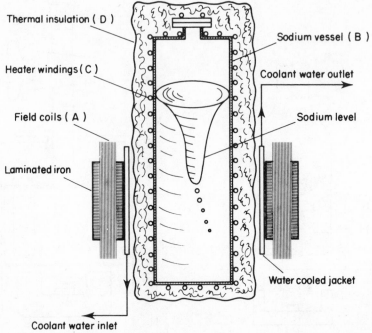

Thermal insulation (D)

Sodium vessel (B)

Heater windings (C)

Coolant water outlet

Field coils (A)

Sodium level

Laminated iron

Water cooled jacket

Coolant water inlet

Figure 2.4   Apparatus employing a rotating magnetic field for stirring liquid sodium

two-pole induction motor was used. Field coils (with 10 cm inside diameter) carrying a current of 4 A surrounded a 5 cm outside diameter sodium vessel B, resistance heater windings C and thermal insulation D. B can be fabricated from any non-magnetic material, and contained about 230 g of sodium. In a typical experiment, sodium was stirred for 200 h at 700°C, and the stirring rate could be altered at will by varying the voltage applied to the field coils.

This technique of stirring has now been generally superseded by the electromagnetic pump, which employs the principle that if a conductor carrying a current is placed in a magnetic field at right angles to it, there will be a mechanical force on the conductor in a direction perpendicular to both the magnetic field and the direction of the current. The direction of the force (i.e. forwards or backwards through the plane of the current and the magnetic field) is determined by Fleming's left-hand rule, and the existence of the force was demonstrated in the early days of physics by 'Barlow's Wheel'. If the conductor is liquid sodium, the mechanical force moves the sodium, and when the pump is incorporated into a reaction vessel very efficient stirring can be achieved.[20]

A typical pump used for laboratory work is illustrated in Figure 2.5. The duct AB carrying the liquid metal may be of glass or metal, and is about 5 mm internal diameter. It is usually flattened somewhat where it passes between the poles of the magnet; this improves the uniformity of the magnetic field. If

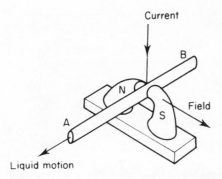

Figure 2.5 Laboratory version of an
electromagnetic pump

the duct is made from glass, two tungsten electrodes are sealed into the top and bottom of the duct where it passes through the magnet; if it is made from metal, electrical leads are attached at the same positions, and the duct is insulated from the magnet by thin mica sheets. Convenient field strengths of the magnet are between 0.2 and 0.6 tesla. When direct current (up up 40 A at 1–2 V) is applied in the direction shown in Figure 2.5, the liquid metal moves in the direction B to A.

Figure 2.6 shows how the electromagnetic pump P can be incorporated into a steel reaction vessel.[21] A cylindrical reservoir (10 × 5 cm) contains the bulk of the liquid metal, and is attached by means of a glass-to-metal seal S to a vacuum frame which enables the vessel to be evacuated, or filled with argon or some reagent gas. The electromagnetic pump is incorporated at P; liquid metal is drawn from the bulk, passes through the side tube, and is returned to the reservoir as a jet of metal J so that a fresh metal surface is continually generated. Various modifications can be made to this basic design which permit the study of liquid metal–gas reactions, rates of solution, and reactions within the liquid metal medium.[22–24] These will be described later. With the use of a steel assembly, the temperature range has been extended to 500°C by enclosing the entire apparatus in an air oven. The magnet material loses its magnetism at about 550°C, but this can be overcome by local cooling of the pole pieces.

Similar apparatus can be made from glass where the liquid metal has a low melting point. Figure 2.7 shows a Pyrex reaction vessel which could be used to study reaction of a sodium–potassium alloy with gases, in which the jet is arranged in a vertical direction to give a fountain of the liquid alloy. High flow rates are easily achieved by increasing the strength of the magnetic field in the pump, and a magnet of 0.5 tesla enables a jet of the liquid alloy to be ejected vertically to a height of up to 50 cm at a rate of 0.6 litres per minute. The metal in the fountain is bright and completely free from surface contamination, and has provided an impressive demonstration of liquid alkali metal chemistry on many occasions.[25,26]

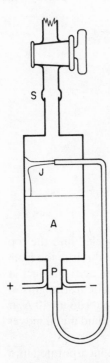

Figure 2.6   Reaction vessel employing an
electromagnetic pump

Extensive use of electromagnetic pumps has been made for circulating sodium in metal-cooled nuclear reactors and associated experimental sodium loops. These are large-capacity pumps requiring correspondingly high currents and powerful electromagnets and are now being superseded, in the fast reactors, by mechanical (centrifugal) pumps.[27] Nevertheless, they will continue to be used in experimental loops, and are important components in the liquid metal laboratory.

2.5   Cleaning and recovery of apparatus

At the end of an experiment, the operator has the task of cleaning and recovering equipment which may be both intricate and costly; when the apparatus has contained one of the alkali metals, this can present major problems. For a systematic review of cleaning agents, and their modes of action, the reader is referred to the 1967 monograph by Mausteller, Tepper and Rodgers.[28] This deals with industrial-scale as well as small-scale processes, and present remarks will be restricted to those aspects shown by recent work to be particularly relevant to laboratory handling.

The bulk of the liquid metal can be poured from the apparatus under an atmosphere of argon, and it is the removal of the residues which must be treated with care. With sodium, this is relatively straightforward. Steel apparatus may be washed with n-propanol, which reacts at a suitably slow rate; the rate can be increased by addition of small amounts of ethanol or

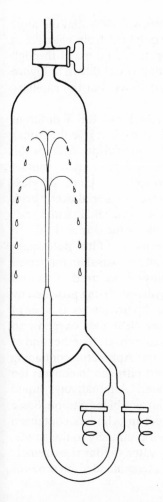

Figure 2.7   Sodium–potassium fountain

water to the propanol. When the reaction is complete, the apparatus may then be washed with water and dried in a stream of dry nitrogen.

Glass apparatus can be treated similarly, except where it appears that a high concentration of sodium propoxide may form in regions where there is a larger amount of sodium residue. The propoxide is sufficiently alkaline to attack the glass on prolonged contact. A typical example is the slow disintegration of the sintered glass plates in bubblers used for gas purfication (Figure 2.1) when cleaned with propanol. In such a case glacial acetic acid, which gives neutral reaction products, may be used in place of propanol.

Lithium reacts with propanol or acetic acid more slowly than does sodium, so that the same cleansing techniques can be employed.

It is always difficult to remove lithium or sodium residues from capillary tubes. The first propanol which comes into contact with the metal reacts to give hydrogen; this, and the plug of solid product which is formed, effectively separate the metal from the propanol. Later chapters will show that the most valuable information in liquid metal chemistry is obtained from changes in electrical resistivity, which is measured by passing the liquid metal through capillary tubes inserted in the side tube of the reaction vessel shown in Figure 2.6. The vessel itself can be recovered as described above, but the capillary tubes have to be cut away and replaced.

Potassium, rubidium and caesium (and NaK alloy) call for a different approach. Whereas lithium and sodium in contact with oxygen give the oxides Li_2O and Na_2O_2, the heavier alkali metals form the superoxides MO_2 as the stable products, and the superoxides can give rise to explosions in contact with organic compounds. In the author's laboratory,[29] a piece of glass apparatus containing residual potassium was immersed in glacial acetic acid, as for sodium. The reaction was more vigorous, giving occasional small bursts of flame. In an attempt to moderate the reaction, the acetic acid was diluted with xylene. A sintered glass plate which had been used to filter potassium, and which had been exposed to air briefly during transfer, was then immersed in the xylene–acetic acid mixture; a violent explosion occurred. This was believed to be a hydrogen–oxygen explosion, the hydrogen being provided by the $K-CH_3COOH$ reaction, and the oxygen by the superoxide. A number of such explosions have been reported, and it is now clear that oxygen and organic compounds must not be allowed to come into contact, together and at the same time, with potassium, rubidium or caesium. Apparatus containing potassium can be cleaned by first flushing with moist nitrogen to convert the metal to hydroxide, and then washing with water. Alternatively, liquid ammonia may be used, when the metal is washed out as a blue solution. Since the ammonia available commercially under pressure in cylinders contains a little water, the metal in the wash liquid is converted eventually to the amide, then the hydroxide. The reaction of caesium with water vapour is extremely vigorous, and washing with liquid ammonia is the recommended method for removing caesium residues.

Chapter 3

Chemistry of purification methods

3.1 Introduction

It is convenient to consider purification processes in two stages; firstly, the removal of gross (usually surface) contaminants where these exist, and secondly the removal of dissolved impurities from the clean metal. The relative importance of these two stages is determined by the form in which the raw metal is normally available to the laboratory, the quantities involved, and the nature of the experiment to be carried out.

3.2 Preliminary purification

Lithium is available in lumps of about 50 g which carry a coating of oxide, nitride and carbonate, and the surface layer can be cut away with a knife, with some difficulty; if this is done in a glove box charged with pure argon, the metal then remains clean and lustrous and in this form can be transferred (also in the glove box) to whatever apparatus is to be used in the second purification stage. The metal is also available in 0.5–1.0 kg ingots, and it is preferable (and easier) to break down these ingots in a glove box, using a hacksaw. Sodium can be given the same preliminary treatment; it is softer and more easily cut.

However, because much of the research work on liquid alkali metals has been concerned with sodium, and because sodium is usually used in larger quantities, much repetitive handling of crude sodium can be avoided if the laboratory is equipped with a sodium loop, from which a supply of partially purified liquid sodium is constantly available. A typical loop is shown in schematic form in Figure 3.1. It consists essentially of a steel tank T (capacity about 12 litres) holding the bulk of the sodium, a system of pipework for manipulating the liquid, and a wellhead W from which liquid metal can be drawn by a pipette. Sodium (from which surface layers have been removed) is added to the tank via the filling valve F, and the tank is heated to 400°C. Calcium (the main impurity) combines with oxygen in solution and calcium oxide is precipitated; this reduces the calcium content. With F closed, the tank is pressurized with argon, and the pipework filled with sodium via the valve

25

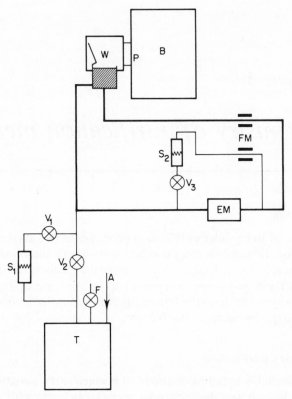

Figure 3.1   The sodium loop

V_1, and the stainless steel sinter plate S_1, which removes precipitate formed in the tank. When the wellhead is filled, V_1 is closed, and W is then connected to the tank through valve V_2. The liquid sodium is then circulated round the circuit (shown in Figure 3.1 by a heavy line) by means of an electromagnetic pump EM and the circuit as well as tank T is maintained at 160–170°C by means of heating coils. By control of the valve V_3, a part of the circulating sodium passes continually through a bypass which incorporates a stainless steel sinter S_2, and the rate of flow in the main and bypass circuits is monitored by flow meters FM. S_2 is maintained at a temperature of 110°–120°C, i.e. just a little above the melting point of sodium. The solubility of non-metal compounds (e.g. sodium oxide, sodium hydride) decreases as the temperature is lowered (see Chapter 5), so that filter S_2 lowers the concentration of an impurity to its saturation level at that temperature. With constant operation, the surface of the sodium in W becomes bright. Access to the wellhead is provided by means of a gas-tight port P, from a steel glove box B which can be evacuated or filled with argon. Transfer of sodium from W to other apparatus can then be effected within box B.

Potassium is not normally handled in litre quantities in the laboratory, so that there is little experience of potassium loops (corresponding to those of sodium) which can be quoted. The metal is usually purified as required for each individual experiment. It is supplied in lumps which carry a thick layer of the yellow superoxide. This can easily be cut away in an argon-filled glove box, and if the argon is sufficiently pure the remaining metal has a bright, lustrous surface. However, in spite of the appearance of the solid, a film of impurity appears on the surface as soon as the metal is melted; the liquid metal should therefore be degassed at 200°C and 10^{-5} torr, and the surface film removed by decantation. These operations can be carried out in glass apparatus. The surface film wets glass more readily than does the liquid metal, which aids separation by decantation.

Preliminary purification is not usually necessary for rubidium or caesium. These metals are produced by heating the chloride with excess of calcium chips at 700–800°C under vacuum, and condensing the evolved rubidium or caesium vapour. Because of the intense chemical reactivity of these metals, they are collected in glass ampoules or steel cylinders sealed under an inert gas, and supplied as such.

3.3 Dissolved impurities

Tables giving the content of dissolved impurities quoted by various suppliers have been given by Hatterer,[30] Mausteller, Tepper and Rodgers[28] and Thorley and Raine.[31] From these tables it is clear that considerable variations exist, and manufacturers often provide several grades of metal. In the present context it is unnecessary to include lists of these initial impurity levels; any values quoted below are approximate and intended only to indicate the order of impurity concentration. This section will summarize the types of impurity involved, their relative magnitudes, and some of the reasons why the major impurities can vary from one alkali metal to another. Details of solubilities will not be given here, as they are the subject of Chapter 5. Analytical values for total transition metal and other heavy metal contents are usually of the order of 50–100 parts per million (ppm). These metals are introduced as a result of corrosion of structural metals during production. The main non-metallic impurities are hydrogen and oxygen (from contact with atmosphere and moisture), carbon (from structural steels) and nitrogen, which is only important in the case of lithium. Traces of chlorine are also introduced during the production of lithium, sodium and potassium.

Lithium is produced by electrolysis of the molten LiCl–KCl eutectic mixture, and carbon can be introduced as a result of the use of graphite anodes. Sodium and potassium occur together with lithium in its minerals, so they are present also in commercial lithium. Potassium is only slightly miscible with lithium, with the result that the potassium content (~70 ppm) is lower than that of sodium (~150 ppm). Lithium is a good solvent for all the non-metals mentioned above, particularly nitrogen, and it is only with great difficulty that they are reduced below the 100 ppm level.

The main metallic impurity in sodium is calcium, which results from the production of sodium by electrolysis of molten $NaCl–CaCl_2$ mixtures. The initial calcium content of sodium as it is first formed in the Downs cell is high, but on cooling the liquid metal some calcium separates, and the calcium content of the sodium is reduced to about 500 ppm. Calcium is also an undesirable impurity because it can increase the oxygen and hydrogen content of sodium above their solubility values in pure sodium. The main alkali metal impurity is potassium (~100 ppm).

Compared with sodium, the content of alkaline earth and other metals in potassium is relatively small; this is because potassium is produced from KCl by reduction with sodium metal. The main metallic impurity is sodium, and values as high as 2 per cent are quoted. However, better grades of potassium are readily produced by redistillation. The production of rubidium and caesium by condensation from the vapour state means that they are, in effect, already purified by distillation during production. Small amounts of sodium and potassium can follow rubidium or caesium through the distillation process, but perhaps the most interesting feature of caesium chemistry is the very high solubility of oxygen.

3.4 Removal of dissolved impurities

There are three main methods of purification of the clean metals; these involve filtration, gettering or distillation. As in the previous chapter, the techniques described will be those applicable to laboratory work. Industrial-scale techniques, based on the same principles, are referred to in many of the references quoted.

Filtration

With glass apparatus, the metal can be filtered by passing it through a filter tube containing a plug of glass wool (as a coarse filter), and then through a glass sinter plate of appropriate pore size. With steel apparatus, filter tubes containing sintered steel plates are used. Simple filtration at a constant temperature will merely remove suspended foreign matter, though clearly the concentration of an impurity after filtering will not exceed its solubility at the filtration temperature. Other steps must therefore be taken, in conjunction with filtration, to lower the impurity level.

The method most commonly used in the laboratory and on the industrial scale takes advantage of the fact that solubilities in the alkali metals generally decrease with decreasing temperature; this is illustrated in Table 3.1 and data for other solutes will be given in Chapter 5. It is clear, in this case, that if sodium can be filtered at a temperature near to its melting point, the oxygen content will be thereby reduced to a few parts per million. This is the function

Table 3.1 The solubility of oxygen in liquid sodium[32]

Temperature (°C)	Solubility (wt ppm)
100	2
200	22
300	103
400	460
500	1420
600	3340

of the filter S_2 in the sodium loop shown in Figure 3.1, and also of the 'cold traps' used in the metal coolant circuits in fast nuclear reactors. Supersaturation is not a phenomenon normally encountered in liquid alkali metal solutions, and this aids the filtration method. However, the impurity which is thrown out of solution on cooling may be slow to nucleate and to aggregate to a particle size suitable for filtration. For sodium oxide in sodium, sintered plates having pore sizes of 15 μm are quoted,[31] but appropriate pore sizes will depend on the impurity, the metal and the experimental conditions, and must be determined by trial and error.

A second method by which a soluble impurity can be rendered insoluble, and thus filterable, is to add another element which will form an insoluble compound with the impurity. This can be illustrated using the removal of calcium from liquid sodium as an example. The obvious requirement for the removal of a metallic impurity by oxygen is that its oxide should be more thermodynamically stable than that of sodium. Table 3.2 gives values of the free energy of formation $-\Delta G_f^\circ$ for a number of metal oxides at 327°C, which is in the temperature range at which these reactions are likely to be carried out. Using these values, we see that

$$Na_2O + Ca \rightarrow CaO + 2Na \ (-240 \ kJ/g \ atom \ O)$$

so that the forward reaction is favourable, and addition of oxygen or sodium oxide to a solution of calcium in sodium will cause precipitation of calcium oxide. The method has been used to remove calcium from commercial grade

Table 3.2 Values of $-\Delta G_f^\circ$ for metal oxides at 327°C (kJ/g atom oxygen)

Oxide	$-\Delta G_f^\circ$	Oxide	$-\Delta G_f^\circ$	Oxide	$-\Delta G_f^\circ$
Li_2O	497	MgO	543	Y_2O_3	574
Na_2O	333	CaO	573	TiO_2	419
K_2O	273	SrO	530	ZrO_2	489
Rb_2O	248	BaO	500	ThO_2	556
Cs_2O	224	Fe_2O_3	231	UO_2	490
		Cr_2O_3	326	Al_2O_3	496
		NiO	184		

sodium, and can even be used to reduce the sodium content of liquid potassium. The reaction itself is not slow within the liquid metal medium, but nucleation and aggregation of the precipitate may be slow. Aggregation occurs more quickly at higher temperatures, so that the liquid metal with its added oxygen is held first at 300–400°C for some time, then cooled for filtration. Any excess oxygen, in the form of sodium oxide, will then be filtered off at the same time. This technique (sometimes termed 'slagging' in industry) has not found much favour as a laboratory method. Carbon and nitrogen can also be used as reagents in special cases, but addition of controlled amounts of these reagents can be difficult and the process is less necessary as purer grades of the alkali metals become commercially available.

Gettering

This is a general term for the removal of impurities (usually non-metals) by converting them into more stable compounds which are insoluble in the liquid metal. The elements (usually metals) which are added for this purpose are termed 'getters', and may be soluble or insoluble in the liquid metal.

The removal of oxygen from sodium by means of added calcium, which was discussed earlier, is a typical example of the use of a soluble getter. In fact, soluble getters are not in frequent use, partly because of the filtration which is also necessary. An insoluble getter, on the other hand, can be immersed in the liquid metal, held there until it has collected the impurity, and then withdrawn. This section is concerned with the choice of suitable metals for use as insoluble getters.

In the general equation

$$\underset{\text{liquid}}{M_1[X]} + \underset{\text{solid}}{M_2} \rightarrow \underset{\text{liquid}}{M_1} + \underset{\text{solid}}{M_2X} \quad (-\Delta G)$$

the efficiency with which the getter M_2 can remove impurity X from the liquid metal M_1 will be dictated, in the first place, by the magnitude of the free energy change associated with this reaction; if ΔG is large and negative, M_2 can be considered for use as a getter. Taking oxygen as a typical impurity, we can examine, as a first step, the relative values for the free energies of formation of the oxides listed in Table 3.2. The greater the difference between the $-\Delta G$ values for the oxide of metal M_2 and the oxide of the liquid metal M_1, the greater will be the potential value of M_2 as a getter. It becomes immediately obvious that all the metals listed in the third column of Table 3.2 have potential use as getters for sodium, potassium, rubidium and caesium, but that only yttrium and thorium are likely to remove oxygen from liquid lithium. Similar comparisons of the free energies of formation of the hydrides, nitrides and carbides of the metals can be made. These values may be obtained from the Ellingham diagrams which follow; as expected, the $-\Delta G$ values for possible insoluble getters do not fall into the same order for each non-metal impurity, so that an efficient getter with respect to one non-metal may not be able to remove others. It may therefore be necessary to use more than one metal for gettering an alkali metal, or to employ an alloy of suitable metals.

Direct comparison of $-\Delta G°$ values does not, however, provide the whole picture. We are usually concerned with the gettering of non-metals from unsaturated solutions in an alkali metal, and allowance must be made for the fact that the chemical potential (and the free energy) of the solute, when in an unsaturated solution, is not the same as in the solid alkali metal compound. This is most readily illustrated in terms of actual solutions. Pulham and Down[33] have determined the free energies of hydrogen, nitrogen and oxygen dissolved in liquid lithium, and the lithium–hydrogen system will be taken as an example. Assuming Henry's law to apply to these dilute solutions, then

$$a_{LiH} = k\, c_{LiH}$$

where a and c represent activities and concentrations respectively. When the hydrogen concentration has been increased to its saturation value $(c_{LiH})_{sat}$, addition of further hydrogen will cause precipitation of solid hydride LiH. At this stage, the chemical potential of hydrogen in solution equals that in the solid LiH, and its activity is unity. It follows that

$$a_{LiH} = c_{LiH}/(c_{LiH})_{sat}$$

The free energy of formation of LiH, $\Delta G°$, is related to the free energy of unsaturated solutions $(\Delta G°)'$ by

$$(\Delta G°)' = \Delta G° + RT \ln a_{LiH}$$

so that we may now write

$$(\Delta G°)' = \Delta G° + RT \ln \left[\frac{c_{LiH}}{(c_{LiH})_{sat}} \right] \tag{3.1}$$

Knowing the solubility of hydrogen in liquid lithium, it is then possible to calculate $(\Delta G°)'$ values for solutions of hydrogen in lithium at any given temperature and concentration. The results are shown in Figure 3.2, which illustrates the variation in $\Delta G°$ with temperature for a number of metal hydrides by full lines. The broken lines represent the 'isopleths' calculated from equation 3.1, and it is immediately clear that the $(\Delta G°)'$ values for the unsaturated solutions are much more negative than $\Delta G°$ for lithium hydride.

Amongst the gettering metals shown in Figure 3.2, only yttrium and lanthanum are able to remove hydrogen from liquid lithium. Where gettering can occur, the difference between the $\Delta G°$ values for hydrogen in the getter and in the liquid metal diminishes as the solution becomes more dilute, so that gettering becomes progressively more difficult, as might be expected. It is instructive to follow changes in the condition of the solution along one of the isopleths (e.g. line AB, Figure 3.2), remembering that the solubility of LiH decreases as temperature decreases. At the highest temperature (A), the solubility is high and the solution is unsaturated. As we decrease the

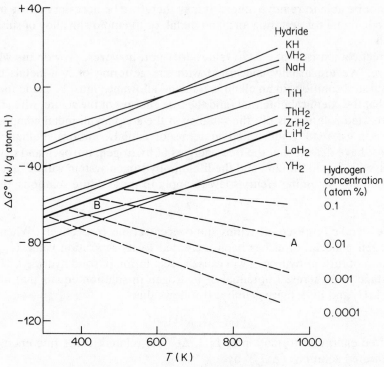

Figure 3.2 Free energy–temperature diagram for hydrides, showing isopleths for unsaturated Li–H solutions

temperature from A towards B, solubility decreases and moves nearer to the actual hydrogen concentration of the solution. At the temperature represented by point B, the solubility equals the hydrogen concentration in solution and the solution is saturated. For the saturated solution the activity a_{LiH} is unity and $(\Delta G^{\circ})'$ now equals ΔG°. The points at which the isopleths meet the line for LiH (solid) therefore represent the temperatures at which each of the solutions is saturated.

Figure 3.3 shows the corresponding Ellingham diagram for oxides. Because of the relatively large negative value for lithium oxide, only a few metals (e.g. Y, La, Ta) will remove oxygen from lithium. The isopleths have the same significance; their slope merely reflects the relative magnitude of the two terms on the right-hand side of equation 3.1. The Ellingham diagram for the nitrides is shown in Figure 3.4. The relative positions of the various nitride lines are in sharp contrast to those for hydrides and oxides, and many metals (e.g. V, Nb, Ta, Ti, Y, Th, Zr, Hf) now have potential as insoluble getters for nitrogen.

In spite of the attempts to move nearer to an understanding of the behaviour of insoluble getters, gettering remains a complex science, and it is still not possible with any certainty to predict the performance of a potential

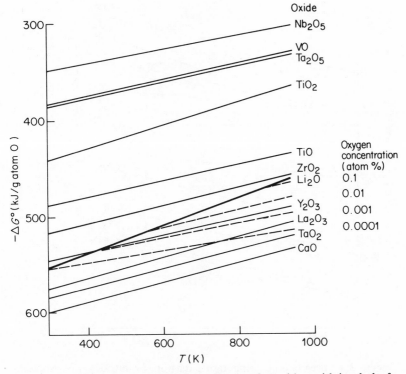

Figure 3.3   Free energy–temperature diagram for oxides, with isopleths for unsaturated Li–O solutions

getter. A number of additional factors have to be considered. For example, the data in Figures 3.2–3.4 ignore the kinetic aspects. Ideally, the concentration to which the getter can reduce the impurity is indicated by the point at which the getter line crosses the isopleth. The final impurity concentration is therefore dependent on temperature, and the diagrams show that the higher the temperature, the less effective is the getter. However, the rate at which gettering takes place, which involves the rate of diffusion of non-metal into the getter, is more rapid at higher temperatures, and surface films have less of an inhibiting effect. This kinetic factor has therefore to be set against the thermodynamic aspects. Again, the Ellingham diagrams offer a comparison between the simple, stoichiometric, binary compounds of non-metal with getter and alkali metal. In fact, diffusion of non-metal into the solid getter may never reach a stage at which a stoichiometric compound is formed, and if such stoichiometry is actually achieved, it may not represent the maximum valency of the metal. Reference to the TaO_2 and Ta_2O_5 values in Figure 3.3, the dihydrides in Figure 3.2 and the mononitrides in Figure 3.4 will illustrate this. Furthermore, the eventual stable product of reaction between getter and impurity may not be a binary compound but a ternary

34

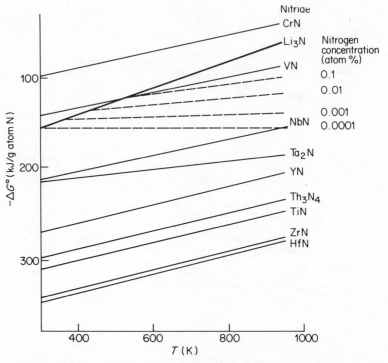

Figure 3.4   Free energy–temperature diagram for nitrides, with isopleths
for unsaturated Li–N solutions

compound between getter, alkali metal and non-metal. For example, when
niobium (or tantalum) is immersed in liquid sodium containing oxygen in
excess of 5 ppm, the following reaction occurs at temperatures over 300°C:

$$Nb + Na(O) \rightarrow Nb(O) + Na_3NbO_4$$

The product is the ternary oxide Na_3NbO_4, formed on the surface of a solid
solution of oxygen in niobium.[34] When chromium metal is immersed in liquid
lithium containing more than 200 ppm of dissolved nitrogen, the ternary
nitride Li_9CrN_5 is formed as a surface layer.[35] Further complications are
introduced if the liquid metal contains more than one dissolved non-metal; in
the chromium example, dissolved oxygen does not interfere with the
formation of the ternary nitride since the $-\Delta G°$ values for Li_2O are greater
than the values for ternary Li–Cr–O compounds; but any carbon present can
interfere owing to the formation of the very stable carbide $Cr_{23}C_6$.

The use of insoluble getters on an industrial scale is described
elsewhere,[28,31] and is termed 'hot-trapping'. In the laboratory, a getter can be
introduced using the type of apparatus shown in Figure 3.5, which is a
modification of the reaction vessel shown in Figure 2.6. The liquid metal is
circulated by the electromagnetic pump P around a steel capillary loop which

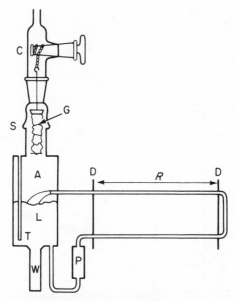

Figure 3.5   Purification by gettering

is spanned by two metal discs D between which the electrical resistance is measured. The resistance is remarkably sensitive to dissolved impurities, which can therefore be monitored continuously by this means. A constant temperature is essential, and this is controlled by a thermocouple inserted at T. Pieces of the getter (e.g. yttrium sponge) are threaded on to a stainless steel rod, which is supported by a steel chain. This is wound around a glass spindle C, which can be rotated through a glass joint, and the whole getter support is joined to the liquid metal container by a glass-to-metal seal S. The getter can then be lowered into the liquid metal, or raised out of it, at will, or it can be isolated from the system in the well W.

If the rate of equilibration with the getter is slow, it may happen that the quantity of liquid metal flowing through the capillary tube is not sufficient to give adequate stirring within the main vessel A, and in this case two loops can be employed, as shown in Figure 3.6.[36] The main pump P_1, which is incorporated in wider tubing, is used to provide rapid circulation, and the orifice can enter A either above or below the metal surface. The second pump P_2 feeds the liquid metal through the capillary (internal diameter of the order of 1.5 mm) for continuous resistance measurement.

Some typical results are shown in Figure 3.7, which illustrates the resistivity changes observed when solutions of hydrogen and deuterium, each at 1.00 mol % concentration in liquid lithium at 400°C, were gettered using yttrium sponge.[36] Resistivity can be converted directly to solute concentration using the appropriate calibration, so that ordinate axes also show concentrations of hydrogen or deuterium present at any given time. When yttrium was added to

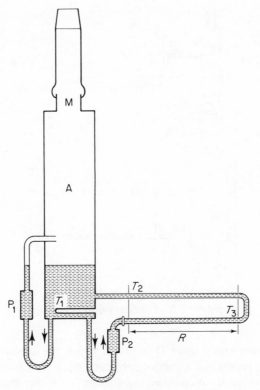

Figure 3.6 Apparatus employing two liquid
metal loops

the hydrogen solution, hydrogen was steadily removed (section AB). The yttrium was then withdrawn from the liquid metal for a 2-day period (section BC), during which time the hydrogen concentration remained constant. The getter was then re-immersed, and hydrogen was again removed until after 23 days its concentration was near to zero. Similar behaviour was observed with the deuterium solutions, and the resistivity–time curve shows a 1-day (FG) and a 2-day (HJ) period during which the getter was withdrawn, and no change in deuterium concentration took place. Again, the deuterium concentration was eventually reduced to near zero.

In experiments using other types of apparatus, it is often possible to maintain an impurity at very low concentration by suitable choice of container materials. For example, in crucible experiments used to study the reactivity of sodium with various materials, the oxygen content of sodium is kept low by the use of zirconium crucibles. One of the best getters for carbon dissolved in liquid sodium is stainless steel (e.g. austenitic type 316 steel), and therefore liquid sodium in a stainless steel container always has a low carbon content.[37] In other containers, plates of stainless steel can be added to maintain a low carbon level.

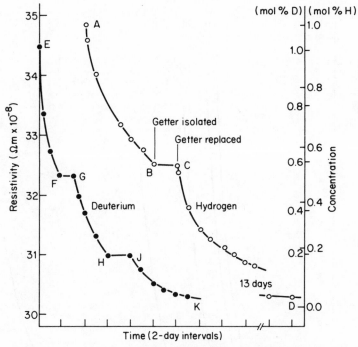

Figure 3.7   Gettering of hydrogen and deuterium from liquid
lithium at 400°C, using yttrium sponge

Distillation

Because of the relatively high vapour pressures of the alkali metals,
distillation is not experimentally difficult, and is an acceptable method for
purification of the metals. Distillation under reduced pressure lowers the
operating temperature, and thereby reduces the risk of contamination of the
distillate by materials from which the condenser is constructed.

Figure 3.8 shows the variation in vapour pressure of the metals with
temperature. In the case of caesium, for example, the vapour pressure is
sufficiently high that the type of equipment used for the distillation of
molecular liquids is applicable to the alkali metal also, and an apparatus
constructed from Pyrex glass which has been used often and successfully[38] is
illustrated in Figure 3.9. The caesium to be purified is contained initially in a
flask which can be inverted and attached at the ground glass joint A. The
caesium is then melted, and about 150 ml are allowed to drain into flask C
through the wide-bore, sparsely greased tap B, which is then closed. The
condenser tube D is about 200 mm long and 20 mm diameter, and wrapped
with heating tapes for the first 150 mm of its length; the remaining 50 mm is
exposed to the atmosphere. Flask C is heated by means of an electrical
heating mantle, and the caesium is found to distil easily at 190°C under a

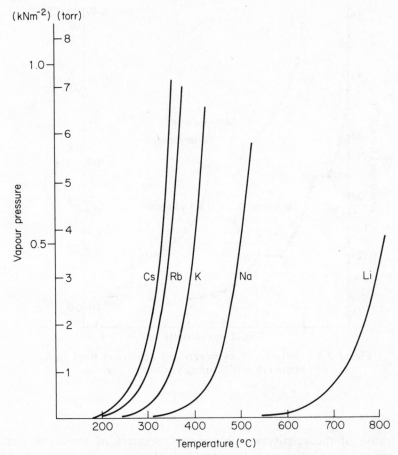

Figure 3.8 Vapour pressures of the alkali metals

pressure of about 10^{-3} torr. The vapour condenses at the lower end of the condenser and the liquid collects in flask E. The arrangement of taps shown in Figure 3.9 is to allow the subsequent separation of the purified caesium in E from the distillation apparatus under an argon atmosphere. These are Teflon-type taps, but the Teflon gaskets are replaced by Viton since the alkali metals in general, and caesium in particular, react readily with fluorine-containing plastics.

The ease with which purification of an alkali metal can be achieved by distillation depends, of course, on the difference between the vapour pressure of the metal and that of the impurity or its decomposition products. Removal of transition metals and alkaline earth metals from the alkali metals by distillation is virtually complete. The efficiency with which non-metals can be removed will depend on the metal itself, and the non-metals which form the major impurities. However, experimental conditions can often be varied to

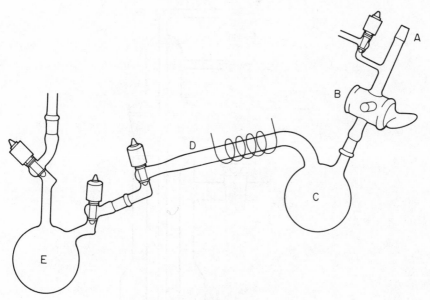

Figure 3.9   Distillation of caesium

improve efficiency. In the case of caesium, the high solubility of oxygen renders filtration an unacceptable purification method, whereas distillation can be efficient. It has been suggested[39] that oxygen can be carried over into the distillate in the form of carbon dioxide (produced by dissociation of carbonate impurity) or carbon monoxide (from the reduction of any oxides present by carbon impurity). Both these processes require the liquid metal to be boiled to dryness, and this would in any case be avoided. Although nitrogen is quite insoluble in liquid caesium, caesium amide is a possible impurity, but this compound decomposes into its constituent elements at about 270°C, and distillation is carried out below this temperature. If the distilland contains caesium hydride it is possible for hydrogen to recombine with the caesium in the receiver, since at the distillation temperature of 190° the hydrogen dissociation pressure is about 1 torr. This reabsorption of hydrogen is minimized by maintaining a good vacuum during distillation, and condensing the distillate at as low a temperature as is practicable.

The above discussion on caesium distillation is also relevant to rubidium, and distillation can be used to provide the quantities of pure metal likely to be required in experimental work. With potassium, sodium and lithium the vapour pressures become progressively lower, the distillation temperatures become higher, and the distillation of appreciable volumes of these metals in glass apparatus is not practicable. The separation of alkali metals by distillation is utilized in the manufacture of potassium, but for laboratory work large volumes of these three metals are more suitably purified by filtration and gettering. There are circumstances, however, in which

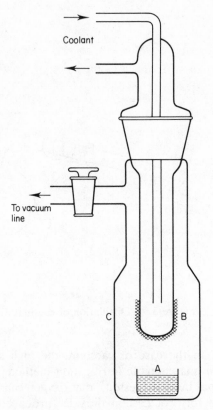

Figure 3.10 Distillation of small
quantities of sodium or potassium

distillation still has a part to play—for example in the purification of small
quantities, and in the removal of liquid metal from a reaction mixture.

With these less volatile metals, it is necessary to reduce the vapour distance
and to employ a larger temperature differential between distilland and
distillate. A simple form of apparatus is illustrated in Figure 3.10. This is
constructed in Pyrex glass, and can be used for potassium and sodium but not,
of course, for lithium. The metal to be distilled is contained in an open steel
vessel A, and a 'cold finger' B is positioned a few centimetres above the metal
surface. This finger can be cooled by circulation of a cooling liquid such as a
light oil; cold water would be an ideal coolant if it were not for the obvious
hazard involved. It is often more convenient to fill the cold finger with pieces
of solid carbon dioxide, though this needs constant replenishing during the
distillation. Container C, which is initially filled with pure nitrogen or argon
cover gas, is heated by an electric heating mantle. With sodium, distillation at
about 300°C and 10^{-3} torr gives crystals of sodium on the cold finger within
several hours, though some discoloration of the glass also occurs. With

potassium, a temperature of 200–250°C is suitable, and no glass discolouration is normally observed.

For work at higher temperatures, or with liquid lithium, it is necessary to employ steel apparatus, and a versatile form of equipment is illustrated in Figure 3.11. The crucible A containing the liquid metal is placed in a stainless steel vessel B, consisting of a flanged cylinder of about 100 mm diameter and 200 mm deep. The O-ring seal inset into the top surface is protected from high temperatures by the cooling coils C. Inserted into this cylinder is a cold trap D which could be bolted on to the flange on B to give a gas-tight seal. The trap can be filled with a coolant (water is now acceptable), or a cooling coil. D also carries a vacuum connection E which allows the space between vessels B and D to be evacuated at will. For distillation, that part of B below the flange F can be heated in an electric furnace.

With sodium, this apparatus is used primarily for studying reactions between substances immersed in the liquid sodium medium in crucible A. Such reactions can be studied at temperatures up to 600°C under an argon

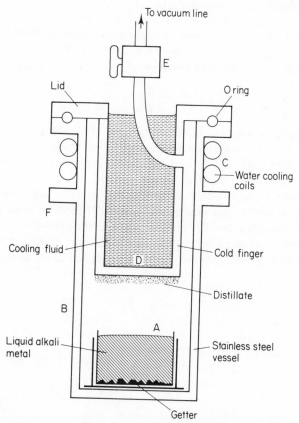

Figure 3.11   Purification by high-temperature distillation

atmosphere, and the sodium subsequently distilled on to the cold trap at 300°C and 10^{-3} torr. This equipment is particularly useful for the distillation of lithium, which is carried out at 600°C and 10^{-5} torr. The purity of the distilled lithium will depend on the extent to which impurities can be carried through the vapour phase, as we have discussed above for caesium distillation. Heavy metals will not be transferred to the distillate, but other alkali metals will distil with the lithium, and fractional distillation becomes necessary if appreciable quantities of these are present. Non-metals (hydrogen, oxygen, carbon and nitrogen) may well be carried over. Of these, nitrogen is the most serious contaminant because of its high solubility and because it is largely responsible for corrosion by liquid lithium. Lithium nitride (Li_3N) undergoes some decomposition below the distillation temperature of 600°C, and the reaction of nitrogen with distilled lithium is rapid, so that distillation in itself will not remove nitrogen from lithium. However, using this technique it is possible to combine gettering with distillation to produce a pure lithium distillate; this is achieved by placing a suitable mixture of gettering metals in crucible A (Figure 3.11) together with the impure lithium, and distilling from this mixture. Figure 3.7 shows that yttrium should getter hydrogen from liquid lithium, and Table 3.2 indicates that yttrium will also getter oxygen. Titanium and zirconium have been found to be effective getters for both nitrogen and carbon, which is in accord with the relevant values for the free energies of formation[40] listed below:

$-\Delta G^{\circ}_{298}$ kJ/g atom N	TiN: 309	ZrN: 337	Li_3N: 156
$-\Delta G^{\circ}_{298}$ kJ/g atom C	TiC: 181	ZrC: 182	Li_2C_2: 89

A mixed getter consisting of titanium 55%, zirconium 35% and yttrium 10% by weight, all in sponge form, has been found very effective for simultaneous removal of all the non-metals mentioned above. Hoffman[41] has given analytical values for non-metals remaining in liquid lithium after distillation under these conditions. The mixed getter is added to the impure lithium in the crucible, and maintained at 650°C for four days in an argon atmosphere to complete the gettering process. During this time cooling water is not passed through the cold trap, so that premature distillation is avoided. The vessel is then evacuated, the cold trap cooled, and the lithium distilled.

Chapter 4

Species formed by dissolved elements

4.1 Introduction

In previous chapters reference has been made to the fact that various elements (both metallic and non-metallic) will dissolve in the liquid alkali metals, and that solubilities can vary widely; but no attempt was made to define the nature of such solutions or the dissolved species formed within the metallic medium. In this chapter an attempt will be made to do just this, but it must be emphasized from the outset that there is as yet no ideal or precise theory of solution in liquid metals. The areas of interest covered by chemists and physicists have not yet overlapped sufficiently to enable the extensive and excellent work on the physics of the metallic state to be applied usefully to chemical problems. Much of the physics is concerned with pure metals, and theories are expressed in terms of mathematical models which are not usually applicable directly to the chemistry of solutions in metals. In contrast, the chemist is primarily concerned with the behaviour of solutes dissolved in a metallic environment, and in the correlation and interpretation of results he must necessarily accept concepts and chemical terminology which the physicist might well regard as too shallow or imprecise.

Let us consider where the problems lie by reference to solutions in molecular liquids, e.g. water. These liquids consist of molecules, and it is the properties of the individual molecules which dominate the character of the chemistry which can be carried out in that particular medium. The classical concepts of acid, base etc. arise directly from the self-dissociation of the water molecule, and its polar nature determines the extent of solvation of a dissolved species, and hence its solubility. Bulk properties of such molecular liquids are relatively unimportant, and since the ionic theory was put forward by Arrhenius over one hundred years ago, it has become possible to develop a satisfactory picture of the nature of solutions in such liquids.

In contrast, a solute dissolved in a liquid metal is immersed in a sea of electrons, and it is the nature of these electrons, their number, their energy levels and the extent to which they are, or are not, localized on the atoms of the metal medium which dominate the chemical behaviour of dissolved substances. For instance, is it possible to consider the existence of the OH^- group in a solution of sodium hydroxide in liquid sodium if sufficient suitable electrons are available to allow separation of OH^- into O^{2-} and H^-? Again, will oxygen dissolve in caesium as O_2, O^- or O^{2-} species, and will nitrogen give solutions consisting of N_2, N^-, N^{2-} or N^{3-} in liquid lithium? The approach to such problems is restricted by the lack of obvious physical techniques for the study of dissolved species. In (say) aqueous solutions, a whole range of spectroscopic methods is available, but by their very nature these methods are not applicable to metallic-type solutions. Some methods (e.g. photo-electron spectroscopy) may prove useful in the future, but many of the conclusions on the nature of dissolved species have still to be arrived at by indirect methods such as kinetics, phase equilibria, electrical conductivities, electrochemical measurements and the like.

4.2 The metallic state

In order to interpret the chemical reactions of, and between, species dissolved in a liquid metal it is first necessary to arrive at some physical picture of the reaction medium itself, and there are two limits between which any metal must fall. On the one hand, we can consider the liquid metal as consisting of a collection of atoms, with the electrons localized on each atom. At the other limit the electrons (usually the valency electrons) may be separated from the atoms; this is the 'free electron' model, free electrons being defined as conduction electrons which have mean free paths which are long in comparison with their wavelengths.

A measure of the extent to which the free electron model is applicable to any given metal is given by studies of the Hall effect.[42-44] When a current flows through a stationary conductor perpendicular to the direction of an applied magnetic field, a small electric potential transverse to the current flow may be generated; this is the Hall effect, and it arises from the bending of the electron paths in the magnetic field. Measurement of the potential can be used to determine a Hall coefficient R, which for free electrons has the value

$$R = -1/ne$$

where n is the number of conduction electrons per unit volume of metal and e is the electron charge. This may then be compared with R_0, the value of the Hall coefficient which would be computed if all the valency electrons were free. Table 4.1 gives some values for various metals, collected from References 42–45. The determinations are difficult to carry out, and the degree of error is about ±3 per cent on most values. The alkali metals (including lithium) exhibit near-free-electron behaviour in the solids also, so that for the liquid alkali metals the free electron model would appear to be the correct one. Taking the Hall coefficient measurements together with other observations, it could be that the alkali metals are, in fact, the only group in the

Table 4.1   Hall coefficients for some liquid metals

Metal	Na	Rb	Cs	Zn	Cd	Hg	Ga	In	Tl
Valency electrons	1	1	1	2	2	2	3	3	3
R/R_0	1.01	0.99	1.00	1.01	0.99	0.99	0.95	0.94	0.77

periodic table to which the free electron model can suitably be applied; it becomes less adequate as the group number and the period number increase.

Another method for determining the concentration of conduction electrons in a liquid metal involves a study of the optical properties of the metal, in which the polarization of the beam which is reflected when plane polarized light is incident on the metal surface is analysed. The results yield the real and imaginary parts of a complex refractive index, which can be compared with a formula given by Drude on the basis of the free electron model.[42] This will not be discussed further since measurements do not appear to have been published on the alkali metals to date, perhaps for purely technical reasons. It is relevant to note, however, that where metals have been studied by both methods, the results are in agreement.

In discussing chemistry in the liquid alkali metal media we shall therefore deal in terms of the free electron model, and regard liquid sodium, for example, as consisting of cores of Na^+ (i.e. Na atoms minus one valency electron) and an equal number of free electrons. The electrons, which occupy the s orbitals of the isolated atom, combine together in the metallic state to give continuous molecular orbitals, which taken together represent the conduction band of the metal. Figure 4.1 shows the distribution of electrons over the energy range of the conduction band, and N_E represents the extent to which the various energy levels are occupied. The s band itself is a symmetrical semicircle, and electrons enter the band from the lowest energy available. An

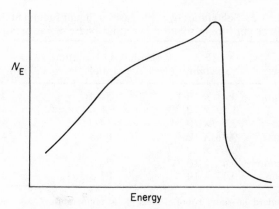

Figure 4.1   Conduction band in liquid sodium

alkali metal supplies only half the number of electrons required to fill the band, so that only the low-energy half of the band is filled, and this gives rise to the profile shown in Figure 4.1. There is therefore a fairly sharp cut-off in the occupancy of the highest energy levels, and it is the electrons at this Fermi level which are important for the chemistry, as well as for the electrical and thermal conductivity of the metals. These curves are much the same for the solid or liquid states. It is one of the basic concepts in liquid metal chemistry that (apart from the noble gases) the formation of a solution involves electron donation to, or withdrawal from, this conduction band. The energy range and to some extent the profile of the band is unique to each alkali metal, and so is the effect of temperature, so that each metal is highly selective in its solvent properties, and in the nature of reactions which are possible in that particular medium.[46]

4.3 The noble gases

These elements provide one of the few examples of solutes which undergo negligible electronic interaction with the liquid metal. Under these circumstances, solubility is very small indeed. Most of the available data relate to the solubilities of krypton and xenon in liquid sodium,[47] since these gases are produced as fission products in the fast reactor. Helium and argon[48,50,51] have also been studied because of their possible uses as cover gases.

Solubility values are listed in Table 4.2. They have been measured with sufficient accuracy despite their very low values, to establish that solubility decreases with increasing size of the noble gas atom. The value for neon should not be regarded as contradicting this rule; most of the other values in Table 4.2 were determined at the Argonne national laboratory, whereas the neon value was determined by independent Japanese workers using a different technique.[49] Results obtained in Germany[48] support the Argonne values for argon, and though they do not agree closely with the helium values, they

Table 4.2 Solubilities of noble gases in liquid sodium at 300°C (moles per litre in solution per atmosphere of gas pressure)

Gas	Solubility	Enthalpy of solution* (kJ mol^{-1})	Reference
Helium	3.4×10^{-7}	54.4	47
	1.1×10^{-7}	68.6	48
Neon	3.7×10^{-7}	54.9	49
Argon	0.48×10^{-8}	80.8	47
	0.65×10^{-8}	83.7	48
Krypton	0.4×10^{-8}	83.7	50
Xenon	0.17×10^{-8}	81.3	51

*Standard gas state: 1 mol of ideal gas at temperature T, and pressure 1 atm.

Table 4.3   Solubility of helium in liquid lithium and liquid potassium
(moles per litre per atmosphere)

Solvent	Solubility	Temperature (°C)
Lithium	3.9×10^{-7}	650
	4.7×10^{-7}	760
	5.6×10^{-7}	870
Sodium	3.4×10^{-7}	300
Potassium	5.9×10^{-5}	482
	11.8×10^{-5}	538
	12.7×10^{-5}	593
	17.1×10^{-5}	704

support the order of solubility. All the solutions obey Henry's law to at least 9 atmospheres pressure. The enthalpies of solution (Table 4.2) are positive, and the solubilities increase with increasing temperature, though to a small extent only; the temperature dependence becomes more marked with increasing atomic weight of the noble gas.

All these properties are consistent with a simple concept in which the solubility represents the number of noble gas atoms which can be accommodated in holes of appropriate size between the metal atoms or cores. An interesting model has been developed in which the difference between two work terms, i.e. the work required to create a cavity in the metal of a size appropriate to accommodate the gas atom, and the attractive work done on the gas atom in the cavity when it is polarized by the surrounding solvent, determines the excess free energy of solution of the gas.[52]

It would be satisfying to be able to add to the above information a set of solubility values for a single noble gas in each of the alkali metals, at the same temperature. Unfortunately, such a survey does not appear to have been carried out. However, some early (1965) values for helium in liquid lithium and potassium[53] would appear to be still acceptable, and are quoted in Table 4.3. In each case the solutions obey Henry's law, the solubilities increase with temperature and solubilities in potassium are much greater than in lithium. This is to be expected since holes between metal solvent atoms will become larger and more accessible to helium atoms as the size of the metal atom increases, so that the results are consistent with the general principles outlined above. Note that the value quoted for sodium (Table 4.2) is for a much lower temperature.

4.4   Diatomic molecules

An immediate distinction can be drawn between those solutes which undergo some electronic interaction with the metal concerned, and those which do not, and a useful guideline is provided by the simple inorganic

48

chemistry of the binary systems. Where the two elements are known to form a stable compound, the non-metal is likely to have an appreciable solubility in the liquid metal; where they do not, solubility is likely to be negligible, as in the case of the noble gases. For example, lithium (solid or liquid) reacts readily with nitrogen to form the stable product Li_3N, and nitrogen also shows appreciable solubility in liquid lithium. In contrast, sodium and nitrogen do not yield a corresponding nitride, and nitrogen has negligible solubility in liquid sodium.

The behaviour of diatomic molecules will be discussed by reference to the chemistry of solutions of hydrogen, oxygen and nitrogen; research on these solutions has been stimulated by the needs of the nuclear industry, and a fairly clear picture of the nature of dissolved species of the elements is now emerging.

Hydrogen

Pressure measurements   Solutions of hydrogen in the alkali metals possess a unique advantage in that they have a readily measurable vapour pressure of hydrogen in equilibrium with the solution, and this can be utilized to study the nature of the solution. A typical pressure–concentration curve for solutions of hydrogen in sodium[54] is shown in Figure 4.2.

Consider a closed vessel containing liquid sodium under a vacuum. When some hydrogen is admitted, part will dissolve and part will remain in the

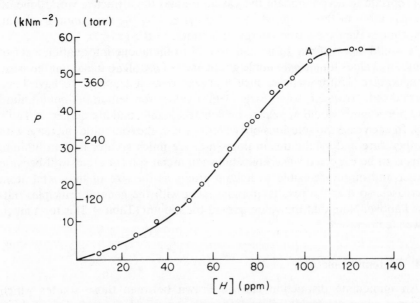

Figure 4.2   Solutions of hydrogen in sodium at 404°C. (From A. C. Whittingham, *J. Nuclear Mat.*, 1976, **60**, 119)

Table 4.4   Plateau pressures over saturated solutions of hydrogen in sodium

Temperature ($^\circ$C)	Plateau pressure (torr)	(kNm^{-2})	Hydrogen solubility (ppm)	(atom % H)	Sievert's constant k_s (atom % H m/$(kN)^{1/2}$)
337	44	5.87	33.5	0.0763	0.0285
359	94.5	12.60	53.5	0.122	0.0300
385	241	32.20	83.0	0.189	0.0273
404	420	56.00	110.0	0.250	0.0293

vapour phase, and Figure 4.2 shows the distribution at 404°C when equilibrium is reached; equilibrium requires times of the order of one or two hours, depending on temperature and pressure. For example, a solution containing 0.1 atom % hydrogen (44 ppm) has an equilibrium vapour pressure of 90 torr ($12kNm^{-2}$). If some hydrogen is now withdrawn from the system, the concentration in solution is decreased and a new and lower equilibrium pressure is set up; if hydrogen is added to the system, both the concentration and equilibrium pressure will increase. The sodium surface remains bright, and no solid sodium hydride is present. The equilibrium pressure increases with increasing dissolved hydrogen concentration until the liquid sodium is saturated with hydrogen (at point A, Figure 4.2). Addition of further hydrogen results in the precipitation of solid sodium hydride; the curve reaches a plateau, and no further increase in equilibrium hydrogen pressure can be achieved at that temperature. Table 4.4 shows the variation of this plateau pressure, and the hydrogen solubility in sodium, with temperature. This plateau pressure reaches one atmosphere at about 425°C (compare 850°C for hydrogen in lithium, and 390°C for hydrogen in caesium).

These results may be used to determine whether hydrogen dissolves as a diatomic or monatomic unit. If hydrogen dissolves according to the equilibrium.

$$H_2(gas) \leftrightharpoons 2H(solution) \qquad (4.1)$$

then the system should obey Sievert's law, expressed as

$$[H] = k_s P_E^{1/2}$$

where $[H]$ is the concentration of dissolved hydrogen, P_E is the equilibrium hydrogen pressure, and k_s is the Sievert's constant. The results in Figure 4.2 are shown in Figure 4.3 in terms of $[H]$ against $P_E^{1/2}$, and it is clear that $[H]$ is directly proportional to $P_E^{1/2}$ over most of the concentration range. These results have been published by Whittingham,[54] but there are many previous publications listed in his paper which leave no doubt that the sodium–hydrogen system does indeed obey Sievert's law over a wide range of temperatures.

Some values for the Sievert's constant are included in Table 4.4. As the hydrogen in solution approaches the saturation value, some deviation from

50

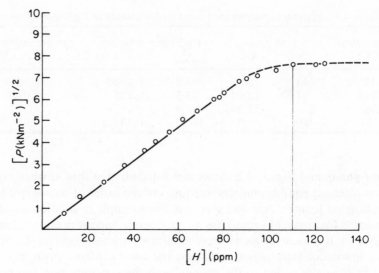

Figure 4.3 Sievert plot for solutions of hydrogen in sodium at 404°C.
(From A. C. Whittingham, *J. Nuclear Mat.*, 1976, **60**, 119)

Sievert's law occurs; this is to be expected, since any hydrogen–hydrogen interactions which might then arise will cause the activity to differ appreciably from the concentration of dissolved hydrogen.

This, and other evidence mentioned below, leave no doubt that when hydrogen enters the liquid metal it does so as the monatomic species, and it only remains to consider whether this species is the hydrogen atom, or whether the atom gains or loses an electron, i.e. whether it should be represented as H, H^+ or H^-. The answer is not immediately obvious since hydrogen in some transition metals may take any of these forms. No definite experimental proof is available, but all the chemistry of hydrogen dissolved in the alkali metals is consistent with its presence in solution as the H^- ion. Thus, the addition of hydrogen beyond the solubility limit causes precipitation of the ionic solid Na^+H^-, and other metals dissolved in sodium are separated by precipitation of their hydrides on addition of hydrogen. In all the following discussions we shall therefore regard the solution of hydrogen as involving the process

$$H_2\,(\text{gas}) \; \underset{-2e}{\overset{+2e}{\rightleftharpoons}} \; 2H^-\,(\text{solution}) \tag{4.2}$$

in which the hydrogen molecule is first adsorbed, then dissociates to atoms. Electrons from the conduction band of the medium then convert H atoms to H^- ions, which dissolve as such.

There are two points concerning terminology which should be clarified at this stage. Firstly, it is common to speak of these solutions as 'solutions of hydrogen' rather than 'solutions of sodium hydride'. This arises largely because the solutions, which are very dilute, are prepared easily, and with highest accuracy, by addition of measured volumes of hydrogen gas to sodium. The use of this terminology is also encouraged by the fact that the process 4.2 above is reversible, so that the solute can be recovered, by evacuating, as hydrogen gas. Needless to say, solutions prepared by addition of the same amount of hydrogen in the form of gas or as solid sodium hydride are identical in all respects. This common parlance extends to solutions of other non-metals, and carries no implication as to the species in solution.

The second point is a more fundamental one. For want of an alternative expression, we use the term 'ion' to describe the H^- species in solution in a liquid metal. This term is traditionally associated with a charged species in an electrically neutral molecular medium, and thus has various characteristic properties associated with it such as migration in a potential gradient. It does not follow, however, that H^- in an environment of free electrons will necessarily possess any of these properties, and in liquid metal chemistry implies only that an electron is largely localized on the H atom. The extent of this localization cannot be defined precisely and will vary with the liquid metal involved. However, considering the low work functions of the alkali metals, electron localization on solute atoms will be more complete in the alkali metals than in any other metallic solvents, and in interpreting their chemistry we shall assume that such localization is indeed complete.

Some information on the mechanism of hydrogen absorption by sodium is available from kinetic measurements. Studies of the rate at which equilibrium is achieved have shown that the rate of hydrogen absorption is directly proportional to the pressure, i.e.

$$\frac{-\mathrm{d}P}{\mathrm{d}t} = k_1 P$$

where $-\mathrm{d}P/\mathrm{d}t$ is the rate of change of hydrogen pressure at pressure P, and k_1 is the rate constant. The reaction is therefore first-order with respect to hydrogen pressure, and the rate constant is independent of the dissolved hydrogen concentration, irrespective of whether the sodium is saturated with hydrogen. This indicates that the rate-determining step involves molecular hydrogen, and thus the process.

$$H_2 \xrightarrow{\ k_1\ } 2H_{ads} \xrightarrow[2e]{\ k_2\ } 2H^- \tag{4.3}$$

occurs at the metal surface. It is not yet clear whether electron transfer plays any integral role in the rate-determining step, i.e. whether k_2 has any kinetic significance. First-order kinetics has also been observed for the evolution of hydrogen from sodium. This cannot be regarded as due to the converse of 4.3, as this would lead to second-order kinetics, and it has been suggested[54] that the rate-determining step in hydrogen evolution is of the type

$$H^-_{solution} \rightarrow H^-_{adsorbed}$$

First-order kinetics has also been observed for both absorption and evolution of hydrogen in a sodium (22%)–potassium(78%) mixture at 380°C.[55]

Using techniques described in earlier chapters by means of which a gas can be brought into contact with a clean metal surface, it has now been possible to obtain reaction rates of hydrogen with each of the alkali metals (except rubidium), and from these to determine activation energies for each reaction. In each case, first-order kinetics have been found. The first-order rate constant k_1 is related to the absolute temperature T by the Arrhenius expression

$$k_1 = A e^{-E/RT}$$

where E is the activation energy and A is the pre-exponential factor. From this expression

$$\ln k_1 = \frac{-E}{RT} + C$$

where C is a constant, so that a plot of $\ln k_1$ against $1/T$ gives a straight line of slope E/R.

Reaction rates vary widely, and relative values (referred to the rate for sodium) are given in Table 4.5. These rates correlate with the activation energies E, though the correlation is not precise because of variations in the pre-exponential factor A. Nevertheless, there is no reason to doubt that in all cases H^- is produced in solution, and that the same reaction mechanism operates. As expected from the rates for the pure metals and the E values, addition of lithium to sodium progressively increases the rate of reaction with hydrogen, and the effect is appreciable for the addition of as little as 1.1 atom % lithium in sodium. Addition of barium increases the reaction rate to a much greater extent than does lithium.

Cryoscopy   An elegant proof that hydrogen dissolves in liquid lithium as the monatomic species has been achieved by measurement of the depression of freezing point.[20,60,61] Liquid lithium was purified by gettering with yttrium sponge to give a melting point of 180.49°C. The liquid was circulated by an electromagnetic pump in a steel vessel, small amounts of hydrogen added, and the freezing points determined from cooling curves using very slow

Table 4.5   Activation energies and relative rates for the reactions of hydrogen with the liquid alkali metals

Metal	Relative reaction rates at 250°C	Activation energy E (kJ mol^{-1})	Reference
Li	76	52.8	58
Na	1	72.4	54
K	6.7	66.5	57
Cs	59	42.4	56
Na–Li (5%)	≫Na	64.1	59
Na–Ba (5%)	≫Na	45.0	59

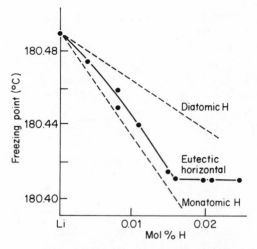

Figure 4.4   Depression of the freezing point of
liquid lithium by dissolved hydrogen

cooling rates of about 15°C per hour. The results are shown in Figure 4.4; it will be noted that the maximum freezing point variation is only 0.08°C, and the self-consistency of the results is a testimony to the skill and care of the experimenters. Assuming an ideal solution, the freezing point depression is related to solute concentration by the expression

$$T_m - T = \frac{RT_mT(x - x_s)}{\Delta H_f} \qquad (4.4)$$

where T_m and T are the freezing points of pure lithium and the solution, R is the gas constant, and ΔH_f is the latent heat of fusion of lithium (2.93 kJ mol.$^{-1}$); x is the total (mole fraction) hydrogen concentration, and x_s the solubility of hydrogen in the solid metal which separates. Calculated values for the depression $T_m - T$ will depend on whether x is expressed in [H] or [H$_2$] units, and theoretical lines which could be followed by monatomic or diatomic species are shown as broken lines in Figure 4.4, assuming the solid solubility x_s to be negligible. The experimental results lie close to the monatomic [H] line, and the inclusion of a small value for x_s will raise this line even closer to the experimental results. If the deviation between the monatomic [H] line and the experimental depressions is regarded as due entirely to solid solubility, the above expression can be used to calculate x_s. On this assumption, the solubility of hydrogen in solid lithium at 180.4°C is 0.002 mol %.[60]

Oxygen

Oxygen dissolves in all the alkali metals. The metal oxides are more thermally stable than the hydrides, with the result that the vapour pressure of

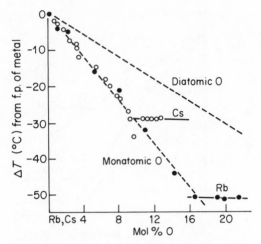

Figure 4.5 Depression of the freezing point of
liquid caesium and liquid rubidium by dissolved
oxygen

oxygen over solutions of oxygen in the liquid alkali metals is negligible. Methods for studying atomicity in solution based on isothermal pressure measurements which have been so useful for hydrogen solutions are not therefore available for oxygen solutions, but sufficient evidence is now available to indicate that oxygen is present in solution as O^{2-} ions. Because of the high solubility of oxygen in rubidium and caesium, freezing point depressions can be measured with precision; results for these two metals[62,63] are shown in Figure 4.5, together with the calculated values for monatomic and diatomic solute species, and leave no doubt that oxygen is monatomic in solution. The chemistry of oxygen solutions in lithium, sodium and potassium is entirely consistent with the presence of monatomic oxygen in these metals too.

A valuable series of measurements of Hall coefficients on solutions of oxygen in caesium has been carried out by Kendall[64] to determine the change in the number of electrons in the conduction band of liquid caesium on addition of oxygen. The results, carried out at temperatures of 25–40°C and 2.8–16.2 mole % oxygen, agreed closely with the free electron values obtained on the assumption that two electrons are removed from the conduction band of liquid caesium by each oxygen atom dissolved, and provided unique evidence that the oxygen is in the O^{2-} ion state in solution. The question as to whether the O^{2-} ions were discrete, or were associated with two caesium ions in the form of Cs_2O molecules, was resolved by nuclear magnetic resonance. If O^{2-} ions separate, then all caesium ions are in a similar environment and only one caesium resonance would be expected; if Cs_2O molecules are present, two resonances should exist. In fact, only one resonance was observed. For solutions of oxygen in the alkali metals, we may therefore write

$$O_2(gas) \xrightarrow{4e} 2O^{2-} (solution)$$

Theoretical solvation models[65,66] are consistent with the presence of O^{2-} in solution, and models involving calculation of the screening effect of conduction electrons on dissolved non-metals also predict that O^{2-} ions are stable.[67] In the course of the Hall coefficient experiments it was observed that the characteristic golden colour of pure caesium darkened on addition of oxygen, and at the highest concentrations became dark bronze in appearance. It would be interesting to know where the energy levels of the electrons removed from the conduction band occur in relation to the Fermi surface, and what new unoccupied levels are introduced by added oxygen.

With molecular liquids, direct evidence concerning the sign and magnitude of the charge on a dissolved species is readily available from the extent and the direction of migration of the solute under an applied potential; movement to the anode or cathode implies a negative or positive charge respectively, and characteristic transport numbers can be assigned. When an electric field is applied to a liquid metal solution, some electromigration again occurs, but the interpretation of the results is by no means so clear in liquid metal solutions.

Many theories have been advanced to explain electromigration, and have been reviewed by Wilson,[43] Verhoeven[68] and Epstein.[69] Most theories are based on the assumption that two forces act on the solute element. One component of the resultant force is that exerted by the electric field on the ions in the medium, and the second component is an 'electron drag' resulting from electron–ion collisions. The direction of electromigration is determined by the relative magnitude of these two forces.

Experimental work has been concerned almost entirely with metal mixtures, and we shall refer to electromigration again later in this context. One of the very few measurements on a solution of a non-metal was reported by Nevsorov.[70] A direct current was passed through sodium containing a known quantity of oxygen, at 300°C and at 600°C, and the oxygen concentration was found by analysis to have increased at the anode and decreased at the cathode. The quantity of oxygen transferred, per ampere hour, was constant and in agreement with a theoretically derived value for the O^{2-} ion. There is no reason to doubt the experimental observations, but in view of the many uncertainties surrounding the interpretation of electromigration in liquid metals, it would be premature to accept these results as providing direct evidence for the O^{2-} ion in solutions of oxygen.

Nitrogen

Unlike for hydrogen and oxygen, there is no direct evidence as to the nature of the species formed on solution of nitrogen. However, on cooling a solution of nitrogen in lithium, the salt Li_3N crystallizes, and the extensive chemistry of nitrogen solutions (discussed in later chapters) is entirely consistent with the solution process

$$N_2 \, (gas) \xrightarrow{\quad\quad} 2N_{adsorbed} \xrightarrow{\quad 6e \quad} 2N^{3-}_{solution}$$

in liquid lithium.

The cryoscopy of solutions of nitrogen in lithium does not yield direct information on atomicity because of the appreciable solubility of nitrogen in solid lithium. Figure 4.6 shows the partial phase diagram for dilute solutions of nitrogen where a eutectic has now been found to exist.[20,72] If we apply the ideal solution of equation 4.4, ignoring solid solubility, the depression is inadequate even for a monatomic solute, because the hypo-eutectic liquidus is displaced as a result of appreciable solid solubility. Using equation 4.4 the extent of solid solubility can be calculated, and this is shown as region α in Figure 4.6; at the eutectic temperature it is about one third of the nitrogen concentration at the eutectic.

So far as the pure alkali metals are concerned, we need to consider lithium only, since nitrogen is virtually insoluble in sodium, potassium, rubidium or caesium. This can be rationalized by reference to Born–Haber cycles for the preparation of Li_3N and the hypothetical Na_3N (Figure 4.7), remembering the general rule that a non-metal which forms no known compound with a metal is unlikely to dissolve in that metal. The value for the enthalpy of formation of Li_3N is consistent with the ready reaction of nitrogen with lithium. In the sodium cycle, the value of ΔH° appears to be very small but negative if we use the quoted value of the lattice energy of Na_3N, which is calculated from the Kapustinskii equation. From similar cycles for the heavier alkali metals, ΔH° becomes large and positive because of the rapid decrease in the calculated value for the lattice energy. We would not therefore expect reaction of

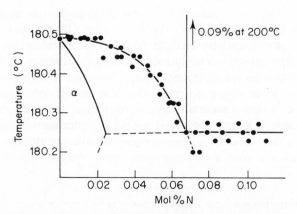

Figure 4.6   Freezing point depression and solid solubility α in the lithium–nitrogen system

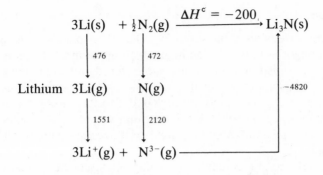

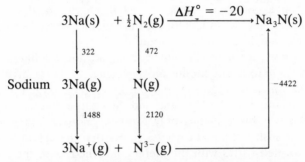

Figure 4.7 Born–Haber cycles for the formation of Li_3N
and Na_3N (kJ units)

nitrogen with potassium, rubidium or caesium, but these cycles indicated that reaction with sodium, and hence solution in liquid sodium, might just be possible. The nitrogen–sodium system has been examined with great care because of the potential value of nitrogen as a cover gas, and measurements using ^{15}N-enriched nitrogen have confirmed that nitrogen is virtually completely insoluble in sodium.[71] The isotopic composition of the minute amount of gas recovered from solution indicates that nitrogen dissolves as the diatomic N_2 species, and the difficulty in accommodating this dumb-bell shaped molecule in holes in the liquid matrix accounts for a solubility $(0.6 \times 10^{-10}$ mol/litre/atm), even lower than that of the noble gases.

The scope of nitrogen chemistry is not so restricted as would appear from the above, however, since nitrogen does dissolve, as the N^{3-} species, in mixtures of lithium with any of the other alkali metals. Again, the group II metals calcium, strontium and barium form nitrides M_3N_2 of high stability and give mixtures with sodium which will dissolve nitrogen. Many instructive experiments which have been fundamental in the development of liquid alkali metal chemistry have been performed using mixtures of sodium with lithium, calcium, strontium or barium, and these will be discussed in later chapters.

58

4.5 Carbon

The natural extension of the series H^-, O^{2-}, N^{3-} would lead us to consider that carbon might exist in solution in an alkali metal in the form of C^{4-} ions. In traditional chemistry this unit is not encountered under ambient conditions because the localization of four negative charges on a single, isolated atom would lead to an unacceptably high charge density. However, in the free electron environment in which a carbon atom would exist in solution in an alkali metal, the possibility that C^{4-} might exist becomes more feasible. There is another general concept which is relevant here. We have chosen to define the liquid alkali metal medium as consisting of closed-shell metal cores plus free electrons; non-metallic solutes such as H^-, O^{2-} and N^{3-} have collected into their orbitals sufficient electrons to form closed shells also, thus extending the free electron model for the pure metal also to the solutions. The existence of C^{4-} would extend this model.

However, a new factor arises with carbon, i.e. the readiness with which carbon catenates to form polyatomic units, which is so characteristic of carbon chemistry and which distinguishes it from hydrogen, oxygen and nitrogen. The simplest such unit is the C_2^{2-} ion, and the charge density on C^{4-} can be reduced in this way by spreading only two charges over a larger ionic unit. The general chemistry of metal carbides shows that the extent to which C or C_2 units are formed varies with the metal concerned; with the alkali metals the only known binary compounds have the formula M_2C_2 and contain the C_2^{2-} (acetylide) ion, and these arguments point to the possibility that carbon in solution in a liquid alkali metal exists as the C_2^{2-} ion, or that both monomeric and dimeric units are present in some sort of equilibrium.

Much careful research is being devoted to this problem. It has practical importance because carbon transport from one part of a metal containment vessel to another by solution in a liquid metal (e.g. in a sodium circuit) can modify the properties of structural components. Experiments which lead to definitive results are difficult to plan, and this is aggravated by the very small values of carbon solubility, particularly in sodium. Considerations of dissolved species are closely involved with solubility determinations, so that solubility values will be included at this stage (Table 4.6). Results are only available for lithium and sodium, but the very small values for sodium

Table 4.6   Solubility of carbon in liquid lithium and sodium

Temperature (°C)	Liquid lithium		Liquid sodium	
	(atom ppm)	(wt ppm)	(atom ppm)	(wt ppm)
200	2	3	0.38×10^{-6}	0.2×10^{-6}
400	66	114	10.7×10^{-2}	5.6×10^{-2}
600	460	800	10.5	5.5
800	1550	2700	184	96

compared with lithium would suggest that solubilities of carbon in potassium, rubidium and caesium are negligible.

The solubility of carbon in liquid lithium is represented by the equation[73]

$$\ln S = -1.100 - 5750/T$$

where S is the atom fraction of carbon in solution; some specific values derived from this equation are given in Table 4.6. Solutions were prepared by addition of Li_2C_2 to lithium, and solubilities determined by direct sampling of the solution. Samples were hydrolysed, and carbon content determined from the evolved acetylene. When carbon is recovered from liquid lithium it separates as the compound Li_2C_2, and there seems to be no reason to doubt that carbon dissolved in lithium is in the form of C_2^{2-} ions. At very high temperatures, or at high dilution, it is possible that some dissociation of the C_2^{2-} species into monomeric carbon units may occur, but at present there is no evidence for this.

The state of carbon in sodium is a more difficult problem, and has been studied by a number of research groups.[74-79] All authors agree that present conclusions are tentative, but the available evidence is now sufficient to imply that the general conclusions are correct. It is difficult, in a short account, to do justice to the elegance of the techniques involved, but the approach has been as follows.

In view of the very low solubility of carbon in sodium (Table 4.6), the activity of carbon in solution (a) may be expressed, assuming Henry's law, as

$$a_{Na} = \frac{x_{Na}}{{}_1S_{Na}} \qquad (4.5)$$

where x_{Na} is the concentration of carbon in sodium at temperature T and ${}_1S_{Na}$ is the saturation solubility of carbon in sodium at that temperature. This equation assumes that carbon is present in sodium as the monatomic species. If, however, carbon is present as a diatomic species, then Henry's law becomes

$$a_{Na} = \left(\frac{x_{Na}}{{}_2S_{Na}}\right)^{1/2} \qquad (4.6)$$

where ${}_2S_{Na}$ refers to the saturation solubility of dimeric carbon. If both monatomic and diatomic species are present in solution, equations 4.5 and 4.6 may be combined, so that

$$x_{Na} = {}_1S_{Na}a_{Na} + {}_2S_{Na}a_{Na}^2 \qquad (4.7)$$

or

$$x_{Na}/a_{Na} = {}_1S_{Na} + {}_2S_{Na}a_{Na}$$

If, now, values for concentration x_{Na} and activity a_{Na} can be found, x_{Na}/a_{Na} can be plotted against a_{Na}. The form of this graph should then give a measure of the relative significance of ${}_1S_{Na}$ and ${}_2S_{Na}$.

60

Values for x_{Na} and a_{Na} have been determined[74-77] by taking advantage of the principle that when two homogeneous phases containing the same solute reach equilibrium, the activity of the solute is the same in each phase. Sodium was therefore contained in a sealed nickel ampule, which was then exposed to carbon sources (e.g. $CO-CO_2$ mixtures) outside the ampule. Carbon diffuses throughout the system by penetration of the ampule walls, and at equilibrium

$$a_{Na} = a_{Ni} = x_{Ni}/S_{Ni}$$

where a_{Ni}, x_{Ni} and S_{Ni} represent activity, concentration and saturation solubility of carbon in metallic nickel. S_{Ni} is a known quantity. The concentrations of carbon in the ampule and in the sodium were determined by combustion and measurement of the CO_2 produced; the use of carbon-14 in this analysis leads to higher precision.[75]

Figure 4.8 shows a plot of x_{Na}/a_{Na} against a_{Na}, obtained for a series of carbon concentrations at 710°C and at 760°C.[75] If equation 4.5 applied, the results should lie along a line parallel to the a axis, whereas if equation 4.6 applied the results should lie on a straight line passing through the origin. In fact, the slope of the line $(_2S_{Na})$ is much greater than $_1S_{Na}$, which would appear as an intercept on the ordinate axis, implying that the dimeric species is dominant over a wide concentration range. As the measurements did not extend below an a value of 0.3, the possibility remained that a small intercept might exist. This research group concluded that some small degree of dissociation to the monomeric species might occur, particularly at low carbon concentration and high temperatures.

Data such as that in Figure 4.8 may be extrapolated to unit activity to give the saturation solubility S_{Na}. This has been done assuming carbon in solution to be monomeric[74] and dimeric.[75,76] Thompson[78] believes that the best

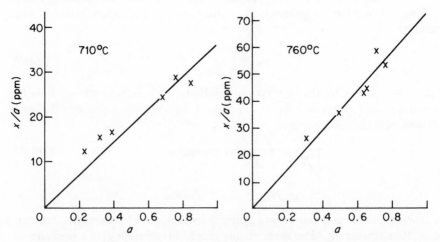

Figure 4.8   Concentration–activity relationship for carbon in liquid sodium

expression for the solubility of carbon in sodium is an average of these results, which gives

$$\log_{10} S_{Na} = 7.45 - 5858/T \tag{4.9}$$

where S is expressed in weight parts per million. This equation is derived from results over the temperature range 500–980°C, and had been used (with some extrapolation) to give the solubility values quoted in Table 4.6.

A more recent study[78] of the variation of activity with temperature has given further support for the presence of dimeric species. Experimental values for carbon activity, at various temperatures, for a carbon solution originally saturated at 500°C (determined using nickel ampoules) clustered around the broken line in Figure 4.9. We may also write equation 4.7 in the general form

$$x_{Na} = k_1 a_{Na} + k_2 a_{Na}^2 \tag{4.10}$$

where k_2/k_1 is the ratio of dimer to monomer concentrations. If k_2/k_1 is given the value n, then by solving equation 4.10 we may express the carbon activity in terms of concentration x_{Na}, the saturation solubility S_{Na} and n. Since S_{Na} varies with temperature, it is then possible to draw theoretical plots of a_{Na}

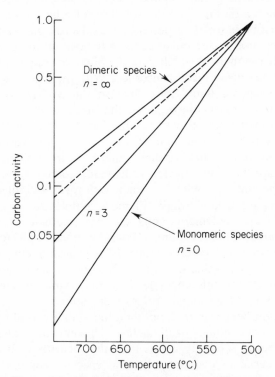

Figure 4.9   Carbon species in liquid sodium, from activity–temperature measurements

against temperature for various values of n. The calculated lines for $n = \infty$ (dimeric species) and $n = 0$ (monomeric species) are also given in Figure 4.9, and it is deduced that a dimer/monomer ratio of the order of 8 to 10 is feasible.

The predominance of the dimeric species also receives support from several theoretical studies. Johnson et al.[79] determined thermodynamic properties for the compound Na_2C_2, and from the solubility of this compound in sodium they derived an equation for the solubility of graphite in sodium. Only by assuming that carbon dissolved as C_2 were they able to obtain agreement with solubilities reported by other workers. Thompson[80] has derived as solvation model for solutions of non-metals in liquid sodium from which the solvation enthalpy can be calculated. This model has been successful for many solutions, but only by regarding carbon as the acetylide ion in sodium solution could satisfactory agreement be obtained between calculated and experimental solvation enthalpies. The dimer C_2 in solution has also been found to be more stable, by about 150 kJ mol^{-1}, relative to the monomer.[81]

4.6 Other non-metals

At the time of writing, there appear to be no published researches giving specific information on the nature of solutions of other non-metals, and any speculation must be based on experience with the four non-metals discussed above. The most useful clue might eventually turn out to be the properties, and particularly the thermal stability, of the binary compound formed by the non-metal with the alkali metal concerned. With the halogens, for example, Bredig[82] has examined liquid miscibility in the M–MX systems at high temperatures, but little is known about the dilute solutions of halogens in alkali metals formed at lower temperatures. The halogens might be expected to give monomeric X^- ions in solution, by analogy with hydrogen solutions; there is a possibility that iodine might form polymeric units in solution—perhaps the I_3^- unit in view of its readiness to form tri-iodides with alkali metals. Sulphur and selenium might well form solutions in which the S^{2-} and Se^{2-} ions are in equilibrium with more complex species, remembering that catenation and polymeric units are characteristic of much of the chemistry of these elements. Similar considerations apply to solutions of phosphorus and arsenic. Electrical resistivity measurements have recently been applied in the determination of the solubilities of silicon and germanium in liquid lithium,[83] and this might lead in due course to definition of the species in solution. Alkali metal borides were unknown before 1963; LiB_4 and KB_6 have now been reported, but their low stability would suggest a very low solubility in the liquid alkali metals. The limited evidence available indicates that boron solubility is less than that of carbon. Boron solutions would almost certainly contain polymeric species, but the experimental problems in investigating boron solutions would seem to be almost insuperable.

4.7 Metallic solutes

These solutes fall into two classes. The first is represented by the alkaline earth metals, all of which have some solubility in the alkali metals (see Chapter 5). Solubility increases with atomic weight of the solute, and freezing point depressions caused by calcium, strontium and barium in liquid lithium, for instance, indicate that these metals dissolve as monomeric species. The chemistry of the group II metal is not inhibited by solution in an alkali metal. Solutions of barium in liquid sodium have been used in many studies of alkali metal chemistry, and often behave as though sodium was no more than an inert diluent for barium. For example, solutions of barium in sodium react readily with nitrogen, and more rapidly with hydrogen than does sodium alone. In such reactions, using stirred solutions, reaction is not inhibited by formation of surface films which would form on solid barium. No conduction band profiles are available for such solutions, but in interpreting the chemistry of alkaline earth metal solutions, it has been satisfactory to regard such solutions as containing M^{2+} closed-shell cores, with two electrons per atom added to the conduction band of the alkali metal medium. A new conduction band will be formed, in which the energy level of the highest-energy electrons will be at least as high as that of the alkali metal medium. The group II metals are probably the only metal solutes which behave in this manner.

The second class of metallic solutes is represented by the less electropositive metals, where electrons from the conduction band can be regarded as localized on the solute atom to an extent which varies with the electropositive nature of the solute. The chemistry of the solution may then be quite different from the chemistry of the individual metals. For example, liquid lithium readily absorbs nitrogen to form the stoichiometric nitride Li_3N, but lithium–mercury solutions containing more than 50 atom % mercury show no reactivity towards nitrogen. Moreover, in the 0–50 atom % mercury range, it is only that amount of lithium in excess of the 1:1 NaHg composition which reacts with nitrogen.[84] There is no evidence for the presence of discrete NaHg molecules in the solution, and we may consider that the affinity of the mercury atom for the valency electron of the lithium atom is sufficient to lower the energy level of the latter in the conduction band below that required for reaction with nitrogen. This affinity involves only one electron per mercury atom, so that lithium which is present beyond the composition LiHg is, in effect, dispersed in a chemically inert medium. The term 'localization of electrons' will be used in interpretation of all effects such as this. In extreme cases, such as the crystalline non-metallic compound CsAu, transfer of the alkali metal valency electron to the transition metal atom is virtually complete, and the product is essentially a salt Cs^+Au^-. This compound, a white solid, exists in the CsCl cubic structure and shows n-type semiconductor behaviour; the compounds KAu and NaAu show increasing metallic character. However, there are many examples, typified by the lithium–mercury solutions, where the electronegativity difference is not so great, where electron transfer is not complete, and the metal mixture retains its metallic character. Mixtures of the alkali metals with antimony and bismuth have been widely studied in this context.

Sodium amalgam finds wide use as a laboratory reagent. Here the electronegativity difference is not so great as for Li–Hg mixtures, but electron localization again modifies the chemistry of the amalgam. Three properties are plotted against composition in Figure 4.10. The electrical resistivity of sodium (curve A) increases rapidly on addition of mercury, reaching the pure mercury value at about 35–40 atom % mercury. The high melting point of the compound $NaHg_2$ (phase diagram B) indicates considerable stability, and the fact that the resistivity curve of the liquid shows a sharp minimum at this composition also would indicate that this is the dominant ratio in the liquid. The rates at which sodium amalgams of various compositions react with glacial acetic acid are also shown in Figure 4.10 (curve C). Because of the relatively low temperature which is required for these experiments, several (and varying) phases were present in the amalgam as composition changed, so that high precision is not claimed. However, it is quite clear that the sodium–acid reaction rate is reduced vastly by addition of mercury; this rate is much greater than any dilution factor, and reaches a minimum at about the composition $NaHg_2$. The reaction rate of liquid sodium with hydrogen is

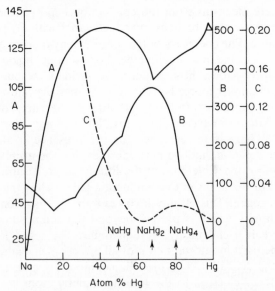

Figure 4.10   Correlation of physical and chemical properties of sodium amalgams. (A) Electrical resistivity at 350°C ($\mu\Omega$ cm). (B) Solid–liquid phase diagram (scale °C). (C) Rate of reaction with glacial acetic acid (g/s Na at 20°C, using 2 ml amalgam in 50 ml of acid)

Table 4.7   The influence of added mercury on the rate of reaction of liquid sodium
with hydrogen

Atom % Hg in Na	0	4	8	12	16	20
First-order rate constant k_1	2.57	1.79	1.21	0.80	0.50	0.32

influenced similarly by mercury. In experiments carried out at 200°C,[59] the reaction rate of hydrogen with sodium was proportional to pressure (i.e. $-dP/dT = k_1 P$) for solutions containing up to 21 atom % mercury, but the first-order rate constants k_1, decreased profoundly with increasing mercury concentration. Some values for k_1 are given in Table 4.7, which show that for solutions containing 20 atom % mercury the reaction is slower by almost an order of magnitude than that for pure sodium. A solution of 50 atom % mercury in sodium was found to be quite unreactive towards hydrogen, so that the affinity of mercury for the conduction band electrons of sodium would appear to decrease the availability of electrons of sufficient energy to dissociate the hydrogen molecule.

Mention has already been made of the electromigration of oxygen in liquid sodium, and this technique has been applied extensively to the study of liquid metal mixtures. Epstein[69] has reviewed published experimental work, and has given a detailed discussion of the possible physical factors which can influence the direction of electromigration in liquid metal mixtures. A high proportion of published experimental work is concerned with mixtures of the less electropositive metals (e.g. Pb, Hg, Bi, Ga, In, Sn, Sb) from which no simple picture emerges. However, where mixtures of these metals with the alkali metals are concerned, the general concepts of electron localization discussed above does provide an acceptable interpretation which allows a correlation of the physical and chemical properties of these solutions. Thus, taking a series of dilute solutions of the solutes Ag, Cd, In, Sn, Sb in liquid sodium, all these solutes were found to migrate to the anode, whereas the same solutes dissolved in mercury or bismuth migrated to the cathode, and the migration velocities in sodium were up to fifty times greater than those in mercury. Similar results were observed for solutions of these solutes in various sodium–potassium alloys. These results are discussed by Epstein in terms of atomic mass, electron drag parameters, diffusion coefficients etc., but the interpretations are made without any recognition that the chemistry of these solutions is very different from that of the component metals. From the purely chemical point of view, the electromigration results may be regarded as supporting the general concept of electron localization on the less electropositive metal, as defined above.

Chapter 5

Solubilities and analytical methods

5.1 Introduction

Solubility values over a temperature range are required in order to determine thermodynamic quantities for the solutions, to assess the possible scope of experiments using particular solutions, and (in the case of the transition metals) to examine the suitability of a metal for use as containing material for the liquid metal. However, it is not the intention of this chapter to present a detailed catalogue of solubility values; rather, the general trend in solubilities across the periodic table will be discussed, and it is hoped that the references given will provide the reader with access to the source of more detailed information.

Consideration of trends in solvent power is severely limited by the patchy nature of available data. The most extensive literature relates to liquid sodium; information on liquid lithium is increasing as interest in the fusion reactor develops, but only occasionally have solubilities in potassium, rubidium or caesium been measured. Solubility values for the non-metal solutes hydrogen, oxygen, and nitrogen have been measured with satisfactory accuracy because of the technological importance of these solutions, but quoted values for metallic solutes are still subject to change, and binary phase diagrams are continually being re-examined and modified, particularly in the alkali metal-rich region. This applies particularly, of course, where the solubilities are very small, and this aspect will be discussed in more detail under the 'transition metals' heading.

Another general observation relates to trends which might be expected when the solutes lie within the same group in the periodic table. The properties of elements in the same group, or their corresponding compounds, often show a regular gradation from one element to the next, but this is not necessarily the case so far as solubilities in the alkali metals (particularly with metal solutes) is concerned. The solubilities are given by the liquidus lines in the binary phase diagrams, and the shape of the phase diagram is determined by many factors, predominant amongst which is the stability of the solid phase. The greater this stability (usually as a result of solid compound formation), the more readily will precipitation occur, and the lower will be the

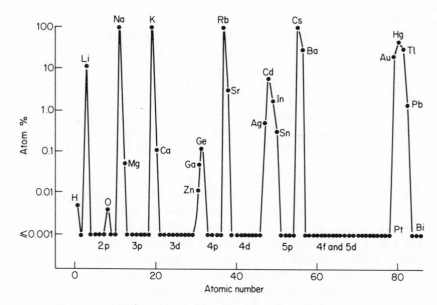

Figure 5.1   Solubilities in liquid sodium at 200°C

solubility in the liquid. The stability of binary solids does not necessarily change regularly as the solute element is replaced by another within the same group, so that regular changes in solubilities do not always occur. This will be seen especially in solute elements in the p-block of the periodic table.

Figure 5.1 gives an overall picture of the relation between the atomic number of a solute and its solubility in liquid sodium at 200°C, from which some generalizations appear. Many non-metals have small solubilities. With the metallic solutes, there is an obvious periodicity with an alkali metal at the head of each peak, so that throughout the periodic table the solubility is intimately related to the electronic structure of the solute metal. The group II metals have high solubilities. The solubility of the early transition metals is near-zero, but solubilities increase towards the end of each transition period, with zinc, cadmium and mercury at (or near) the head of a set of secondary peaks. Thereafter, solubilities decrease as we pass through the p-block groups along any given period.

Solutes will be considered in the order in which they occur in the periodic table, i.e. hydrogen, the group II metals, the transition metals, and the p-block elements taken group by group. Where appropriate, reference will also be made to methods by which solubilities have been determined.

5.2 Hydrogen

Alkali metal–hydrogen solutions have been studied for some years, by a variety of techniques. These include thermal and differential thermal analysis,[60,97] physicochemical analysis[98] and the diffusion-type[99] and electrochemical hydrogen meters.[100,101] Solubilities may also be determined by a direct sampling technique, and the values for deuterium quoted in Table 5.2 were obtained by this method. Analysis of the samples was achieved by degassing the sample at 850°C for ten days in a fused silica tube with metal liner connected to a Toepler pump; the evolved gas was collected in a known volume and its deuterium content determined by mass spectrometry.[113] Most of the recent data, however, have been derived from measurement of either the electrical resistance of solutions,[102] or of hydrogen pressures in equilibrium with dissolved hydrogen as a function of hydrogen concentration.[54,103-107]

Resistance techniques are equally applicable to all the alkali-metal solvents, and will be discussed in more detail in a later chapter; they rely on the fact that resistivity increases progessively with increasing solute concentration. The increase caused by the addition of most non-metal solutes to the liquid alkali metals is comparatively high, thereby enhancing the accuracy of the technique. The resistivity increases almost linearly with hydrogen concentration to the saturation point; further addition of gas leads to the formation of hydride, but no more hydrogen dissolves and the resistivity remains constant. The change in resistivity on saturation is unmistakeable.

Vapour pressure techniques depend on the fact that the hydrogen pressure in equilibrium with a solution increases with hydrogen content of the metal. Saturation is taken to be the concentration of hydrogen in solution corresponding to the point on the equilibrium pressure–composition isotherm where the pressure becomes invariant. The method is less precise than the resistance method, because there is more difficulty in assessing the exact concentration at which the equilibrium pressure becomes invariant. Vapour pressure measurements also involve a number of inherent problems. The equilibrium pressures of hydrogen above the alkali metal hydrides, and the corresponding solutions, increase in the order[30]

$$LiH \ll NaH \sim KH < CsH < RbH$$

reaching one atmosphere at about 850°C for LiH, 425°C for NaH, and 390°C for CsH.[108] Pressure measurements cannot be easily applied to Li–H solutions at low temperatures because of the low hydrogen equilibrium pressures generated below 700°C. The high pressures generated above solutions in the heavier alkali metals (Na–Cs), coupled with the relatively high vapour pressures of the metals themselves, necessitate the use of isothermal closed vessels in all experiments; otherwise, metal hydride collects in the colder parts of the apparatus, thus invalidating the concentration analysis. Figure 5.2 shows equilibrium pressure–composition isotherms for sodium–hydrogen solutions at four temperatures, as determined by Whittingham.[54,61] At 404°C the isotherm AB has the shape required by Sievert's law, except at the higher concentrations, and beyond point C the pressure becomes invariant. At the lower temperatures the shape of the curves is the same, but equilibrium

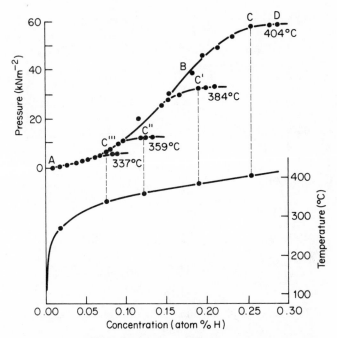

Figure 5.2 Hydrogen equilibrium pressure–concentration isotherms for sodium–hydrogen solutions, and the solubility of hydrogen in liquid sodium

pressures are lower. Points C, C', C'' and C''' represent points on the hypereutectic liquidus of the Na–NaH phase diagram, which is shown in the lower part of Figure 5.2.

Data for the heavier (Na–Cs) alkali metal–hydrogen systems are restricted to dilute solutions for reasons discussed above, and the complete M–MH phase diagram has been determined only for lithium–hydrogen solutions (Figure 5.3). That part of the diagram above about 10 atom % hydrogen has been known for some years,[97] but has been extended only recently to dilute hydrogen concentration by means of the resistivity technique.[61] Lithium hydride is the only alkali metal hydride which melts at normal working pressures before dissociating completely; the phase diagram exhibits a liquid miscibility gap which extends over a considerable proportion of the composition range (19.0 to 49.5 atom % H at 685°C), and is reminiscent of alkali metal–alkali metal halide systems.[114]

Over the temperature ranges used in the determinations, all solutions give a straight line Arrhenius plot of solubility against $1/T$, and the solubility (S, atom %) can be represented by the general equation

$$\log_{10}S = A - B/T$$

Coefficients A and B for each solution are collected in Table 5.1, and some solubility values calculated from these equations are given in Table 5.2. These

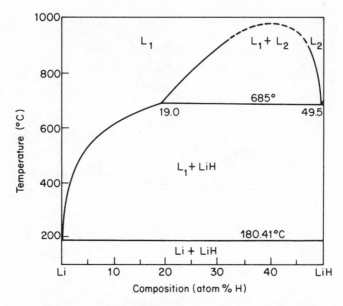

Figure 5.3   Lithium–lithium hydride phase diagram

values sometimes involve extrapolation of the solubility equations beyond the experimental limits; values at the highest and lowest temperatures may therefore be less precise, and the original references should be consulted if solubility at a particular temperature is required. This applies particularly to caesium, about which there is only one published report,[116] describing measurements carried out between 300°C and 400°C. Again, there is only one set of data for potassium solutions;[110] these show an abnormally high temperature coefficient compared with the other solutions, so that the solubility values at the high temperatures may be somewhat suspect. Allowing for these uncertainties, there is no doubt that hydrogen solubilities show a steady decrease from caesium to sodium, with disproportionately high solubilities in lithium. No measurements for rubidium solutions have been reported, but the values in Table 5.2 are sufficiently consistent to allow an estimate to be made of the solubility of hydrogen in liquid rubidium. Values for Na–K alloy (78 wt % K) are included in Table 5.2 because of the many applications of this low-melting alloy. Solubility values lie close to those of

Table 5.1   Coefficients in the hydrogen solubility equations[61,113,115]

	Li		Na	K	Cs[116]	Na–K
	(H)	(D)				
A	3.523	3.347	3.830	4.305	1.584	3.886
B	−2308	−2207	−3023	−2994	−900	−2629

Table 5.2   Solubility of hydrogen in the liquid metals[61,113,115] (atom % hydrogen or deuterium)

Temperature (°C)	Li		Na	K	Cs[116]	Na–K
	(H)	(D)				
100	—	—	0.00005	0.0002	0.148	0.0007
200	0.044	0.048	0.0028	0.0095	0.480	0.0213
300	0.313	0.313	0.036	0.120	1.03	0.200
400	1.24	1.17	0.220	0.720	1.77	0.955
500	3.44	3.11	0.832	2.70	2.63	3.06
600	7.57	6.68	2.33	7.50	3.57	7.50

potassium, so that mixing these alkali metals would not appear to modify their solvent powers. The lithium–deuterium and lithium–tritium systems are involved in the development of the thermonuclear reactor, and solubility values for deuterium in lithium are also given in Table 5.2. They are very similar to, but slightly lower than, the hydrogen values. Extrapolation of hydrogen and deuterium solubilities to the lithium–tritium system, and extension of the observed high-temperature phase relationships for all these systems, indicates that solubility will decrease in the order LiH > LiD > LiT, but that these will be only marginal quantitative differences.[117]

5.3   The group II metals

Figure 5.4 shows liquidus curves from binary diagrams for mixtures of sodium with calcium, strontium and barium, from which solubilities in liquid sodium are available. The sodium–barium curve[86,87] shows a eutectic at 4.5 atom % Ba, and an inflexion at about 200°C due to the stability at the composition Na_4Ba in the solid. The very high solubility of barium in sodium over the lower temperature range is one reason for the wide use of these solutions in the study of reactions in solution. The Na–Sr liquidus[88] (Figure 5.4) illustrates the much lower solubility of strontium, compared with barium, at corresponding temperatures. Sodium and strontium give a simple eutectic system, but the eutectic composition is reduced to 0.65 atom % strontium; there are no detectable intermetallic compounds, and strontium, like barium, shows complete miscibility with sodium in the liquid state at temperatures above their melting points. The calcium–sodium system is quite different.[89,90] The highest concentration of calcium which can be achieved is 4.1 atom % at 710°C, at which point the system separates into two immiscible liquid phases containing 4.1 and 77.9 atom % calcium. This propensity for phase separation is consistent with the very low concentration of calcium available in liquid sodium; it may also be responsible for some of the peculiar surface properties which these solutions exhibit, since calcium is known to concentrate at an interface.

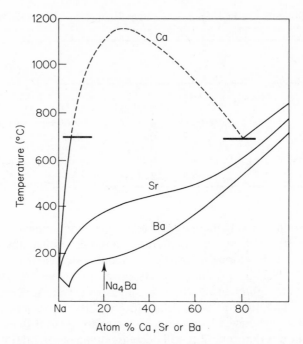

Figure 5.4   Liquidus curves for Na–Ca, Na–Sr and
Na–Ba mixtures

Magnesium and beryllium give phase systems with sodium resembling that for sodium–calcium, but the immiscibility range becomes progressively larger. Quoted values for the maximum solubilities for magnesium in sodium are 7.2[92] and 1.4[91] atom % magnesium at 638°C, but recent experience indicates that both values are much too large. The solubility of beryllium in liquid sodium is so small that beryllium has been tested for use in fuel assemblies for liquid-metal cooled reactors, where it is in intimate contact with flowing sodium.[93]

The solubilities of the group II metals in liquid lithium illustrate once again that liquid lithium is often a much better solvent than liquid sodium. Values published for the solubility of beryllium (0.23 atom % at 732°C, 1.08 atom % at 1016°C and 0.17 atom % at 1000°C) show a wide scatter, but the significant feature is that the beryllium content of saturated lithium is now measurable. Solubilities for the heavier group II metals in lithium are taken from published phase diagrams, and listed in Table 5.3. Magnesium and calcium have high solubilities, and there is no liquid immiscibility region in the phase diagrams. With increasing atomic number the phase diagrams move towards a simple eutectic pattern, and the solubility of barium in liquid lithium is quite similar to that in liquid sodium. No reliable data are available for the solubilities of group II metals in liquid potassium, rubidium and caesium. If we continue the trend suggested by the lithium and sodium values,

Table 5.3 Solubilities of group II metals in liquid lithium (atom %)

	200°C	300°C	400°C	500°C
Mg	4	18	30	43
Ca	18	48	59	68
Sr	40	58	70	81
Ba	30	45	56	66

beryllium and magnesium should be virtually insoluble, with calcium, strontium and barium moving towards lower solubility.

5.4 Transition metals

Published investigations on solubility of the early transition metals have been concerned very largely with liquid sodium, and some solubility values for metals in the first eight of the ten transition metal groups in sodium are quoted in Table 5.4. The literature up to 1970 has been surveyed in a critical review by Claar[118] and some later data and references are given by Thompson.[119,120] All the solubilities are extremely small (often less than one part per million over most of the available temperature range); and although Table 5.4 gives what have been recommended as 'best' values, it should be emphasized that values quoted by different research groups may vary by a factor of 10, or even 100. No studies on scandium, titanium and vanadium are known, but there is no reason to doubt that their solubilities are also very small. The problems involved in the determination of these very low solubility values are prodigious, and would surely never have been undertaken had it not been for the fact that these are the metals used for containment of the liquid alkali metals and that mass transfer from one part of a flowing liquid metal system to another must involve some solubility of the container metal, however small. Active isotopes of some of these metals can also be produced under reactor conditions; for example, manganese and cobalt can be produced by the reactions ^{54}Fe(n,p) ^{54}Mn and ^{58}Ni(n,p) ^{58}Co, and may thus give rise to additional decontamination problems.

The scatter of results is due as much to the problems involved in obtaining a representative sample as in the analytical methods themselves. Thus, there are difficulties in establishing true equilibrium conditions, and in maintaining the quality of the sodium (particularly its oxygen content) over the long periods of time necessary to achieve maximum solubility of the transition metal. Furthermore, metals dissolved in liquid sodium are known to adsorb on to the surface of dissimilar metals (for example, nickel is an effective getter for manganese dissolved in liquid sodium), so that the containers in which the sodium sample is transferred, and the capsule in which it is collected, have to be considered carefully from this point of view; it has sometimes been necessary to analyse for the solute metal both in the sodium sample and on

Table 5.4   Solubilities of the early transition metals in liquid sodium and liquid lithium (atom ppm)

	In sodium		In lithium	
	600°C	800°C	800°C	appm N in Li
Sc	—	—	—	—
Ti	—	—	1	55
V	—	—	—	—
Nb	0.03	0.22	1	38
			12	70
			34	260
Cr	0.05	4.0	7	150
			18	790
Mo	0.01	0.09	—	—
Mn	0.4	1.7	—	—
Fe	0.4	2.5	7	90
Co	~1	~1	—	—
Ni	0.7	1.5	400	146
			3520	220

the walls of its container. Another problem feature is the difficulty in removing all particulate matter from the sample by filtration. However, the major complication arises from the fact that, with many of the transition metals, the solubility values are dependent on oxygen concentration in the sodium, and many early results are scattered because the influence of dissolved oxygen was not realized, and the oxygen level was not controlled.[121]

The influence of oxygen is usually rationalized in terms of the formation of a ternary oxide which is more soluble in liquid sodium than is the transition metal itself. In most cases the formulae proposed for these ternary oxides when in solution are no more than speculation, but there does seem to be some firmer evidence that, in the case of iron, the ternary oxide Na_4FeO_3 may well be responsible.[122,123] This compound has been extensively reported, and is formed most readily at temperatures above 550°C and at oxygen levels near saturation. Similarly, the compounds $NaNiO_2$, $NaCrO_2$, $NaMnO_2$ and $Na_4Mn_2O_5$ may well be involved in enhancing the solubility of these metals in liquid sodium. It is interesting that carbon, hydrogen and nitrogen appear to have relatively little influence on metal solubility in sodium compared with the effect of oxygen. This is presumably a reflection of the lower stability of ternary carbides, hydrides and nitrides in liquid sodium, compared with the ternary oxides.

Nevertheless, the mechanism by which oxygen influences the solubility of transition metals in liquid sodium is still open to question. It is easy to assume (and may well be the case) that anions such as FeO_3^{4-}, NiO_2^-, CrO_2^- or MnO_2^- are formed in solution and have a higher solubility than the metal itself. However, it might then be expected that the increase in metal solubility would be related to the available oxygen in solution, whereas experimental results do not bear this out (see Chapter 17); Thompson[119] reports that at 700°C, each iron atom requires about 100 oxygen atoms for increased dissolution in sodium. There is an alternative concept which on present evidence is equally acceptable, and has analogies with solutions in molecular liquids. For example, liquid N_2O_4 (an oxidizer fuel in space propulsion) dissolves about 1 ppm of iron, but this rises readily beyond 10 ppm on addition of trace amounts of nitric acid. Being a polar molecule, nitric acid increases the dielectric constant of the medium, introduces an irregularity in charge density throughout the medium, and increases its solvent powers. The addition of oxygen to sodium may also distort the electron distribution within the liquid to a sufficient extent to enhance the solvent powers of the whole medium, without the need to assume any direct association of oxygen with dissolved iron atoms.

The traditional wet methods of analysis are now accepted as being too imprecise for these dilute solutions, and new analytical tools have been developed, such as atomic absorption spectrometry and spark-source mass spectrometry. Radiochemical methods are also finding increasing application. For example, for iron solubility the liquid sodium is allowed to come to equilibrium in iron crucibles which have been previously irradiated to ^{59}Fe, and the sodium samples then analysed for ^{59}Fe by activation analysis. Similarly ^{51}Cr, ^{54}Mn, ^{60}Co and ^{63}Ni have been used in analyses for these metals in solution. When the influence of oxygen was realized, it became necessary to work with sodium in which the oxygen content was known more precisely, especially at the lower oxygen concentrations. Much use has been made of the fact that uranium is able to getter the oxygen in sodium to below 1 ppm by weight; known quantities of oxygen can then be added as Na_2O or Na_2O_2. These experiments have led to the observation that low concentrations of oxygen are not very effective, and only have a pronounced influence on metal solubility when the oxygen content exceeds about 20 per cent of its saturation value at that temperature.

Information on the solubility of the early transition metals in liquid lithium is meagre. The few values quoted[16,124,125,128] would indicate that solubility is somewhat greater than in liquid sodium, but by a factor of only about five, so that the increased solubility in lithium is little more than the scatter in analytical values. Only in the case of nickel is the solubility in lithium much higher than in sodium. The values quoted in Table 5.4 are taken from the review by Anthrop.[128] Solubilities in lithium, as in sodium, are influenced profoundly by dissolved non-metals; available evidence on lithium solutions relates to dissolved nitrogen only, but there is no doubt that other non-metals (especially oxygen) also have an effect. The emphasis on nitrogen arises because nitrogen dissolves only in lithium, because it is difficult to remove dissolved nitrogen in order to study the effect of oxygen alone, and because nitrogen is the solute primarily responsible for corrosion by liquid lithium (see Chapter 17).

Available results (Table 5.4) are sufficient to show some relation between transition metal solubility and nitrogen concentration. If we regard the enhanced solubility as arising from an increase in the solvent powers of liquid lithium brought about by dissolved nitrogen, then the mechanism is the same as already discussed for oxygen in liquid sodium. If we regard the mechanism as involving solution of a ternary nitride, then we must consider why it is that when lithium containing both oxygen and nitrogen reacts with a transition metal, the ternary nitride is formed in preference to the ternary oxide. This arises because the stability of lithium oxide (in terms of its free energy of formation) is much greater than that of lithium nitride ($-\Delta G_{298}$ for Li_2O and Li_3N are 561 and 129 kJ/mol respectively). Oxygen therefore tends to remain in solution rather than react to form a ternary oxide, whereas dissolved nitrogen reacts with the transition metal to form a ternary nitride of higher thermodynamic stability. The compounds Li_3FeN_2, Li_5TiN_3, Li_7MN_4 (where M is V, Nb or Ta) and Li_9MN_5 (where M is Cr, Mo or W) have been considered in this connection.[129] This is an acceptable explanation of the formation of ternary nitrides as corrosion products when transition metals are corroded by liquid lithium; but it does not necessarily follow that the enhanced solubility is due to the presence of ternary nitrides as distinct entities in solution.

No values for the solubilities of these metals in liquid potassium, rubidium or caesium have been reported, but are presumed to be negligible; steel is an entirely satisfactory material for the containment of these liquids.

Towards the end of the transition block of metals, solubilities in the liquid alkali metals increase rapidly, and it is now possible to quote values at temperatures as low as 200°C (Table 5.5). Nickel has been mentioned above; the solubilities of palladium and platinum have not been measured, but a piece of platinum foil disintegrates rapidly when immersed in liquid sodium, and it seems likely that these two metals have appreciable solubilities, at least in liquid lithium and sodium. In the copper and zinc groups, solubility increases rapidly with increasing atomic number. Over the temperature range 200–600°C, the solubility of silver in liquid lithium is about ten times that in liquid sodium. Phase diagrams are available for binary systems of gold with each of the five alkali metals[94-96], and it is interesting that over the

Table 5.5 Solubilities of the later transition metals in liquid sodium
(atom % at 200°C)

Ni	Cu	Zn
2×10^{-6}	3.7×10^{-6}	0.011
Pd appreciable	Ag 0.27	Cd 4.9
Pt appreciable	Au 20	Hg 45

200–500°C range the solubility of gold in each of the five liquid metals changes very little (e.g. at 400°C, solubilities in Li, Na, K, Rb and Cs are 20, 33, 35, 35, 28 atom % respectively).

5.5 The p-block elements

In these groups there is a general trend towards higher solubilities in the liquid alkali metals as the character of the solute changes from non-metallic to metallic with increasing atomic number, but for reasons mentioned at the beginning of this chapter the trend is not a regular one. A survey of the phase diagrams for M–X systems (M is an alkali metal and X a p-block element) reveals that interest in the past has centred on the X-rich regions (the Na–S system is a typical example), and precise information on the metal-rich region is very sparse indeed. Only in the case of the group IV elements has a comparative study of solubilities been made[126,127] and this group will therefore be discussed first.

The group IV elements

Methods for determining the solubility of carbon have been discussed in the preceding chapter. For the heavier group IV elements, simultaneous measurements of temperature and resistance were made as the solutions cooled and the temperature at which saturation occurred was shown by abrupt changes in resistance, or in the rate of heat loss. Some interpolated solubility values are given in Table 5.6. We see that there is an increase in solubility from carbon to lead in both lithium and sodium, which is generally attributed to an increase in metallic character, and to the balance between the sol-

Table 5.6 Solubility of group IV elements in liquid lithium[126,127] and liquid sodium[130,131]

Temperature (°C)	Solubility (atom %)				
	C	Si	Ge	Sn	Pb
In lithium					
250°	0.0006	0.06	0.07	0.03	0.08
300°	0.001	0.19	0.22	0.08	0.25
350°	0.003	0.49	0.56	0.21	0.63
375°	0.005	0.75	0.85	0.32	0.95
In sodium					
250	4×10^{-8}	<0.01	0.20	1.3	4.0
300	3×10^{-7}	<0.01	0.35	2.7	8.1
350	2×10^{-6}	<0.01	0.65	6.2	12.0
375	5×10^{-6}	<0.01	1.0	10.2	16.0

ute–solvent electronegativity difference and the difference in their atomic size. However, tin has an anomalously low solubility in liquid lithium. The extent of the increase from carbon to lead is much greater in sodium than in lithium; the solubility of the heavier elements (Si–Pb) in lithium increases only slightly, whereas solubility in sodium increases profoundly, reaching a very high value for lead. Unfortunately there are no published results for solubilities in liquid potassium, rubidium or caesium. The solubility of carbon is smaller, by several orders of magnitude, than that of the heavier members of the group.

The factors which contribute to these effects can best be understood in terms of the simple Born–Haber cycle below; the Na–Ge system is used for simplicity, but the magnitude of the various terms would need to be modified where the precipitating phase involves other than 1:1 stoichiometry:

$$Na\mathrm{Ge(s)} \xrightarrow{\ H_{soln}\ } Na(l) + Ge(l, in\ Na)$$

$$\uparrow \Delta H° \qquad\qquad \uparrow H_{solute}$$

$$Na(s) + Ge(s) \xrightarrow{\ \Delta H_{fus}\ } Na(l) + Ge(s)$$

Here $\Delta H°$ is the standard enthalpy of formation of the precipitating phase and ΔH_{fus} the enthalpy of fusion of sodium (2.60 kJ g atom^{-1}). H_{soln} and H_{solute} are the partial molar enthalpies of solution of the element relative to the precipitating phase, and of the solute element alone.

Values for H_{soln} and H_{solute}, and the composition of the precipitating phases, are given in Table 5.7. Cycles for silicon and germanium cannot be completed since $\Delta H°$ is unknown for NaSi and NaGe. Carbon dissolves endothermically in both solvents, whereas all the heavier elements dissolve exothermically; this, and the relatively large values for H_{soln} for carbon, can be related to its low solubility and its particularly large bond energy. The high solubility of lead in liquid sodium has been noted, and sodium–group IV element phase

Table 5.7 Enthalpies of solution (kJ g atom^{-1}) for group IV elements in liquid lithium[127] and liquid sodium[130]

	In lithium			In sodium		
Solute	Precipitating phase	H_{soln}	H_{solute}	Precipitating phase	H_{soln}	H_{solute}
Carbon	Li_2C_2	47.8	15	Na_2C_2	105	112
Silicon	$Li_{22}Si_5$	56.3	−104	NaSi	—	—
Germanium	$Li_{22}Ge_5$	55.1	−113	NaGe	44	—
Tin	$Li_{22}Sn_5$	56.5	−126	$Na_{15}Sn_4$	47	−50
Lead	$Li_{22}Pb_5$	55.9	−115	$Na_{15}Pb_4$	30	−64

diagrams show a eutectic in the sodium-rich region only in the case of lead. This generates disproportionately high solubilities at low temperatures, and gives rise to the abnormal value for H_{soln}

The major factor influencing solubility is the stability of the solid phase in equilibrium with the solution, so that solubility trends can be profitably discussed only in the case of systems having precipitating phases of the same stoichiometry. For this reason, general trends in the sodium solutions would not be easy to define because of the several precipitating phases involved, even if all experimental data were available. The lithium solutions are better in this respect since the precipitating phase $Li_{22}X_5$ is common to all elements except carbon, and we are justified in regarding the low solubility of tin as anomalous. An interesting correlation can therefore be drawn between the solubility and the anomalously high enthalpy of formation of $Li_{22}Sn_5$; where X = Si, Ge, Sn or Pb, the enthalpy of formation $-\Delta H°$ of $Li_{22}X_5$ is 735, 775, 845 and 791 kJ mol^{-1} respectively.

The group III, V and VI elements

Table 5.8 presents a survey of the published literature on solubilities of elements in groups III, V and VI. It is evident that there are insufficient data available to allow any general conclusion to be made regarding solubility trends within any one group. The values shown in parentheses are obtained from extrapolations to the alkali metal axis in published phase diagrams,[94-96]

Table 5.8 Solubilities of elements of group III, V and VI in the liquid alkali metals at 200°C (atom %)

	Li	Na	K	Rb	Cs
Group III					
B	0	0	0	0	0
Al	(3.5)	0.001	0	0	0
Ga	—	(0.1)	0	—	—
In	2	3	—	—	—
Tl	3	32	10	—	—
Group V					
N	0.086	0	0	0	0
P	—	—	—	—	—
As	—	—	(2)	—	—
Sb	—	0.0005	2	(0.1)	(0.1)
Bi	(2)	0.0005	(2.5)	—	(4.5)
Group VI					
O	0.0009	0.004	0.67	24.2	25.1
S	—	—	—	—	—
Se	—	—	—	—	—
Te	—	(1.5)	—	—	—

(0 = very slight or no solubility; − = no information).

Table 5.9   Solubilities of nitrogen and oxygen in the liquid alkali metals (atom %)

Temperature (°C)	Nitrogen in lithium	Oxygen in:				
		Li	Na	K	Rb	Cs
50	—	—	—	—	—	19.7
100	—	—	0.0004	0.24	21.0	20.3
150	—	—	0.0015	0.44	23.3	22.7
200	0.086	0.0009	0.0039	0.67	24.2	25.1
300	0.416	0.0099	0.0185	1.46	26.4	26.2
400	1.450	0.051	0.0534	3.70	31.2	27.6

and experience has shown that such values are usually too high. Results in parentheses form a large proportion of the total results shown in Table 5.8, so that reliable values for solubilities in this area of the periodic table are scarce indeed. Furthermore, the precipitating phases would appear to vary widely within each group. In group III, large solubility is observed in the heaviest member of the group (compare lead in group IV), and this is associated with the presence of a eutectic which appears in the phase diagrams for thallium at 1 and 7 atom % thallium in lithium and sodium respectively.

The only elements which have been studied in detail are nitrogen and oxygen, because of the important role which these elements have played in alkali metal chemistry, and solubility values over a wide temperature range are available (Table 5.9).[102,109] Nitrogen dissolves only in liquid lithium, and solubility increases with temperature, giving a value of 39.2 kJ mol^{-1} for the partial molar enthalpy of solution, H_{soln}, with respect to the precipitating phase Li_3N. The practical importance of these values lies in the purification of the metal; the nitrogen content of lithium can be reduced to 0.08 atom % by filtration just above the melting point (180.5°C). Oxygen is less soluble in lithium than is nitrogen (H_{soln} is 52.5 kJ mol^{-1} with respect to Li_2O), and also has a low solubility in sodium. Impurity oxygen can, therefore, also be removed from both lithium and sodium by filtration near the melting point. Oxygen solubility increases from lithium to caesium, and with potassium, rubidium and caesium is so high that, even allowing for the lower melting points of the metals, filtration becomes an ineffective method of purification. The high solubilities of oxygen in liquid rubidium and caesium are quite remarkable. The molar enthalpies of formation $-\Delta H_{298}$ of the monoxides Li_2O, Na_2O, K_2O, Rb_2O and Cs_2O are 597, 421, 362, 330 and 318 kJ mol^{-1} respectively, so that as the thermodynamic stability of the monoxides decreases there is an increasing tendency for oxygen to remain in solution. However, in the case of rubidium and caesium there are further precipitating phases (Rb_3O, and Cs_7O, C_4O, $Cs_{11}O_3$) which are likely to have lower stabilities than the monoxides and which will further enhance the oxygen content of the liquid metal at equilibrium with these solid phases. This is illustrated in the phase diagrams for the Rb–O (Figure 5.5) and Cs–O (Figure 5.6) systems.

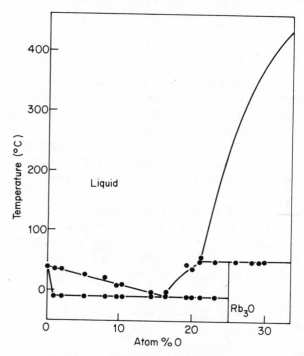

Figure 5.5   Rubidium–oxygen phase diagram[132,133]

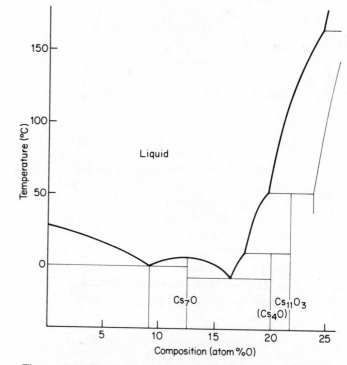

Figure 5.6   Caesium–oxygen phase diagram.[133] (This is a simplified form of the diagram published by A. Simon, *Z. Anorg. Chem.*, 1973, **395**, 310)

The halogens

When a halogen reacts with, and dissolves in, an alkali metal, the halide ion is produced in solution, so that a solution of (say) chlorine in sodium is identical with a solution of the salt NaCl. With each halogen the same simple salt MX forms the precipitating phase, so that solutions of the halogens are not complicated by many of the factors which arise with solutes elsewhere in the periodic table. In consequence, we might now expect to find a greater regularity in the solubility pattern, and indeed this is the case.

Almost the whole of our knowledge on alkali metal–metal halide systems is due to the extensive work of Bredig and co-workers.[134-137] The phase diagrams for mixtures of a liquid metal and the corresponding molten halide are dominated by large miscibility gaps. This is illustrated in Figure 5.7 for metal–metal fluoride systems, and shows that complete miscibility between the two liquids is achieved at progressively lower temperatures as the metal

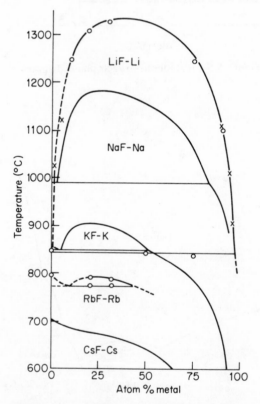

Figure 5.7 Alkali metal–alkali metal fluoride systems. (From M. A. Bredig in *Molten Salt Chemistry* (Ed. M. Blander), Interscience, 1964); reproduced by permission of John Wiley & Sons, Inc.

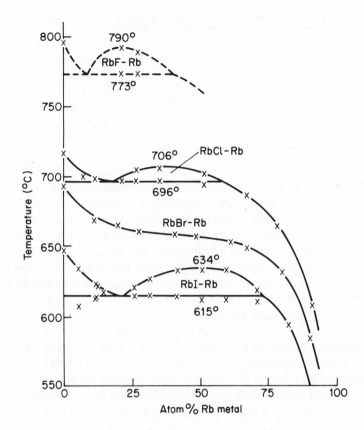

Figure 5.8 Rubidium metal–rubidium halide systems. (From M. A. Bredig in *Molten Salt Chemistry* (Ed. M. Blander), Interscience, 1964); reproduced by permission of John Wiley & Sons, Inc.

changes from lithium to caesium. Again, if the same metal is used and the halide ion changed, there is a steady decrease in the temperature required for complete miscibility as the halide is changed from fluoride through the chloride and bromide to iodide. This is illustrated in Figure 5.8 for solutions in liquid rubidium.

Throughout Bredig's work, the emphasis was on the molten salt aspects, so that high temperatures and high salt concentrations were generally employed. Nevertheless, we might anticipate that at lower temperatures (i.e. below the melting points of the salts) where dilute solutions of the halogens in the liquid metals will be involved, the solubility of a given halogen would increase going from liquid lithium to liquid caesium, and that for a given alkali metal the solubilities would increase from fluorine to iodine. Some of Bredig's results extend into this region, and the general position is illustrated in Table 5.10 by reference to solutions of the sodium halides in liquid sodium. The solubility

Table 5.10   Solubility of the halogens in liquid sodium (mol % NaX)

	750°C	650°C	550°C	300°C	100°C
NaF	*0.95*	0.30	0.05	4×10^{-5}	3×10^{-11}
NaCl	*1.8*	*0.75*	*0.12*	9×10^{-4}	7×10^{-8}
NaBr	*3.0*	*0.91*	*0.21*	2×10^{-4} (3×10^{-5})	6×10^{-9} (1×10^{-9})
NaI	—	*0.90*	*0.16*	9×10^{-5} (6×10^{-5})	2×10^{-10} (3×10^{-9})

values in italic in Table 5.10 are obtained by interpolation (or slight extrapolation) from experimental results. They show the expected increase in solubility from fluoride to bromide, but indicate that bromide and iodide have very similar solubilities. These results would bear re-examination because techniques designed to study liquid miscibilities over the full concentration range may not be ideal for systems where the solid halide is in equilibrium with dilute solutions in the liquid metal. On the other hand it may well be the case that in this respect bromine, as in some other aspects of its chemistry, is a little out of step with the other halogens.

It is also desirable to obtain some estimate of the solubility of the halogens at temperatures below the experimental range, i.e. at the lower temperatures at which chemical reactions of the solutions are more likely to be carried out. It should then be possible to make comparisons with the solubilities of other non-metals, and also to assess the value of filtration as a purification method. Solubility is related to temperature by the general relationship

$$\ln x = \frac{S_{\text{soln}}}{R} - \frac{H_{\text{soln}}}{RT}$$

where x is the mole fraction of solute, and S_{soln} and H_{soln} the partial molar entropy and enthalpy of solution. Bredig[137] has calculated values for H_{soln} for each of the sodium halides in liquid sodium; these are 109 ± 8, 84 ± 12, 100 ± 4 and 117 ± 8 kJ mol^{-1} for NaF, NaCl, NaBr and NaI respectively. If we use these values to determine the slope of the linear graph relating $\ln x$ and $1/T$, and use the experimental solubility values to fix the position of the graph on the solubility axis, it is then possible to obtain extrapolated solubility values at lower temperatures, and values obtained in this way are shown in Table 5.10 for 300° and 100°C. These can be no more than rough estimates, but they do indicate the order of magnitude of the solubility at these temperatures. It is clear that the solubility of the halogens in sodium is much smaller than that of hydrogen or oxygen, and also that filtration of sodium at 100°C should reduce the halogen content to a mere trace.

These extrapolated values receive support from more recent direct measurements[138] of the solubility of bromine and iodine in liquid sodium; interpolated values are shown in parentheses in Table 5.10. These measurements were made in response to the constant demands from the nuclear energy industry for background information. In a nuclear reactor the fission yields of bromine and iodine are high; these halides are therefore amongst the most important fission products, and a knowledge of their possible solubility in a liquid sodium coolant is highly desirable. In experiments carried out at UKAEA, Dounreay,[138] sodium purified to 10–20 ppm by weight of oxygen was equilibrated with sodium bromide and sodium iodide labelled with 87-Br and 131-I, and the solubility determined by gamma spectrometry and chemical analysis over the temperature range 150–400°C. Both bromide and iodide are reversibly adsorbed on to a steel surface from solution in sodium, and distribution coefficients have been calculated.[138,139] From these coefficients the solubilities of the halides in liquid sodium can be determined, and the results are in good agreement with those determined by direct measurement. The extrapolated and experimental values for the solubility of the bromide and iodide agree within an order of magnitude and the values of H_{soln} obtained from the experimental results (97.5 ± 4.7 and 89.2 ± 2.6 kJ mol^{-1} for NaBr and NaI respectively) are similar to Bredig's values. It is interesting to note, however, that the experimental results indicate that iodide is rather more soluble than the bromide. We may therefore consider the values given in Table 5.10 for the fluoride and chloride also to be good estimates of their solubilities.

Estimates of the solubility of the halogens in the other alkali metals over the lower temperature ranges can be attempted in the same manner, but the extrapolations are less satisfactory. Only in the case of lithium and chlorine have any measurements been reported, and the two sets of data differ greatly. Nakajima et al.[140] used a technique which involved the sampling of saturated solutions of chlorine in liquid lithium, and found the solubility to range from 0.007 to 0.210 mol % LiCl at 294 to 540°C. The results of Konovalov et al.[141] were obtained by neutron-activation analysis of filtered samples, and ranged from 0.001 to 0.011 mol % LiCl over a similar temperature range. The poor agreement prevents any useful comparison being made between the solubilities of chlorine in liquid lithium and liquid sodium.

Chapter 6

Alkali metal mixtures

6.1 Introduction

Because the alkali metals resemble one another in atomic structure, and in the physical and chemical properties of the bulk metals, it is pertinent to enquire to what extent the properties of mixtures of two or more of the liquid metals are means of those of the separate liquids. Such information can have practical as well as academic significance. For example, the eutectic mixture of sodium and potassium is liquid at room temperature, which has many advantages in industry; but are there other properties of the mixture which differ from those of the component metals to the same extent? Again, how far is it necessary to purify a given alkali metal from other alkali metals present as impurities before we can accept its properties as characteristic of that element?

This chapter will survey available information on some of the properties of the mixtures which deviate from values which would be calculated additively from the composition and the corresponding values for the two component metals, and which have some chemical interest. In many aspects the variation from the average of the components can be surprisingly extensive. Physical properties of mixtures have been fairly fully explored, but relatively little is known of their chemical properties.

6.2 Two-component mixtures

Liquid miscibility

The general position is shown in Figure 6.1.[92] The heavier elements sodium to caesium are miscible with one another in all proportions in the liquid state, and only in the case of mixtures with lithium does immiscibility arise. The lithium–sodium system is the only one which can be investigated under atmospheric pressure, and the phase diagram has now been fully defined[142] using a combination of resistance and thermal methods (Figure 6.2). The consolute point occurs at 305°C and at 63 atom % lithium; below this temperature the mixture separates into two liquid phases. The miscibility gap

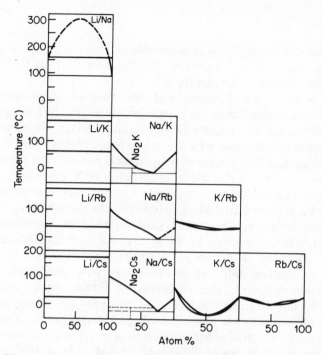

Figure 6.1 Phase diagrams for binary mixtures of the alkali
metals

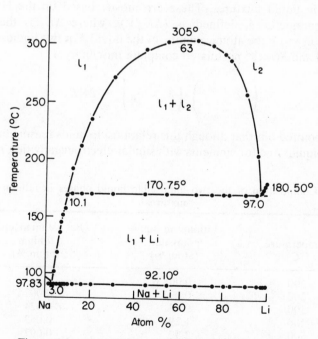

Figure 6.2 Immiscibility in lithium–sodium mixtures

is not symmetrical, and lithium is more soluble in sodium than is sodium in lithium. The immiscibility boundary extends from 10.1 to 97.0 atom % lithium at the monotectic temperature 171°C.

Lithium is even less miscible with the heavier alkali metals. The lithium–potassium phase diagram is presumably similar in form to that for lithium–sodium, but the miscibility gap is much wider. Thus, analysis of filtered samples indicated the presence of two liquid phases above the melting point of lithium, one liquid phase at temperatures between the melting points of the two metals, and no detectable liquid below 63°C (the melting point of potassium).[143] The consolute temperature has not been determined, but is believed to be greater than 1000°C. However, the compositions of the two liquid phases have been measured over a range of temperatures, and interpolated values are given in Table 6.1. As with the lithium–sodium system, the solubility of lithium in potassium is much greater than that of potassium in lithium, but the miscibility is now much less than for lithium–sodium mixtures. The miscibility of lithium with rubidium and caesium has not been studied, but is regarded as negligible in the corresponding temperature range. At higher temperatures (where high vapour pressures are also generated) caesium is reported to be soluble in liquid lithium to the extent of 0.013 and 0.17 atom % at 760° and 970°C respectively.[148]

A number of attempts have been made to devise a theoretical treatment of the liquid state which would enable predictions to be made of miscibility or otherwise, in liquid mixtures. These are mostly based on the Hildebrand solubility parameter δ, defined as $(\Delta E_v/V)^{\frac{1}{2}}$, where ΔE_v is the heat of vapourization and V the atomic volume of the liquid. For two liquids A and B, Hildebrand and Scott[144] postulated complete miscibility if

$$\left(\frac{V_A + V_B}{2}\right) \left(\delta_A - \delta_B\right)^2 < 2RT$$

Mott[145] pointed out that, though this relationship holds reasonably well for non-polar liquids and for elements with similar electronegativities, it does not

Table 6.1 Composition of the two immiscible liquid phases in lithium–potassium mixtures[143]

Temperature (°C)	Lithium in liquid potassium (atom %)	Potassium in liquid lithium (atom %)
100	0.0083	—
200	0.091	0.0074
300	0.43	0.024
400	1.29	0.053
500	2.92	0.097

take into account the extent to which electrochemical interactions interfere with the random distribution of solvent and solute atoms. Mott therefore proposed a modification to the Hildebrand–Scott relation, by introducing a term representing the electronegativity difference between the two metals ($\chi_A - \chi_B$), so that for complete miscibility

$$\left(\frac{V_A + V_B}{2}\right)\left(\delta_A - \delta_B\right)^2 - kn\left(\chi_A - \chi_B\right)^2 < 2RT$$

where n is the maximum number of Pauling bonds which the atoms can form in the liquid state.

However, neither of these approaches has had any particular success so far as the alkali metals are concerned. The Mott modification has minimum significance in this case since the electronegativities of the alkali metals are not greatly different from one another, and the original Hildebrand–Scott approach does not appear to be applicable to these strongly electropositive metals. The four immiscible systems (all of which involve lithium) agree with the Hildebrand rules. However, there are six other mixtures which can be formed from the four heavier alkali metals; all of these are miscible, yet the Hildebrand rules predict miscibility in two cases only.

Later, Kumar[146] suggested that the heat of fusion might be more appropriate than the heat of vapourization in defining the solubility parameter, and many systems were tested in which this substitution was made in the relationships given above. The collected results are still treated in terms of a 'success rate', and theory will need to develop much further before accurate predictions of miscibility or immiscibility in liquid metals can be attempted. At present, the only useful observation that can be made is that immiscibility occurs in lithium-containing systems alone, and that this is consistent with other physical properties of lithium. Thus, the exceptionally small atomic radius, vapour pressure and volume increase on melting, the exceptionally high melting and boiling points, and the exceptionally large ionization energy, density, surface tension and heats of fusion and vapourization, all point to the fact that the mutual attraction between atoms of lithium is much greater than between atoms of lithium and atoms of another alkali metal, so that the observed tendency of liquid lithium to remain as a separate phase is understandable.

An interesting effect, which must have real significance so far as the state of the liquid mixture is concerned, was observed during measurement of the electrical resistivity of lithium–sodium mixtures.[147] The resistivity of all solutions decreases almost linearly with decreasing temperature. However, as the temperature approaches that of the consolute temperature (305°C) at which phase separation occurs, the temperature coefficient of resistivity passes through a sharp peak (Figure 6.3) which occurs at the consolute composition (63 atom % lithium). This is evident at 315°C, and becomes more pronounced at 310°C and 307°C. This property (and maybe others) is

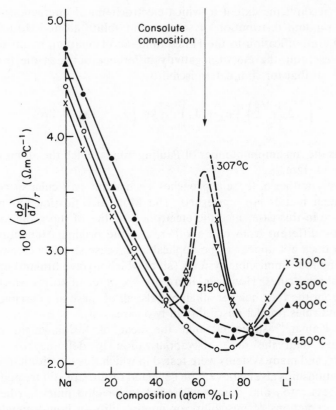

Figure 6.3 Temperature coefficient of resistivity for a sodium–lithium mixture as a function of composition and temperature

therefore sensitive to incipient immiscibility, and reflects a 'state of tension' in the liquid for several degrees before the mixture actually breaks into two phases.

Melting points

The phase diagrams from which melting points may be determined are shown in Figure 6.1, and full-scale diagrams may be consulted elsewhere;[94–96] the main purpose is to indicate the extent to which the melting points of mixtures differ from those of the component metals. The sodium–potassium system has been the most fully explored and a full account of its physical properties was published[2(a)] as a supplement to the *Liquid Metals Handbook* at a time when it seemed possible that the eutectic mixture (NaK) might become the accepted coolant in fast nuclear reactors.

One of the first recorded accounts of the formation of this alloy occurs in a letter which Jocelyn Thorpe wrote to Henry Roscoe in Manchester in 1867 after his arrival at Bunsen's laboratory in Heidelberg. The letter is quoted in Thorpe's obituary notice published in the Journal of the Chemical Society for 1926 (pp. 1036–7). It appears that Thorpe had been asked to include in his luggage samples of quite pure sodium and potassium which had come into Roscoe's hands, and in which Bunsen was known to have considerable interest. After introductory remarks, the letter (abridged by the author) goes on:

'. . . and we prepared ourselves for the alkali metals. The servant was called to unpack the box, when to my consternation he produced a bottle . . . partially filled with naphtha, at the bottom of which were a few tablespoonfuls of a bright shining rather mobile fluid. I had given the Geheimrath such a glowing account of the size of the sticks of the two metals that I was simply speechless with astonishment and felt indeed rather like an imposter . . . "Well", I said at last, "potassium and sodium were certainly put into the bottle before I left home, but what is there now is uncommonly like quicksilver". "No", said Bunsen, who was holding the bottle, "it is not quicksilver. Feel the weight of it!" The fact was that our old friend Heywood, who had been ordered by you to pack the specimens at the time in separate bottles, perceiving that both could be got into one bottle, had . . . placed the two metals together, with the untoward result I have indicated. Chemical combination between solids is not of frequent occurrence. I had no knowledge at the time of the existence of the fluid alloy—nor had Bunsen—which, perhaps, is not very creditable to us, since it is actually mentioned by Davy".'

Small quantities of the liquid alloy are easily prepared in the laboratory. Pieces of solid sodium and potassium are placed in the two limbs of an inverted U tube under pure argon. The U tube is then turned up to bring the two pieces of metal into contact, and if the surface of each metal is bright, the liquid alloy forms within two or three minutes at room temperature. The phase diagram is relatively simple, with a eutectic temperature of $-12.5°C$ and a eutectic composition of 67.8 atom % potassium. The low eutectic temperature relative to the melting points of the pure metals is remarkable, and can be related to their different atomic sizes. There is very little solid solubility, and each pure metal condenses to a body-centred cubic structure in the solid state. An inflexion in the liquidus curve is due to the formation of a 'compound' Na_2K, which melts peritectically at $6.9°C$, and has an ordered hexagonal structure. Presumable the sodium and potassium atoms can pack into this solid structure in the 2:1 ratio with some small degree of stability. Na_2K is formed without significant change in volume and no change in magnetic susceptibility, and its enthalpy of formation is very small; its thermodynamic properties have been studied.[149] However, both X-ray diffraction and magnetic susceptibility measurements on the liquid phase of this composition have established that no Na_2K molecules as such exist in the liquid mixture.[150] The sodium–rubidium and sodium–caesium phase diagrams are similar to the sodium–potassium system, and give eutectic temperatures of $-5°C$ and $-29°C$ respectively. A slight inflexion in the Na–Rb liquidus suggests that there may be compound formation in the solid state, but this has not been investigated, and Na–Cs mixtures are said to yield a compound Na$_2$Cs with peritectic melting point $-8°C$.

The three binary systems formed by potassium, rubidium and caesium give rise to an entirely different type of phase diagram. Due largely to the small differences in metallic radii there is now virtually complete mutual solubility in the solid state, so that the liquidus and solidus curves lie close together. In the K/Cs system, the minimum temperature is $-37°C$. In all three systems crystallization occurs more readily than in the sodium mixtures. With the K/Rb and Rb/Cs mixtures, each of which involve neighbouring elements with similar metallic radii, the minimum liquidus temperatures are $34°C$ and $9°C$, representing falls of only $5°C$ and $19°C$ below the melting points of rubidium and caesium respectively.

Electrical resistivity

The extensive use of resistivity measurements in the study of chemical reactions is the subject of a later chapter; but it is relevant here to note that

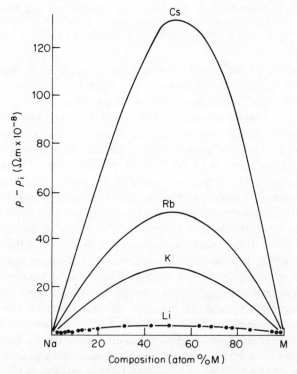

Figure 6.4   Excess resistivity for mixtures of sodium and another alkali metal (M = Li, K, Rb, Cs) as a function of composition. All are at $100°C$, except for lithium at $310°C$

the solution of any other element, metal or non-metal in a liquid alkali metal increases its resistivity, so that the resistivity of any alkali metal mixture is greater than that of either of the pure metals. The curve relating resistivity with composition is not symmetrical because of the different resistivities of the pure metals and is best considered in terms of the 'excess resistivity'. This is the differences between the value determined experimentally and the interpolated resistivity calculated additively from the composition and the values for the pure metals. For mixtures of sodium with alkali metal M, where M is lithium, potassium, rubidium or caesium, the interpolated resistivity ρ_1 is given by

$$\rho_1 = \rho_{Na}x + \rho_M(1 - x)$$

where ρ_{Na} and ρ_M are the resistivities of sodium and metal M respectively, and x is the atom fraction of sodium. The excess resistivity is then $\rho - \rho_1$, where ρ is the experimental resistivity of the mixture. The results for each of the four mixtures with sodium are shown in Figure 6.4.[147] The curves for potassium, rubidium and caesium are derived from data at 100°C, but that for lithium is necessarily at a higher temperature of 310°C (i.e. just above the consolute temperature for sodium–lithium mixtures). This does not invalidate the comparison, since ρ changes by a relatively small amount with temperature. The extent to which excess resistivity changes with difference in atomic size is remarkable; the excess for Na–Li mixtures is very small indeed, whereas that for Na–Cs mixtures is ten times the resistivity of sodium alone (13.5 $\mu\Omega$ cm). Each curve is symmetrical, with its maximum at about 50 atom % composition, which is consistent with the theory for mixtures of similar components where no pronounced interaction is involved.

Surface tension

This brief section deals only with the surface tension of liquid alkali metal mixtures against a vacuum or inert gas; wetting phenomena are not relevant here since they involve a third (solid) phase. Experimental evidence is scanty, but there seems no reason to doubt that the changes in surface tension on mixing the liquid metals are similar in principle to those which are well known for mixtures of molecular liquids such as water and ethyl alcohol. Surface tension, like other properties such as boiling point, vapour pressure etc., reflects the intermolecular attraction between molecules in the liquid.

Hydrogen bonding in water leads to a high boiling point and high surface tension; hydrogen bonding in ethyl alcohol is weaker, so that the boiling point decreases from 100°C to 78°C and the surface tension from 73 to 23 ($Nm^{-1} \times 10^{-3}$) at 20°C. The attraction between molecules in the liquid decreases in the order H_2O–H_2O, H_2O–EtOH, EtOH–EtOH, so that inevitably the alcohol molecules tend to be expelled from the bulk liquid to the surface, which then has a higher alcohol content than the bulk liquid.

94

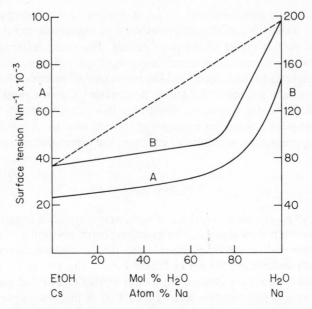

Figure 6.5 Surface tensions of mixtures. (A) Ethyl
alcohol and water at 20°C. (B) Liquid caesium and liquid
sodium at 105°C

The effect on surface tension of the mixture is illustrated in Figure 6.5; on addition of alcohol to water the surface tension falls rapidly, and above 50 per cent alcohol the surface tension is little different from that of the pure alcohol. Some unpublished and preliminary results by Pulham and Meredith[151] for liquid sodium–caesium mixtures at 105°C show that a similar effect is observed with these metals (Figure 6.5). The surface tension of liquid caesium, $74 \, Nm^{-1} \times 10^{-3}$, lies much below that of sodium, $197 \, Nm^{-1} \times 10^{-3}$, so that the concentration of caesium in the surface should be much greater than in the bulk liquid. Consistent with this, the surface tension of sodium falls rapidly on addition of caesium, and in the region 40–100 per cent caesium the surface tension does not differ greatly from that of pure caesium. There may be a significant break in this curve somewhere in the composition region 66–75 atom % Na corresponding to compounds Na_2Cs or Na_3Cs. This merits further investigation; weak compound formation of this sort has previously been considered in connection with the solid state only, and if it can also influence the properties of the liquid state to this extent, it would be highly relevant to the surface chemistry of liquid metals.

Surprisingly, surface tensions over the full composition range have not been reported for sodium–potassium mixtures. However, a value of 103 $Nm^{-1} \times 10^{-3}$ at 200°C is reliably reported[152] for the eutectic composition (67.8 atom % K), and this is virtually identical with the surface tension of pure potassium. Over 30 atom % of sodium can therefore be added to liquid potassium without altering its surface characteristics.

Little further experimental evidence is available. The negative deviation from linearity illustrated in Figure 6.5 would be expected to be greatest for mixtures with liquid lithium. However, in view of the miscibility gaps which arise in these mixtures (discussed earlier), surface tensions over a continuous composition range could only be measured at very high temperatures.

The fact that the composition of the surface in any alkali metal mixture will differ from that in the bulk liquid has important ramifications outside the realm of surface chemistry. We do not know how rapidly the equilibrium composition of a surface is established, and at the surface of a high-velocity jet, for example, non-equilibrium conditions might be detectable. Under most conditions encountered in practice, however, the equilibrium surface may be regarded as being established instantaneously. Many chemical reactions of the liquid alkali metals involve their reaction with gases, and great care must then be exercised in relating reaction rate to bulk metal composition, when it is the surface composition which is the relevant factor.

Ultrasonic absorption

Because the passage of sound through a liquid involves transmission of a pressure wave, it is the mechanical and thermophysical properties of the liquid which determine both the velocity and the absorption of sound. The velocity of propagation of longitudinal waves C is given by the expression[154]

$$C^2 = \frac{1}{\rho} \left(\frac{1}{\beta_s} + \frac{4}{3} G \right) \qquad (6.1)$$

where ρ is the density, G the shear modulus and β_s the adiabatic compressibility of the liquid. Values of C, and corresponding values for the compressibility, are given in Table 6.2 for all alkali metals except lithium.[153,154] The sonic velocities decrease rapidly with increasing atomic weight; the corresponding compressibilities (equation 6.1) increase with increasing atomic volume, which is reasonable. The compressibilities of the alkali metals are an order of magnitude larger than the values for the noble or polyvalent metals, which is consistent with the fact that their interatomic forces are smaller.

Sonic absorption is usually quite small, and frequently negligible, for sound waves in the audio range, and becomes significant only at ultrasonic

Table 6.2 Sonic velocity and compressibility of the heavier alkali metals

	Sonic velocity at the melting point (m s^{-1})	Compressibility (cm^2/dyn)
Na	2526	16.9
K	1890	34.0
Rb	1260	42.7
Cs	967	58.1

frequencies. In liquid metals, both the shear viscosity and the thermal conductivity contribute to the 'classical' sound absorption at ultrasonic frequencies, i.e. $\alpha_{Cl} = \alpha_s + \alpha_T$. The two components are defined[156] as

$$\alpha_s = 8\pi^2 \eta_s f^2/3\rho C^3 \qquad (6.2)$$

$$\alpha_T = 2\pi\beta^2 K T f^2/\rho C J c_p^2 \qquad (6.3)$$

which include an impressive array of mechanical and thermophysical properties; f is the frequency of the sound waves, η_s the shear viscosity, ρ the density, C the sound velocity, β the volume expansion coefficient, K the thermal conductivity, T the absolute temperature, J the mechanical equivalent of heat and c_p the specific heat at constant pressure. In some cases, e.g. for liquid sodium, an absorption value in excess of the classical value is observed; this is defined as

$$\alpha_e = (2\pi^2 f^2/\rho C^3)\, \eta_B \qquad (6.4)$$

where η_B is the bulk viscosity.

Many attempts have been made to relate the ultrasonic absorption with the state of a single liquid metal, and in particular to define the relaxation processes which are responsible for excess absorption. Often these are highly mathematical models which have had only limited success, and discussion of them is outside the scope of this volume. It is apparent from equations 6.2 and 6.3, however, that many physical properties of the metals must be known with high accuracy before excess absorption in a single metal can be given real physical significance.

Turning now to binary mixtures of alkali metals, there are both advantages and disadvantages in the interpretation of results. The various physical constants of the mixtures are known with less accuracy, so that theoretical treatments become more difficult. On the other hand, atomic association in the liquid is now possible, and this gives rise to sonic absorption peaks, at particular compositions of the mixtures, which can be of considerable magnitude. The liquid potassium–rubidium system behaves as an ideal mixture, but ultrasonic absorption measurements on sodium–caesium mixtures suggest substantial association in the liquid.[155] The absorption curves are shown in Figure 6.6. The units of sonic absorption are expressed as α/f^2, and since the expressions for α (equations 6.2, 6.3 and 6.4) include f^2 in the numerator, the results may be regarded as independent of frequency. The peak is not detectable at 250°C, but becomes much more pronounced as thermal agitation diminishes and the temperature is lowered towards the melting point (sodium begins to crystallize from this composition at 70°C). The considerable variation in peak magnitude with temperature suggests very weak association between sodium and caesium atoms, perhaps of an order which is particularly sensitive to ultrasonics. Of considerable interest, however, is the fact that the peak occurs at 75 atom % sodium, indicating a

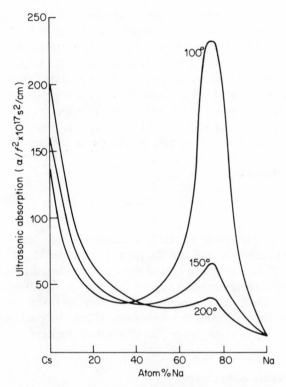

Figure 6.6 Influence of temperature and composition on sound absorption in liquid sodium–caesium mixtures

weak 'compound' of stoichiometry Na_3Cs in the liquid, in contrast to the compound Na_2Cs which has been suggested from thermal studies.

In the present state of the art, it is still open to question whether sonic absorption measurements are likely to make any major contribution to the chemistry of the liquid alkali metals. Absorption peaks of the type discussed above reflect the behaviour of the whole medium, and it is not yet known to what extent a non-metal present in dilute solution can influence the absorption, even though the interatomic and interionic forces introduced into the liquid metal are much greater.

Whether or not this proves to be the case, it is already clear that ultrasonics can make a major contribution to liquid metal technology; using ultrasonics, it is now possible to 'see through' liquid sodium, and to 'photograph' objects which are completely immersed in liquid sodium. The development of the 'under sodium viewer' is a remarkable achievement, and its application in the Prototype Fast Reactor at Dounreay has recently been described.[484] The technique depends on a pulse-echo principle but operates at a much higher frequency than that used in ship sonar systems. A short pulse of sound (at 5

MHz for only 200 ns) is emitted from a transmitter, and the echo detected by a receiver; transmitter and receiver can be combined in a single device, the transducer. The pulse-echo time gives a measure of the distance between the transducer and the target. To produce a picture of the target, the ultrasonic beam is moved over the target by systematic movement of the transducer, and the stored echoes assembled by computer to form a picture of the target. During 1982 the ultrasonic viewer succeeded in producing pictures of the top of a reactor core immersed in sodium, and the potential value of the technique in the study of displacements or distortions of reactor core components without the need to remove the core from the well of liquid sodium is considerable.

Other properties

Many other properties vary linearly with composition; they include density, viscosity, compressibility and specific heat. The properties of sodium–potassium eutectic (NaK) alloy are best known, and are listed in detail in Reference 152. The thermal conductivity of this alloy is significantly lower than that of either of the liquid components. The chemical reactions of alkali metal mixtures with other elements or compounds will not be dealt with here, since they are discussed at appropriate places in other chapters.

6.3 Multicomponent mixtures

Our knowledge of mixtures containing more than two alkali metals depends almost entirely on the work of Tepper, King and Greer[157] on three-component systems. The remarkable feature is the extremely low melting points which are available. Because of its immiscibility with the heavier alkali metals, lithium is not suitable as a major component of a low-melting alloy. Of the other four metals, there are four ternary systems which are possible, and the lowest melting points which can be achieved are as follows:

Na–K–Rb	−25°C
Na–Rb–Cs	−37°C
K–Rb–Cs	−38°C
Na–K–Cs	−78°C

These are claimed to be the lowest melting points of any known metallic systems and the alkali metals can therefore be obtained as liquids at temperatures as low as are normally available with organic solvents. The melting points corresponding to various compositions of the Na–K–Cs system are shown in Figure 6.7; the lowest melting point, −78°C, is obtained at a eutectic composition of Na 28, K 57.5 and Cs 14.5 atom %. Experiments have also been performed to ascertain whether the addition of a fourth component to the Na–K–Cs eutectic could lower the melting point even further. It appears,

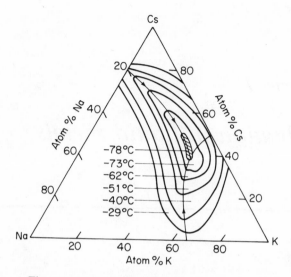

Figure 6.7 Sodium–potassium–caesium system

however, that addition of low-melting metals such as Tl, In, Ba, Hg or Ga has little effect on the melting point of the eutectic, and addition of rubidium causes no further reduction of the melting point.

The same authors also measured the vapour pressure, density, compressibility and viscosity of the Na–K–Cs eutectic, but these show no remarkable features. The experimental vapour pressures lie close to the computed Raoult law vapour pressures calculated from vapour pressures for the individual elements. The density falls linearly with increasing temperature, the values being near to those which would be calculated from an ideal mixture of the three components, and the viscosity of the eutectic lies among the values for the three pure elements.

Chapter 7

Solvation in liquid metals

7.1 Introduction

Research on the alkali metals carried out during the twenty years up to about 1968 covered such practical aspects as solubilities, reactions with gases, the identity of precipitates and so forth. By 1968, a general pattern covering this area of chemistry was beginning to emerge, and sufficient experimental evidence had been accumulated to support a theoretical interpretation of some of the observed behaviour, such as the nature of dissolved non-metals. However, some of the most fundamental questions remained unanswered. Thus, why do some non-metals, and some salts, dissolve in the liquid alkali metals? When chlorine or sodium chloride dissolve in liquid sodium, what are the energy terms involved? In an electronic environment, is it possible that species may be stable which would not merit consideration in solution in molecular liquids? Do dissolved species interact with one another, and if so, by what mechanism? When precipitation of a solid occurs from liquid metal solution, what are the factors which govern the onset of precipitation? In the presence of an excess of electrons covering an energy range, is there any reason why polynuclear species containing covalent bands should be stable? For instance, when hydrogen and oxygen are dissolved together in liquid sodium, is it possible to consider the existence of the OH^{-1} group, or must solutions of non-metals invariably be monatomic?

In considering such questions, some properties of solutes such as lattice or bond energies, enthalpies of solution and other energy terms, which are usually included in Born–Haber cycles, apply equally well to metal as to other solutions, but the one factor which did not receive serious consideration was solvation in liquid metal solutions. With hindsight, solvation should have been considered at a much earlier stage; but the reluctance of inorganic chemists to think in these terms is understandable, since solvation had traditionally been associated with solutions in molecular liquids.

7.2 Comparison with solvation in molecular liquids

Some of the general principles governing solvation in molecular liquids will be discussed briefly at this stage in order to outline the background against which solvation in liquid metals had to be approached, and to enable comparisons to be made. Although the approach had to be different, some of the results are surprisingly similar. The solvation of a metal ion M^{n+} in aqueous solution arises because of the need to minimize the high charge density associated with an isolated ion in an electrically neutral liquid. The solvated ion becomes a coordination complex (Figure 7.1). The important point here is that solution is only appreciable in polar liquids, where the solvent molecules possess a dipole; the atom which carried the negative charge (the O atom in the case of water) is attracted to the M^{n+} ion, and a number of solvent molecules (often six) orientate themselves around the solute ion, with the atom carrying the negative charge directed towards the M^{n+} ion. By this means this ion is surrounded by a sheath of negative charges, and the local charge density is reduced. Similarly, when negative ions are solvated, the positive end of the solvent dipole will now be directed towards the dissolved anion. In all cases where solvation by molecular liquids is concerned, the number of solvating molecules, and the shape of the complex, is determined largely by the extent to which the atomic orbitals of the solute ion are occupied by electrons, and the degree of overlap of these orbitals with those of the solvating molecules.

Since solvation occurs as a result of attraction between the solute ion and the dipolar solvent molecules, energy is released in the process, and when a salt dissolves in water the combined solvation energies of cation and anion exceed the lattice energy of the solid salt by an amount which is represented by the free energy of solution. The combined solvation enthalpy can be measured directly, but there is no purely thermochemical way by which this

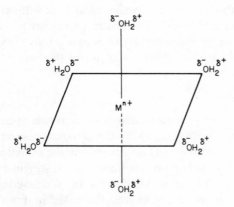

Figure 7.1 Solvation of an M^{n+} ion in
water

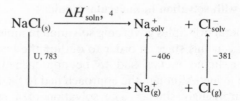

Figure 7.2 Enthalpy terms for solution of
sodium chloride in water

can be separated into its two parts. However, by comparing the solvation (in this case hydration) enthalpies of a series of related salts it is possible to allocate solvation enthalpies to individual ions.[158] As would be expected, the enthalpy of hydration increases with the charge on the cation (for Na^+, Mg^{2+} and Al^{3+} the values are 406, 1921 and 4665 kJ g ion^{-1} respectively) and for the same numerical charge, the enthalpies of hydration of cations and anions are of the same order (for Na^+ and Cl^-, values are 406 and 381 kJ g ion^{-1} respectively). The solution of sodium chloride in water (at 25°C) can therefore be represented by the cycle shown in Figure 7.2. The lattice enthalpy U is almost equal to the sum of the hydration enthalpies of the separate ions, so that the enthalpy of solution ΔH_{soln} is very small indeed. The values of ΔH, ΔG and $T\Delta S$ in this case are actually $+3.8$, -9.2 and $+13.0$ kJ mol^{-1} respectively, so that the entropy term is sufficiently large to convert the small positive ΔH_{soln} change to a negative free energy change, making the solution process favourable.

At first sight, few analogies, would seem possible between a solution of a salt in a molecular liquid and in a liquid metal. However, if we abandon those aspects which can be attributed directly to solvent molecules, the most important feature, i.e. the need to reduce the high charge density in the neighbourhood of an ion which is introduced into the liquid, still remains and is common to all solutions. The description of the state of the solvent metal atoms in the neighbourhood of a foreign solute ion will vary, of course, with the metallic model employed. Thompson[80] has carried out extensive and valuable calculations on the solvation of negative ions dissolved in the alkali metals, and we shall use the model which forms the basis of his calculations.

7.3 Solvation of ions

The calculations apply in particular to non-metals such as the halogens, oxygen, nitrogen and hydrogen, where the electrons responsible for the ionic charge are localized on the ion. (For this reason, the enthalpies so determined will be termed 'ion' solvation enthalpies to distinguish them from the 'inclusive' solvation enthalpies which will be discussed later.) It will be convenient in the first instance to express the model in terms of the solution of sodium chloride in liquid sodium. In the free electron picture of liquid sodium, the metal consists of Na^+ cores and free electrons, but the cores remain electrically neutral by attracting around themselves a screening charge of free electrons. The presence of a Cl⁻ ion will have a strong local effect on the screening charge of free electrons associated with the surrounding sodium cores. This screening charge will tend to be repelled from the space between the Cl⁻ ion and the sodium cores, and the net effect is the establishment of a sphere of partially charged sodium atoms around the Cl⁻ ion. The environment of the Cl⁻ ion can be represented diagrammatically as in Figure 7.3, which may be compared with Figure 7.1. For overall electrical neutrality, the charge on the solute ion is equal to the sum of the partial charges on the surrounding sodium atoms. In the illustration in Figure 7.3, each sodium atom would carry a partial charge of $+1/6$, but this will obviously vary for solute anions having different charges, and for different coordination numbers.

According to Thompson,[80] the solvation energy of an anion in its shell of partially charged metal atoms is the sum of the electrostatic potential between the anion and the surrounding partially charged atoms, and between the partially charged atoms themselves. The solvation energy U_x can be expressed in the simplified form

$$-U_x = \frac{1273\,Z^2}{\sigma_0}\,(1 - I) \tag{7.1}$$

where Z is the formal anion charge, σ_o the equilibrium separation between the anion and the surrounding solvent atoms, and I is a factor depending on the number of solvating atoms and their configuration around the anion.

If we anticipate later discussion on the results of Thompson's calculations, the solution of sodium chloride in liquid sodium can be represented by Figure 7.4, which is to be compared directly with Figure 7.2 (units are again kJ mol⁻¹). The lattice enthalpy is somewhat lower because the values given in

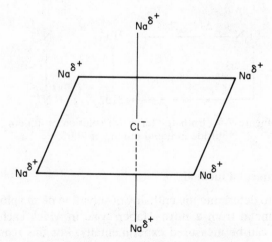

Figure 7.3 Solvation of chloride ion in liquid sodium

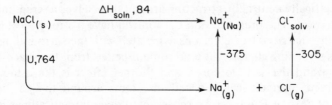

Figure 7.4   Enthalpy terms for solution of sodium chloride in
liquid sodium

Figure 7.4 refer to the higher temperature (750°) necessary to form those solutions. The value quoted for the transfer of $Na^+_{(g)}$ into the liquid metal is not truly an enthalpy of solvation; a change of physical state is involved, and this value can be calculated (see below). The enthalpies of solution of all the alkali metal salts in their corresponding liquid metals are small, so that (as in water) the major part of the lattice enthalpy is distributed, on solution, between the enthalpy changes involved on introducing the cation and the anion into the solution. The interesting feature which appears here is the enthalpy of solvation of Cl^- in liquid sodium, which is of the same order as that obtained for solution of Cl^- in water. Clearly, the phenomenon of solvation is just as important in liquid metal solutions as it is in water. As expected, solvation enthalpies of ions dissolved in the alkali metals also increase with charge on the ion, and this is well illustrated by the various values for the solution of lithium nitride in liquid lithium (Figure 7.5). Because of the high charge on the nitride ion, the lattice enthalpy and N^{3-} solvation enthalpy terms are much larger than for a chloride ion. The enthalpy of solution is very small by comparison, and the entropy change on solution will readily convert this to a negative free energy change.

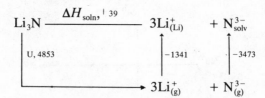

Figure 7.5   Enthalpy terms for solution of lithium
nitride in liquid lithium at 700°C

7.4   The experimental method

It is possible to determine the enthalpy of solvation of an anion dissolved in a liquid alkali metal from a Born–Haber cycle in which each of the other enthalpy terms can be measured experimentally. For this reason, values so determined are termed 'experimental' values. An an example, the cycle used to determine the enthalpy of solvation of the nitride ion in liquid lithium is

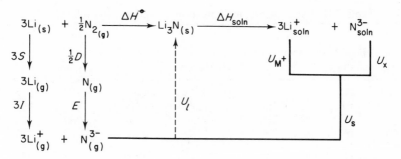

Figure 7.6 Born–Haber cycle for solution of nitrogen in liquid lithium. ΔH° and ΔH_{soln} are enthalpies of formation and solution of Li_3N; U_l and U_s are respectively the lattice enthalpy and the total enthalpy of solvation; S, I and D are sublimation, ionization and dissociation enthalpies; and E is the electron affinity

shown in Figure 7.6, and similar cycles can be used for other non-metals (e.g. hydrogen and oxygen) which give rise to clearly defined ions in solution. The steps leading to the formation of $Li_3N_{(s)}$ are standard literature values. For the solution part of the cycle

$$U_s = \Delta H_{\text{soln}} + U_l \qquad (7.2)$$

and the enthalpy of solution can be obtained from the variation in solubility with temperature, so that the total enthalpy of solvation is also available. U_s is composed of two parts, corresponding to the separate enthalpy values involved in the introduction of the cation, and the anion, into the liquid metal, and designated U_{M+} and U_x respectively. Once the M^+ ion is in solution it becomes indistinguishable from the rest of the metal cores, so this is not actually a solvation process. U_{M+} can, however, be determined from standard known quantities. In order to introduce an ion M^+ into the liquid metal, we can consider that an electron must first be removed from the metal (the work function). This electron combines with the M^+ ion (which involves the ionization enthalpy) and the M atom so formed then condenses into the liquid metal (involving the enthalpy of sublimation). Therefore

$$U_{M+} = -(S + I - \Phi)$$

where Φ is the work function for the metal concerned. In calculating the values shown in Table 7.1, Thompson[80] used work function values of 234, 220, 216, 202 and 175 kJ mol^{-1} for the liquid metals Li, Na, K, Rb and Cs respectively.

When the anion enters the liquid, however, it collects a sheath of positive charges about itself as discussed above, so that U_x is truly a solvation enthalpy and its value can be determined from the expression

$$U_x = U_s - U_{M+} = U_l + \Delta H_{\text{soln}} + S + I - \Phi$$

Table 7.1 Experimental ion solvation enthalpies, and related data

Solvent metal	Dissolved anion	Relevant temperature (°C)	$-U_{M^+}$ (kJ/g ion)	ΔH_{soln} (kJ/mol)	$-U_x$ (kJ/g ion)	Reference
Li	H^-	700	427	44.2	425	102
	Cl^-	700	427	68.4	343	159
	O^{2-}	700	427	55.4	1,960	102
	N^{3-}	700	427	39.2	3,473	102
Na	H^-	300	384	64.8	344	54
	F^-	1100	374	109	410	137
	Cl^-	1000	375	84	305	137
	Br^-	1000	375	97.5	257	138
	I^-	900	376	89.2	225	138
	CN^-	300	384	176	169	80, 159
	O^{2-}	700	377	47.2	1,721	80, 159
K	F^-	1000	286	46	467	159, 160
	Cl^-	900	286	92	316	159, 160
	Br^-	900	286	92	295	159, 160
	I^-	900	286	92	251	159, 160
	H^-	300	286	57.3	362	20
	O^{2-}	500	286	15.1	1,642	20
Rb	Cl^-	300	279	113	286	136
	Br^-	300	279	92	278	136
	I^-	300	279	92	238	136
Cs	F^-	300	268	44	406	159, 161
	Cl^-	300	268	92	280	159, 161
	I^-	300	268	75	243	159, 161
	O^{2-}	300	268	~0	1,580	159

Table 7.1 gives values for the anion solvation enthalpy determined by this method and some of the related data on which the determinations are based. Since a large number of physical constants is involved in each determination (see Figure 7.6), there is inevitably some inaccuracy in the $-U_x$ values, and the original papers should be consulted for detailed assessment of errors. For instance, work function values measured by different techniques do not agree closely, and 'preferred' values have to be chosen. Again, solution enthalpy values are being constantly refined as solubility measurements are repeated. Nevertheless, the values in Table 7.1 (column 6) give a good general picture of the magnitude of ion solvation enthalpies in the liquid alkali metals, and trends which they illustrate should be meaningful.

Thus, in any given solvent metal the solvation enthalpy diminishes as the size of the anion increases; through the series F^-, Cl^-, Br^-, I^- $-U_x$ decreases to about half its original value. On the other hand, for a given dissolved anion the solvation enthalpy does not vary to any great extent with change in solvent metal. The $-U_x$ value for the Cl^- ion changes fairly regularly from 343 kJ g ion^{-1} in lithium to 280 kJ g ion^{-1} in caesium; but results for the other halogen ions do not support any general trend of this nature, and we may conclude that the solvation enthalpies in these systems are determined much more by the solute than the solvent. Increase in charge causes a large increase in the solvation enthalpy; results for several of the metal solvents are available only in the case of oxygen, where $-U_x$ for O^{2-} undergoes a small but regular decrease from 1960 kJ g ion^{-1} in lithium to 1580 kJ g ion^{-1} in caesium. If we ignore variations in the factors σ_0 and I, equation 7.1 indicates that the solvation enthalpy for H^-, O^{2-} and N^{3-} should increase in the order $1^2{:}2^2{:}3^2$, and the values in Table 7.1 for these anions in lithium show that this is broadly true.

7.5 Calculated values of ion solvation enthalpy

The model on which calculations are based has been outlined in section 7.3 above, and we see that U_x can be calculated directly from equation 7.1 if the terms σ_0 and I are known. Both these terms depend on the charge and coordination number of the anion; Thompson's approach[80] has been to assume various configurations for the coordination shell of metal atoms surrounding the anion, to calculate U_x for each configuration, and then to compare the values with the experimental U_x value. By this means, useful information becomes available on the probable configuration shell of the dissolved anion.

The term σ_0, which is the equilibrium distance between the anion and the surrounding solvent atoms, may be approximated to the sum of the hard-sphere radius of the anion and the effective radius of the partially charged solvent atoms. The induction of a charge on the coordinating metal atoms will cause a contraction in their size, and the relation between the effective radius r and the partial charge Ω may be represented as

$$\frac{1}{r} = A\Omega - D$$

The partial charge Ω is obtained by distributing the charge on the anion among the available solvent atoms in the coordination shell. The constants A and D can be evaluated, since when $\Omega = 1$, r is the radius of the cation, and when $\Omega = 0$, r is the radius of the metal atom in the liquid metal. For any given coordination number, the term σ_0 can therefore be evaluated. The term I also depends on coordination number and configuration, and $(1-I)$ is analogous to the Madelung constant in solid ionic lattice enthalpy calculations.

Thompson[80] carried out extensive calculations covering all the halogens, all the alkali metals and a range of configurations, and some of his results are shown in Table 7.2. For the fluoride ion in liquid sodium, the closest agreement between calculated and experimental solvation enthalpies is given

108

Table 7.2 Calculated values of ion solvation enthalpy ($-U_x$) for various coordination shells

Solvent metal	Dissolved anion	Configuration of coordination shell	Coordination number	$-U_x$ (calculated) kJ/g ion	$-U_x$ (experimental) kJ/g ion
Na	F⁻	Octahedral	6	406	410
		Body-centred cubic	8	305	
		Cubo-octahedral	12	272	
Na	Cl⁻	Octahedral	6	339	305
		Body-centred cubic	8	305	
		Square antiprism	8	330	
		Cubo-octahedral	12	272	
Na	Br⁻	Body-centred cubic	8	289	257
		BCC plus 6 nearest neighbours	8 + 6	259	
		Cubo-octahedral	12	272	
Na	I⁻	Body-centred cubic	8	268	225
		BCC plus 6 nearest neighbours	8 + 6	228	
		Cubo-octahedral	12	272	

by an octahedral arrangement of six close-packed, partially charged sodium atoms around the fluoride anion. With the larger chloride ion, the best agreement is obtained on the assumption that the anion is surrounded by a close-packed cubic coordination shell of metal atoms. The bromide ion is presumably too large to fit comfortably into a close-packed cubic coordination shell; the results suggest that the Br^- ion opens the cubic structure so that sodium atoms in the second coordination shell (one atom outside each face of the cube) become affected by the ionic charge. This arrangement appears also to apply suitably to the solvation of the iodide ion in sodium. It must be remembered, however, that these systems are liquid and can display only short-range order, so that no configuration can be assigned rigidly to a solvated anion. It follows that the measured solvation enthalpy may well be an average of a number of preferred configurations.

Similar calculations on solutions in the other alkali metals indicate that, as the size of the solvent atom increases, the smaller is the number able to pack around an anionic sphere of a given radius. For solutions in potassium and rubidium, the octahedral configuration is preferred for chloride and bromide ions, and only with the iodide does cubic coordination arise. In caesium, the octahedral arrangement is preferred even with the iodide ion.

7.6 Solvation of binuclear species

In spite of the electronic environment in a liquid metal, there is now no doubt that anions can associate to give binuclear, or more complex, species in solution. In Chapter 4, evidence was given that carbon in solution exists predominantly as the C_2^{2-} ion. Sodium cyanide dissolves in, and crystallizes from, liquid sodium,[71] and there seems no reason to doubt that the CN^- ion can exist in solution in liquid sodium. Evidence for the stability of other such species in solution will appear in later chapters. In the case of the OH^- ion, for instance, an equilibrium

$$OH^- + 2e \rightleftharpoons H^- + O^{2-}$$

exists in sodium solution, and the change in solvation enthalpy which is involved in dissociation of the hydroxyl ion is one of the factors which determines the extent of the dissociation. It is therefore important to obtain an assessment of the solvation enthalpy of such binuclear species, but this raises a number of problems, particularly where dissociation equilibria occur. The experimental method uses a Born–Haber cycle (e.g. Figure 7.6) which includes the enthalpy of solution derived from a solubility–temperature curve. This is satisfactory where a single solute species is involved, but the solubility curve for sodium hydroxide in sodium, for example, must be a composite curve representing the solubilities of NaOH, NaH and Na_2O. The ΔH_{soln} term is relatively small in such cycles, but can be significant if the solvation enthalpy of the binuclear species is also small.

For potassium,[182] rubidium[183] and caesium,[184] no dissociation occurs when the hydroxide is dissolved in the liquid metal (see Chapter 9), and phase diagrams have been published from which it is possible to obtain some measure of the variation in hydroxide solubility with temperature. From equation 7.2,

$$U_x^{OH-} = (U_l - U_{M^+}) + \Delta H_{soln}^{\circ}$$

Values for the lattice enthalpies for all the hydroxides are known.[188] The enthalpies of solution of KOH, RbOH and CsOH are approximately the same, and fall in the range 17 ± 3 kJ mol^{-1}; this lack of precision is of minor importance in view of the relatively small magnitude of this quantity. The 'experimental' values for the ion solvation enthalpy in the various metal solvents are given in Table 7.3. The values in parentheses are based on extrapolation of the ΔH_{soln}° values to lithium and sodium solutions; this is justifiable since this term makes an almost negligible contribution to the solvation enthalpy, and in any case does not change greatly with change in metal solvent. Again, dissociation of OH$^-$ takes place in both sodium and lithium solutions, so that the quoted values of U_x^{OH-} are those which are relevant to the solution before dissociation takes place. This is particularly relevant to the treatment to be given in Chapter 9 to this dissociation.

Calculation of the solvation enthalpy of binuclear species depends entirely on a knowledge of the effective radius of the solute (see section 7.5). At present we have no direct means of determining the extent to which the internuclear distance in a group such as OH might change when immersed in the electronic environment of a liquid metal, so that calculations of solvation enthalpies are not possible. However, the following argument may make a contribution to solving this difficulty if it proves, by application to other systems, to be generally applicable. The U_x^{OH-} value for sodium hydroxide in sodium is near to that of the F$^-$ ion. Let us assume therefore that OH$^-$, like F$^-$, is octahedrally coordinated in sodium solution, in which case from Thompson's data[80] $(1-I)$ in equation 7.1 is 0.764, and the effective radius of the coordinating sodium atoms is 1.07Å. Substituting the U_x value for OH$^-$ into equation 7.1, we find that $\sigma_0 = 2.16$, so that the hard-sphere radius of OH$^-$ is 1.09Å. This is very near to the measured interatomic O–H distance, 0.98Å, in crystalline sodium hydroxide. Now solid sodium hydroxide, at

Table 7.3 Ion solvation enthalpy for the OH$^-$ and CN$^-$ ions, and related data (kJ mol^{-1})

Solvent metal	Solute	$-U_l$	$-U_{M^+}$	ΔH_{soln}°	$-U_x$
Li	LiOH	925	427	[17]	[481]
Na	NaOH	852	384	[17]	[451]
K	KOH	753	286	17	450
Rb	RbOH	724	279	17	428
Cs	CsOH	676	268	17	391
Na	NaCN	729	384	176	169

higher temperatures, has the rock-salt structure in which the OH^- ion is effectively spherical because of thermal motion, and is surrounded by six sodium ions. The inference here is that when sodium hydroxide is dissolved in liquid sodium, it undergoes very little change in its environment, and its radius is therefore very little influenced. If this concept proves to be a real one, it would simplify the theoretical treatment of many other solutions, such as CN^- and C_2^{2-}, and their interactions. Unfortunately, the simple relation between U_x and ionic radius expressed by equation 7.1 does not apply to these multiple-bonded solute ions; it neglects polar and other short-range forces arising from electron orbital overlap, and a more elaborate treatment will be necessary. An experimental value for U_x for cyanide solutions in sodium is included in Table 7.3. It is much smaller than the lattice energy for NaCN (and U_x for OH^-), as a result of an exceptionally high enthalpy of solution (see Table 7.1). This is the only ΔH_{soln} value at present available for a binuclear, multiple-bonded species, and further speculation must await additional experimental data.

7.7 Inclusive solvation enthalpies

All the above discussion is concerned, by definition, with the solvation of ions, where the electrons responsible for the charge on the solute atom are fully localized on that atom. It will be clear, however, that this treatment can only be applied in a very limited number of systems. There are many elements which dissolve in the alkali metals to give solutions in which the nature of the dissolved species is unknown; this applies particularly to the heavier elements in the p-block of the periodic table. On the other hand there may be experimental evidence to indicate that the solute species is polyatomic in nature, or that electron transfer to the dissolved species is not complete. In such cases the treatment given to ion solvation is again not applicable. If the concept of solvation is to have the wider significance in liquid metal solutions which it already has for solutions in other liquid media, it should not be restricted in its application by a precise knowledge of the nature of the dissolved species, and Hubberstey and co-workers[126,127,162] have proposed an alternative definition of solvation enthalpy termed the 'inclusive' solvation enthalpy. The relation between this and the 'classical' or ion solvation enthalpy (U_s) is shown in Figure 7.7. The inclusive solvation enthalpy U_{incl} is the enthalpy associated with the transfer of the neutral, monatomic, gaseous solute atom into the solution, and is so named because it includes all enthalpy terms arising from this transfer. In Figure 7.7, the final solution is represented merely by the word 'solution', this avoiding any implication regarding the nature of the solution. Values for U_{incl} have been calculated for solutes known to give ions in solution (H, O, N, the halogens), and are compared with the ion solvation enthalpies in Table 7.4. In this connection it is also useful to introduce a term ΔH_{solute}, which is the enthalpy of solution of the element itself. Its relation with the enthalpy of solution of the precipitating phase ΔH_{soln} (LiH in Figure 7.7) is shown in Figure 7.7.

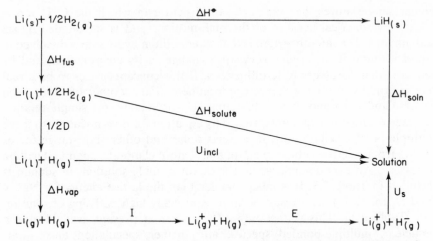

Figure 7.7 Relation between ion and inclusive solvation enthalpies. ΔH°, ΔH_{soln}, U_s, I, E and D have been defined in Figure 7.6; U_{incl} is the inclusive solvation enthalpy; ΔH_{solute} is the partial molar enthalpy of solution of the solute element alone (as opposed to the solution of the precipitating phase); and H_{fus} and H_{vap} are respectively the enthalpies of fusion and vapourization of lithium

The cycle required for the determination of the inclusive solvation enthalpy of a solid solute such as silicon is shown in Figure 7.8. This follows the same principles as for a gaseous solute (Figure 7.7), except that care must now be taken to allow for the (sometimes unusual) stoichiometry of the precipitating

Table 7.4 Some values of inclusive solvation enthalpy (U_{incl} kJ/g atom)

Solvent metal	Solute	$-U_{incl}$	$-U_x$	$-\Delta H_{solute}$
Li	Si	557	—	104
	Ge	490	—	113
	Sn	428	—	126
	Pb	311	—	115
	H	267	427	—
	O	780	1960	—
	N	638	3473	—
Na	Ge	387	—	—
	Sn	306	—	50
	Pb	248	—	64
	H	220	344	—
	O	636	1721	—
K	F	599	467	—
	Cl	478	316	—
	Br	422	295	—
	I	353	251	—

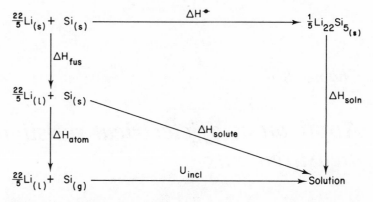

Figure 7.8   Inclusive solvation enthalpy for the solution of silicon in liquid lithium. ΔH_{atom} is the enthalpy of atomisation of silicon; other enthalpy terms are defined in Figures 7.6 and 7.7

phase, and the enthalpy of atomization of the solid solute now replaces the enthalpy of dissociation of the gas. ΔH_{solute} is again the enthalpy of solution of elementary silicon. Values for U_{incl} have now been determined for each of the heavier group IV elements, and are given in Table 7.4; U_x values are not available since the nature of the dissolved species is not known with certainty. U_{incl} for these elements shows a fairly regular decrease with increasing atomic weight, in both liquid lithium and liquid sodium.

The aim of this chapter has been to establish solvation enthalpy as a physical property of solutions in the liquid alkali metals which can be expressed in quantitative terms. The phenomena which give rise to solvation are reflected also in other physical properties of the solutions, and correlation of these various physical properties with solvation enthalpies is clearly desirable; the correlation between solvation and electrical resistivity will be discussed in Chapter 8. Because of its application to a wide range of solutes, it is the inclusive solvation enthalpy which is particularly useful in these correlations.

Application of electrical resistivity measurements

8.1 Introduction

For the study of solutions of single solutes in the alkali metals, and for monitoring the course of interactions which take place between solutes, the measurement of changes in electrical resistivity has proved to be by far the most useful technique. This is understandable, since the conduction band electrons are also those which are involved in solute interactions and in the solution process.

There is now general agreement on the manner in which a solute influences the conduction band electrons. The resistivity ρ of a pure metal is represented by the expression

$$\rho = mV_f/ne^2l \qquad (8.1)$$

where m and e are the mass and charge of the electron, n the number of free electrons per unit volume, V_f the velocity of electrons at the Fermi surface, and l the mean free path of the electron. In the discussion (Chapter 4) on the nature of dissolved species, solute elements were classified according to whether they added electrons to the conduction band (e.g. the group II metals) or withdrew electrons to form ionic species (e.g. H^-, O^{2-}, N^{3-}). We should therefore consider first how far the resistivity of a solution might be determined by variation in the number of conduction electrons. With all other factors in equation 8.1 constant, $\rho_1 = K/n_1$ for pure solvent metal and $\rho_2 = K/n_2$ for a solution, so that $\rho_1/\rho_2 = n_2/n_1$. For an alkali metal, on the free electron model, n_1 is known, and n_2 can be calculated from the concentration of solute. This hypothesis has been tested for a number of solutions and found to be quite untenable, so that variation in the number of conduction electrons on formation of a solution is not, by itself, the major factor responsible for changing the resistivity. All experimental evidence, however, is consistent with the belief that the introduction of a solute into the metal solvent reduces the mean free path of the electrons, i.e. leads to scattering of the conduction electrons. The most important observation is that the resistivity of the pure solvent metals is invariably increased, whether the solute is a metal or non-metal. An extensive theoretical treatment of the electronic theory of metals, as it applies to equation 8.1, has been given by Cusack,[163] and by Faber;[164] in this chapter we are primarily concerned with the extent to which various solutes increase the resistivity, and it will suffice for present purposes if we regard all such increases as reflecting the varying extents to which the solutes scatter the conduction electrons.

The technique whereby reactions in solution are followed from changes in conductivity is a useful and standard method for solutions in molecular liquids, e.g. in aqueous solutions. In such cases the liquid medium is usually a poor conductor, and the conductivity arises largely from the species, usually ionic, in solution. Liquid metal chemistry presents a complete contrast; the liquid medium now has a high conductivity, and the changes caused by solutes are very small. If the results are to be meaningful, the liquid alkali metal must be very pure, and the measurements must be made with high precision. These two aspects can conveniently be discussed at this stage.

Table 8.1 compares the electrical properties of the liquid alkali metals with other good conductors—copper, silver and mercury. The alkali metals are among the most highly conducting metals in the periodic table, though their specific conductivities fall short of the best conductors, copper and silver (column 4); all are more conducting than mercury. Of the alkali metals, sodium has the highest conductivity, and we can best regard lithium as being anomalous in an otherwise steady decrease in conductivity with increasing atomic size.

Table 8.1 also includes the related values for electrical resistivity. In solution chemistry in molecular liquids it is usual to deal in conductivity units, but the scientific literature on metals almost invariably uses resistivity values. In future we shall therefore present results in resistivity units; this has the added advantage that it is the resistance which is actually measured, and all the changes which are brought about on forming solutions in the alkali metals are positive values. The temperature coefficients of resistivity $d\rho/dT$ for the

Table 8.1 Electrical properties of some pure metals at 200°C

Metal	Resistivity ρ $\Omega m \times 10^{-8}$	Temperature coefficient $d\rho/dT$	Specific conductivity $(\Omega^{-1} m^{-1} \times 10^6)$
Li	29.1	0.037	3.44
Na	13.5	0.039	7.41
K	20.6	0.066	4.85
Rb	35.8	0.090	2.79
Cs	56.6	0.119	1.77
Cu	2.90	0.0068	34.5
Ag	2.33	0.0041	42.9
Hg	114	0.13	0.877

liquid alkali metals except lithium do increase slightly with temperature[165] and the values given in Table 8.1 are for 200°C. Resistivity increases more rapidly in the liquids than in the solids, which is exceptional. Liquid lithium is unique among metals in having values of $d\rho/dT$ which decrease as temperature increases.

Electrical resistance R is related to resistivity ρ by the dimensions of the metal sample, i.e.

$$R = \rho l/A$$

where A is the cross-sectional area and l the length of the sample. When measuring the very low resistances of the alkali metals, they are contained in steel capillary tubes in which A is as small, and l as long as is practicable, in order to obtain R values which can be more accurately determined. Early equipment for measuring the resistivity of liquid sodium[165] used a single capillary having internal diameter 1.5 mm and length 120 mm in conjunction with a precision Kelvin–Wheatstone bridge. The filled capillary represents a pair of conductors in parallel, and the resistance of sodium, R_{Na}, is obtained from the relation

$$R_{Na} = R_C R_T/(R_C - R_T)$$

where R_T is the resistance of the cell containing sodium, and R_C the resistance of the empty capillary. Errors can arise, however, from the difficulty in measuring the internal diameter of a narrow tube and the possible variations over its length, and there are inherent problems associated with the electrical insulation of the capillary tube. The apparatus which has proved to be satisfactory is illustrated in Chapter 3 (Figures 3.5 and 3.6). The capillary tube is in the form of a loop bridged by two steel discs 20 cm apart. These are silver-soldered to the capillary and to silver electrical leads; current passes between the discs, and electrical insulation in the main body of the apparatus is not relevant. To obviate any errors arising from measurement of the capillary tube dimensions, the apparatus is calibrated using pure sodium. With sodium circulating through the capillary, the cell is initially heated to 250°C to facilitate wetting of the steel by the liquid metal, and thus to eliminate contact resistances within the capillary. The resistivity of any solution (ρ_s) then introduced into the capillary can be determined from the simple ratio $\rho_s/\rho_{Na} = R_s/R_{Na}$.

The main advantage of this type of apparatus is that it can be used to give a continuous record of changes taking place in the liquid metal passing through the capillary, and is particularly useful for following rates of solution or precipitation, and for detecting when a solution is saturated or when a reaction in solution is complete.[166] The technique has been used, for instance, to determine the solubility of non-metals and their salts, and is applicable to all the alkali metals. Wide temperature ranges are available, and because of the accuracy with which solubilities can be determined, the data are of importance in the contexts of purification, and the study of solvation in liquid metals. Liquid miscibility gaps can also be detected; in the sodium–lithium system, breaks occur in the resistivity–temperature curve, as a given mixture is cooled, at temperatures corresponding to phase separation. The technique has also been used successfully to define liquidus curves and eutectics in many binary systems (notably Li–Na, Na–K, Na–Ba, Na–Sr, and systems involving Li and Na with Si, Ge, Sn and Pb) which are discussed in detail elsewhere in this volume.

Interaction between two solutes can also be followed. Association between dissolved species is reflected in a deviation from additivity in the resistivity, since a composite species will increase the resistivity of the solvent metal to an extent which differs from the sum of the separate species. When there is no association, however, the increase in resistivity of the solution is additive. This provides one of the most important methods for determining both the extent and the stoichiometry of interactions such as occur in Na–H–O, Li–N–H, Na–Ba–N and Na–Ba–N–C solutions. Solvation of dissolved species is also reflected in resistivity changes. It is convenient to use resistivity changes to monitor reaction kinetics; an example was given in Chapter 3 (Figure 3.7) where the rate of hydrogen removal from liquid lithium is monitored by observing resistivity decrease as a function of time.

8.2 Resistivity coefficients

The resistivity of a wide range of solutions of non-metals, their salts, and metals have now been measured, and in all cases it is found that dissolution of the solute increases the resistivity of the liquid alkali metal; this increase is attributed to electron scattering.

The relation between resistivity and concentration is illustrated in Figure 8.1,[175] which is typical of many solutions of non-metals. The remarkable feature is the linear increase in resistivity with concentration, which indicates that each solute atom or ion scatters electrons to the same extent as its predecessor. The slope of this line is entirely characteristic of a given solute in a given liquid metal, and is defined in terms of a resistivity coefficient $d\rho/dx$, which is the increase in resistivity ($\Omega m \times 10^{-8}$) caused by one mole per cent of the solute at a given temperature. In Figure 8.1, for example, the resistivity coefficient for the nitride ion in liquid lithium at 400°C is 7.0×10^{-8} Ωm (mol % N)$^{-1}$. Dissolved nitride has a greater effect on the resistivity of lithium than has hydride; $d\rho/dx$ for the hydride ion is 4.9×10^{-8} Ωm (mol % H)$^{-1}$. Deuteride and hydride effects are indistinguishable although their solubilities are slightly different. The linear relationship continues up to saturation; thereafter, if the solutions are being prepared by addition of nitrogen or hydrogen to liquid lithium, the gas will continue to react, but an insoluble phase of nitride or hydride is formed. No further nitride or hydride enters the solution, and resistivity remains constant.

The change in resistivity at saturation is exceptionally clear and provides, therefore, a precise method of measuring the solubilities of the salts. In very

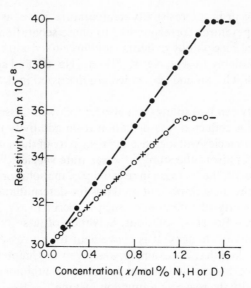

Figure 8.1 Resistivity–concentration isotherms at 400°C for solutions of nitride (solid
circles), hydride (open circles) and deuteride
(+) in liquid lithium

dilute solutions, it appears possible that $d\rho/dx$ may deviate from values
characteristic of most of the concentration range.[175] There is a distinct possibility that this may be due to trace impurities in the solvent metal, or to
absorption of the non-metal by the steel container on prolonged contact.
However, even if these minor deviations are real this does not invalidate the
significance of resistivity coefficients since the concentrations where such
deviations have been observed are below those normally used in the study of
chemical interactions.

Another assumption usually made in this work is that the resistivity coefficients are independent of temperature. Where $d\rho/dx$ has an appreciable magnitude (Table 8.2) and temperature ranges of the order of 100–400°C (Figure
8.2) are concerned, this would seem to be an acceptable assumption. It should
be recorded, however, that variations with temperature can be observed
where the coefficient is very small; thus, solutions of lithium in liquid sodium
have $d\rho/dx$ values of 0.22, 0.14, 0.09 and 0.04 Ωm $\times 10^{-8}$ (mol % Li)$^{-1}$ at
100°, 200°, 300° and 400°C respectively.[147] The application of this technique
to obtain both resistivity and solubility data over a temperature range is
illustrated in Figures 8.2 and 8.3, which show this correlation for nitrogen[20]
and hydrogen[166] respectively in liquid lithium. Taking the Li–N diagram (Figure 8.2), addition of nitrogen at 300°C will show a sharp break in resistivity when saturation is reached, which pinpoints the position of the
hypereutectic liquidus for 300°C, as shown on the lower curve. By raising the
temperature excess nitride will redissolve, and further breaks in the resistivity

Table 8.2 Resistivity coefficients $d\rho/dx$ ($\Omega m \times 10^{-8}$ (mol % solute)$^{-1}$)

Solvent metal	Solute	$d\rho/dx$	Temperature (°C)	Reference	Solvent metal	Solute	$d\rho/dx$	Temperature (°C)	Reference
Na	N^{3-}*	8.8	300	177	Li	N^{3-}	7.0	400	175
	H$^-$	4.5	400	24		H$^-$	4.9	400	175
	O^{2-}	1.8	400	24		D$^-$	4.9	400	175
	Li	0.04	400	147		O^{2-}	2.1	300	176
	K	1.14	300	168		Na	<0.1	400	147
	Rb	3.4	300	168		Al	5.8	420	72
	Cs	4.2	300	168		Si	10.4	400	167
	Sr	0.2	400	171		Ge	11.2	400	167
	Ba	2.5	300	21		Sn	11.3	400	162
	Ge	8.6	400	170		Pb	9.0	400	167
	Sn	11.7	400	170					
	Pb	11.2	300	170	K	F$^-$	3.9	700	178
	Ag	2.8	300	168		Cl$^-$	5.7	700	178
	Au	5.0	300	168		Br$^-$	6.9	740	178
	Cd	5.4	300	168		I$^-$	9.1	700	178
	Hg	4.1	300	172		Na	1.2	300	173
						Rb	0.22	100	174
						Cs	1.61	110	169
					Rb	K	~0	100	174
					Cs	O^{2-}	3.2	30	179
						H$^-$	4.7	300	56
						Na	5.71	110	169
						K	1.01	110	169

*In sodium–barium solutions.

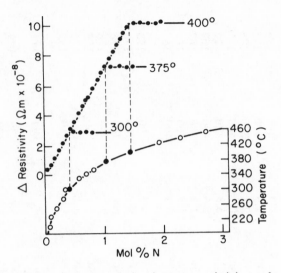

Figure 8.2   Correlation between resistivity and
solubility for solutions of lithium nitride in liquid
lithium

can be observed, thus defining the solubility–temperature curve. A similar
procedure is illustrated in Figure 8.3 for the lithium-hydrogen system.
Both diagrams indicate also that the variation of resistivity coefficient with
temperature is very small, and apparently no greater than the errors involved
in measuring concentration and resistivity.

The same results can be obtained by a method involving progressive
cooling of the solution, and this is often more convenient experimentally. The
two approaches are compared in Figure 8.4 for solutions of silicon in liquid
lithium.[72] Figure 8.4(a) corresponds directly with Figures 8.2 and 8.3. At
407°C, resistivity rises linearly with increasing silicon concentration up to
saturation at 1.17 mol % Si, and thereafter remains constant. At a lower
temperature, the solubility of silicon is less and the resistivity rises to
correspondingly lower values before levelling. The gradient for the
unsaturated solutions gives a coefficient $d\rho/dx$ of 10.4×10^{-8} Ωm (mol %
Si)$^{-1}$; and this is virtually independent of temperature.

The procedure employing cooling of the solution is shown in Figure 8.4(b) and
takes advantage of the fact that when an unsaturated solution is cooled, a
temperature is reached at which the solution is exactly saturated. At that
temperature precipitation begins, and a break is observed in the
resistivity–temperature curve. On cooling a 1.17 mol % silicon solution from
a temperature above 407°C, the resistivity decreases linearly with decreasing
temperature until precipitation occurs at 407°C; this is shown by a filled circle
in Figure 8.4(b). The precipitating phase in the Li–Si system is the compound
$Li_{22}Si_5$. Subsequently, the resistivity decreases nonlinearly as the now
unsaturated solution is gradually depleted of silicon. More dilute solutions

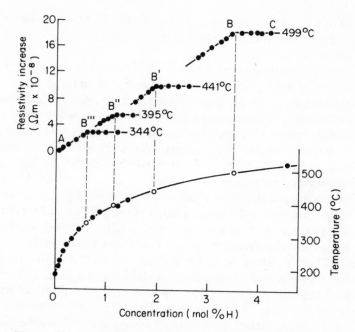

Figure 8.3 Increase in resistivity over that of lithium for lithium–hydrogen solutions, as a function of concentration; and the solubility of hydrogen in liquid lithium

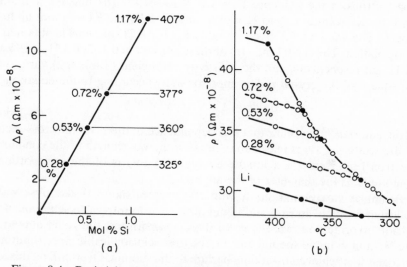

Figure 8.4 Resistivity changes for dissolution and precipitation of $Li_{22}Si_5$ in liquid lithium

(e.g. 0.72, 0.53 and 0.28 mol % Si, Figure 8.4(b) give more extensive linear regions, since precipitation of $Li_{22}Si_5$ occurs at progressively lower temperatures.

The existence of the linear resistivity–concentration relationship implies that an atom of solute added to a near-saturated solution encounters an electronic environment in the medium which is little different from that of the pure metal solvent. With solutions of non-metals this will be the situation, since solubility values are generally very low and even in a saturated solution the metal solvent is in very large excess. Where the solute is another metal, however, a different situation can arise; solubilities can be high, and the solute and solvent may even be miscible in all proportions. Under these circumstances the $\rho–x$ line must eventually bend towards the resistivity of the solute metal. However, if the concentration of the solute metal is restricted to a few atom per cent, a linear $\rho–x$ relation is usually observed, and the slopes can be compared with those obtained for solutions of non-metals.

Table 8.2 gives values of resistivity coefficients for both metallic and non-metallic solutes in the liquid alkali metals, and the list is believed to be complete at the time of writing. There are several generalizations which can be made on the basis of these values. The resistivity coefficients will be determined by many factors, predominant among which should be the charge on the solute ion, the size of the solute atom or ion, the extent to which it is solvated, and the liquid metal used as solvent. Some idea of the size factor can be obtained by comparing similar solutes in the same alkali metal. As early as 1962, Bronstein et al.[178] studied the resistivity of solutions of the potassium halides in liquid potassium; they found that an increase in the size of the anion increases the resistivity, reflecting an increase in the scattering cross-section for electrons in the sequence $F^- < Cl^- < Br^- < I^-$. The linear $\rho–x$ relation holds for KF solutions up to 12 mol %, and for KBr and KI solutions up to 20 mol %. The value of 5.7 given in Table 8.2 for KCl solutions is obtained by interpolation. The crystal radii (r) of these anions are 0.133, 0.181, 0.196 and 0.219 nm respectively, and the resistivity coefficient (Table 8.2) varies with the square of the crystal radius approximately according to the equation

$$d\rho/dx = 1.6 \times 10^{18}r^2 + 1.2$$

Bronstein et al.[178] consider that this dependence of $d\rho/dx$ on the cross-section of the scattering ion is reasonable, because the wavelength of the conduction electron (~0.5 nm for a kinetic energy of 5 eV) is of the same order of magnitude as the diameter of the anions.

A similar variation of $d\rho/dx$ with crystal ionic radius is also observed in solutions in liquid sodium and liquid lithium, with respect to solutions of the nitride, hydride and oxide ions. In these comparisons, the value quoted for the N^{3-} ion in liquid sodium may not be precisely applicable since solution of nitrogen was achieved by adding barium to the sodium. Overlooking this, and accepting 0.145, 0.154 and 0.171 nm as the ionic radii of O^{2-}, H^- and N^{3-}, respectively, we find (Table 8.2) that the resistivity coefficient in liquid sodium, as well as in liquid lithium, increases regularly with increase in ionic radius, despite the fact that the solutes are taken from different groups of the periodic table. This observation (involving solute ions of different charges), taken together with corresponding observations on the halide ions (where charges are unchanged), would suggest that the size of a dissolved species is more important than its charge in determining the resistivity coefficient.

It is also interesting to note that the coefficients seem to be largely independent of the alkali metal used as solvent. At present true comparisons are only possible for the hydride and oxide ions, but in these cases the results are as follows:

	in Li	in Na	in Cs
H^-	4.9	4.5	4.7
O^{2-}	2.1	1.8	3.2

The value for O^{2-} in liquid caesium is an early determination, and would bear re-examination. This observation has practical value; it indicates that where it might be desirable to achieve (say) wider solubility ranges, a mixture of alkali metals may be used as solvent, to give resistivity coefficients which are still meaningful. Some results by Bronstein et al.[178] suggest that this feature may extend outside the alkali metals. For solutions of $BiCl_3$ in liquid bismuth, $d\rho/dx$ is 17.0×10^{-8} Ωm (mol % $BiCl_3$)$^{-1}$. Since each mole of $BiCl_3$ introduces three Cl^- ions into the solution, the value of $d\rho/dx$ for each Cl^- ion in liquid bismuth is 5.7, which is the same as the coefficient for Cl^- in liquid potassium.

Resistivity coefficients for solutions of the alkali metals in one another are relatively small. They tend to be near zero for metals which adjoin one another in the group (e.g. Li in Na and Na in Li), and to increase somewhat as the two metals concerned are further separated in atomic number.

8.3 Correlation with solvation enthalpies

The factors which determine the magnitudes of resistivity coefficients are also the factors responsible for solvation in liquid metal solutions. However, the relative significance of the various factors will depend on the particular physical property concerned, and any correlations which can be identified make a contribution to our understanding of these factors and their relative importance. Hubberstey and Dadd[162] have reported on a correlation between resistivity coefficients and inclusive solvation enthalpies, which is shown in Figure 8.5, and they discuss their results in terms of the charge density of the solute atom, on which both parameters ultimately depend. In this context, however, charge density is difficult to define in any quantitative way. It depends most directly on the charge and the size of the solute species, and for four of the elements (Pb, Sn, Ge and Si) included in Figure 8.5 the charge to be assigned is not known. It is for this reason that the inclusive, rather than the ionic, solvation enthalpy is used in the correlation.

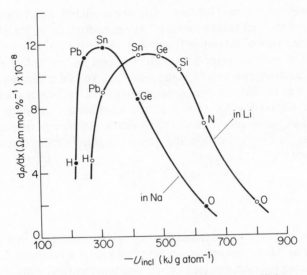

Figure 8.5 Correlation between resistivity coefficients
and solvation enthalpies U_{incl} for solutions in liquid
sodium and liquid lithium

We note, however, that electron localization on the solute is likely to increase in the order Pb < Sn < Ge < Si, as the solute changes from metallic to non-metallic in character, and that the radius to be assigned to the solute ion also changes with the charge. This radius, in turn, determines the number of metal atoms which can cluster around the solute ion in the coordination sphere, so that charge and radius of the solute species are probably the factors which are mostly responsible for the order H < Pb < Sn < Ge < Si < N < O into which the solvation enthalpies fall (Figure 8.5). The variation in resistivity coefficients, however, is more complex. The coefficient is a measure of the effect of the solute species on the electron mean free path in the liquid metal; this increases with the charge on the solute ion, but also with the disruption of the liquid structure in the vicinity of the solvated solute ion so that solvation itself now introduces an additional factor influencing the resistivity coefficient. These two factors can oppose one another, and Hubberstey and Dadd[162] believe that this is responsible for the way in which the resistivity coefficients pass through a maximum (Figure 8.5).

It is of interest that very similar curves are obtained for solutions in liquid lithium and liquid sodium. Corresponding values for the halide ions in liquid potassium also give a smooth plot of $d\rho/dx$ against U_{incl}, but insufficient data are as yet available in the low $d\rho/dx$ range to ascertain whether the maximum also occurs for solutes in liquid potassium.

Chapter 9

Dissociation of hydroxide and amide groups in liquid alkali metal media

9.1 Introduction

Up to this stage, we have been concerned primarily with simple solutions consisting of one solute only, with solubilities, with the nature of species formed in solutions in liquid alkali metals, and with some of the techniques by which the presence of a solute may be detected and its concentration monitored. Some of the most interesting aspects of liquid metal chemistry arise, however, when we consider the possible reactions which can occur between two or more solutes present together in a liquid metal medium. The chapters in this book have been arranged so that concepts such as solvation, resistivity coefficients etc. can be used in the interpretation of solute interactions without further explanation, and so that reference to experimental details can be kept to a minimum. The reactions which have been studied include non-metal–non-metal, metal–non-metal, and metal–metal interactions; we shall not attempt, however, to classify reactions under these headings, choosing rather to deal first with those reactions which have made the greatest contribution to our understanding of reactivity in an electronic medium.

The reaction which has been most extensively studied is probably that between hydrogen and oxygen; this is not only because of its considerable theoretical interest, but also because the results have practical implications. Thus, when water (perhaps inadvertently) is introduced into an excess of liquid sodium (e.g. in a fast nuclear reactor coolant circuit) sodium hydroxide is first formed, and this will dissociate into H^- and O^{2-} ions to an extent which will depend on temperature and concentration. It is a general observation that corrosion of metal containers is greatly enhanced when sodium hydroxide is introduced into the liquid sodium. To take two examples from many such observations, Thorley and Tyzack[180] measured this enhanced corrosion for niobium in liquid sodium; and Oberlin and Saint-Paul[181] have reported that the introduction of sodium hydroxide sets up stress corrosion cracking in the austenitic alloys used in the generator tubes of fast breeder reactors, whereas in sodium polluted by sodium oxide only, the same materials were not affected in this way. It is easy to dismiss this effect as 'hydroxide corrosion', but in view of the formation of hydride ion, and hydrogen gas during hydroxide decomposition, it is essential to understand the equilibria involved before such corrosion phenomena can be interpreted.

Some analogies can be drawn between the reactions of the alkali metals with water and with ammonia. The simple reaction

$$M + NH_3 \longrightarrow MNH_2 + \tfrac{1}{2}H_2$$

is analogous to the reaction

$$M + H_2O \longrightarrow MOH + \tfrac{1}{2}H_2$$

so that in this sense the amide ion NH_2^- is analogous to the hydroxyl ion OH^-. The subsequent dissociation of the NH_2^- ion

$$NH_2^- \xrightarrow{4e} 2H^- + N^{3-}$$

can occur in liquid lithium, and again corresponds to the behaviour of the OH^- ion. However, the analogy is limited by the fact that the heavier metals, sodium to caesium, do not form nitrides. Hydrogen–nitrogen interactions will be discussed in section 9.4 below; because of the higher temperature ranges involved, we are necessarily concerned with reactions of gaseous ammonia, rather than with the well-known alkali metal–liquid ammonia systems.

9.2 Reactions involving separate metal–metal hydroxide phases

When the reaction between the metal and its hydroxide is represented by the equation

$$MOH + 2M \xrightarrow{\Delta H^{\circ}_{reac}} MH + M_2O \qquad (9.1)$$

it is assumed that the reaction takes places between metal and hydroxide in two separate phases, and that the hydride and oxide are also produced as separate phases. The magnitude and sign of the enthalpy change ΔH°_{reac} gives a picture of the variation in reactivity as the alkali metal M is changed. The ΔH° values derived from equation 9.1 do not necessarily give direct evidence of the extent of hydroxide dissociation when the OH^- ion is dissociating while in solution in the single metallic phase to give H^- and O^{2-} which are also in solution; for this purpose we need to examine the equation

$$OH^- + 2Na \xrightarrow{\Delta H^{\circ}_{diss}} H^- + O^{2-} + 2Na^+ \qquad (9.2)$$

which shows that dissociation of OH^- requires the transfer of two electrons from two sodium atoms to be the OH^- ion.

In fact, the decomposition of the OH^- ion when separate phases (equation 9.1) or a single phase (equation 9.2) are involved proceeds along very similar

Table 9.1 Enthalpy changes ΔH°_{reac} (kJ mol^{-1}) for the reaction
MOH + 2M → MH + M$_2$O

Metal	ΔH°_{MOH}	ΔH°_{MH}	$\Delta H^\circ_{M_2O}$	ΔH°_{reac}
Li	−509.4	−90.4	−595.8	−176.8
Na	−426.7	−57.3	−415.9	−46.5
K	−425.8	−56.9	−361.5	+7.4
Rb	−413.8	−54.4	−330.1	+29.3
Cs	−404.7	−56.5	−317.6	+30.6

lines as M is changed from lithium to caesium; this in itself is interesting, since the observations of 'traditional' chemistry, which is usually concerned with ionic solids or molecular liquids, do not often give clear indications of the nature or stoichiometry of interactions occurring in the electronic environment. We should therefore consider first the reaction according to equation 9.1. Enthalpy changes (which are used in the absence of free energy data) are given in Table 9.1, together with the corresponding standard enthalpies of formation of the hydride and oxide.[61] These values suggest that with the heavier metals, potassium, rubidium and caesium, there should be negligible hydroxide decomposition.

The liquidus curves for the three-phase systems are shown in Figure 9.1, and bear a close resemblance to the Bredig M–M halide systems discussed in Chapter 5. The rubidium and caesium curves are extremely simple; the liquidus curve falls smoothly from the melting point of RbOH (382°C) or CsOH (346°C) to that of the metal, so that at temperatures above the hydroxide melting points the liquid hydroxide and metal are miscible in all proportions. With potassium there is a diminishing tendency towards mutual solubility. The liquidus curve for K–KOH mixtures passes through a maximum, with consolute point at K, 36.6 mol % and 610°C, above which temperature the two liquids are again miscible in all proportions; the hydrogen pressure over the liquid mixture is negligible. In none of these diagrams are there regions corresponding to the appearance of any other component, so that no detectable reaction occurs, in agreement with the positive enthalpy change.

The phase diagram for the Na–NaOH system was described in limited detail by Shikov[185] and extended above the consolute point by Maupre.[181,186] It is extremely complicated, and need not be reproduced here; it consists of regions corresponding to solutions of sodium in the hydroxide, and the hydroxide in sodium together with oxide and hydride regions and the solid and liquid solutions formed by these various components. It is clear that in the sodium system an equilibrium exists which is represented by equation 9.1 and is consistent with the small but negative ΔH°_{reac} value for the reaction, and that the position of the equilibrium varies with temperature and composition. The enthalpy change for the Li–LiOH reaction is large and negative. Accordingly, liquid lithium reacts readily with its hydroxide and the forward

128

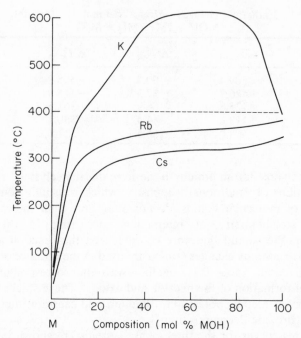

Figure 9.1 Liquidus curves for the K–KOH[182],
Rb–RbOH[183] and Cs–CsOH[184] systems

reaction to give oxide and hydride is virtually complete; a two-component phase diagram (Li + LiOH) is therefore not possible. Thermochemical calculations by Migge[187] also indicate that the compound LiOH cannot coexist with liquid lithium.

9.3 Hydrogen–oxygen interactions in solution

If the interaction between hydrogen and oxygen, when dissolved in the liquid metal (equation 9.2), follows the same pattern as outlined above for the separate phases, we might expect that these solutes would show no mutual attraction in liquid lithium, form OH^- in liquid potassium, rubidium and caesium, and that an equilibrium between OH^-, H^- and O^{2-} would be evident in liquid sodium. That this is indeed the case is supported both by the Born–Haber cycles for the reaction, and by a limited amount of experimental evidence.

A Born–Haber cycle for the dissociation of OH^- in sodium is shown in Figure 9.2. This cycle is built up from a series of steps, each of which might be considered to take place within the solution. The symbols used to denote the separate steps are the same as those which have been defined in Chapter 7. The numerical values assigned to each step are the best values available at the time of writing. Some are known precisely, but others particularly U_{M^+} and

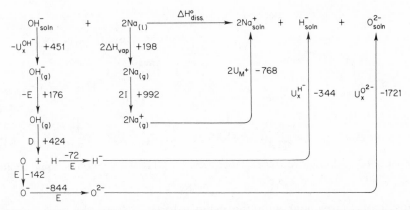

Figure 9.2   Born–Haber cycle for the dissociation of OH⁻ in liquid sodium

the three U_x values) are less precise, since they are determined from other quantities which are themselves less precise. Sufficient data are available to allow similar cycles to be prepared for lithium and potassium, and calculation of the enthalpy of dissociation of OH⁻ leads to the following results:

Solvent	Li	Na	K	
ΔH°_{diss}	−96	+38	+100	(kJ mol⁻¹)

The main conclusion to be drawn from these values is that in the heavier alkali metals the species present is predominantly OH⁻, but that in liquid lithium the species H⁻ and O²⁻ will predominate. The situation changes over at sodium. These are enthalpy values, but the inclusion of an entropy term is unlikely to make a significant difference; the dissociation of OH⁻ should lead to an increase in entropy, but this may be offset by an increase in short-range ordering in the liquid metal. If the equilibrium constant K is represented by [OH⁻]/[H⁻] [O²⁻], then since

$$\Delta G = -RT \ln K$$

a small value of ΔG for the OH⁻ dissociation in liquid sodium indicates a situation in which the quantities of the three species are comparable. Experiments indicate that this is indeed the case, and the ΔH_{diss} value for sodium solutions quoted above is quite near the ΔG value determined from direct measurement of the H⁻ and OH⁻ concentrations in solution (see below).

There is an alternative Born–Haber cycle which can be drawn up which involves the reaction between the elements themselves, the enthalpies of formation of Na₂O, NaH and NaOH, and the enthalpies of solution of these compounds in sodium (Figure 9.3). The preferred experimental values for these various processes have been selected by Ullman and co-workers;[191,192] using these values, this cycle gives a value of +10 kJ for the enthalpy of OH⁻ dissociation, which is again small and positive and in this respect is in

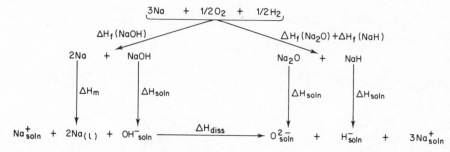

Figure 9.3   Alternative Born–Haber cycle for OH⁻ dissociation in liquid sodium

satisfactory agreement with the value calculated from the cycle given in Figure 9.2. The two cycles are, of course, complementary in that they are both derived from the same original data; but in view of the experimental errors inherent in all the various steps, the approach via two separate cycles gives increased confidence that the $\Delta H^{\circ}_{\text{diss}}$ value for OH⁻ in sodium is indeed small and positive.

All experimental work supports the belief that OH⁻ dissociates in liquid lithium and sodium, but not in potassium, rubidium or caesium. So far as the three heavier metals are concerned, the phase diagrams themselves (Figure 9.1) provide sufficient evidence that no hydride or oxide component is produced. Resistance measurements on liquid lithium at 480°C, in which hydrogen and oxygen were successively added to the liquid metal, indicated that no interaction occurs between these solutes.[189] Conclusive evidence for the interaction between hydrogen and oxygen in liquid sodium was first provided by Pulham and Simm[105] who studied the reaction of water vapour with liquid sodium by measuring the hydrogen pressures produced over the metal as a function of the amount of water added. The initial reaction of water vapour at the metal surface is extremely rapid, and each mole of water was found to liberate half a mole of gaseous hydrogen, according to the initial reaction

$$Na_{(l)} + H_2O_{(g)} \longrightarrow NaOH_{(l)} + \tfrac{1}{2}H_{2(g)}$$

The hydrogen then dissolves at a measurable rate to give an equilibrium pressure which was found to depend on the quantity of sodium hydroxide originally produced in the solution. This equilibrium pressure is proportional to the H⁻ concentration in solution (see Chapter 4). Assuming that this proportionality is not influenced by the other species present in solution, the H⁻ concentration can be determined from the measured hydrogen pressure. From these data, the proportion of hydrogen existing in the solution as H⁻ could be deduced.

The hydrogen present as hydride (plus the small amount in the gas phase) is plotted in Figure 9.4 as a function of the total hydrogen added as water at 400°C. If all oxygen in solution is present as OH⁻, results should follow the

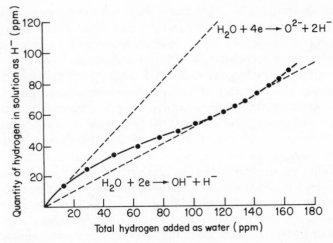

Figure 9.4   Dissociation of water in liquid sodium at 400°C

lower broken line in Figure 9.4; if no association between H^- and O^{2-} occurs, results should follow the upper broken line. The experimental results show a smooth transition from one extreme to the other. Full dissociation of OH^- occurs at very low concentrations; with increasing concentration, however, progressively less hydride is produced, and hydroxide is formed at the expense of hydride and oxide. Limited evidence, derived from equilibrium pressure measurements, also indicates that an attractive interaction occurs between hydrogen and oxygen dissolved in NaK (78 wt % K) alloy, the results being rationalized on the basis of a similar equilibrium.[190] The presence of an equilibrium as indicated by hydrogen pressure measurements is supported by changes in the resistivity of liquid sodium as water vapour is absorbed. Simultaneously with their pressure measurements, Pulham and Simm[105] measured the resistivity of sodium at 400°C. For very small additions of water, the resistivity increase was that expected from the sum of the known resistivity coefficients of H^- and O^{2-} (Chapter 8); the coefficient for H^- (4.5) is much larger than that for O^{2-} (1.8), and at higher concentrations the resistivity increase was reduced as some H^- was withdrawn into the OH^- ion. The hydrogen pressure measurements also gave an indication of the variation in the equilibrium constant with temperature. At 400–450°C, the behaviour was as described above. At 350°C, the pressures were indistinguishable, for a given hydrogen concentration in solution, irrespective of whether the hydrogen was added as the element or introduced as water, thus suggesting that the dissociation of OH^- diminishes with increasing temperature. This may, however, be a solubility effect. The solubility of hydride and oxide is reduced at the lower temperature, and association of H^- and O^{2-} is also reduced at the lower available concentration, as Figure 9.4 illustrates.

In sodium-water heat exchangers which are a feature of both the prototype fast reactors and of many planned commercial designs, the sodium and water circuits are separated by a single metal wall, and the possibility that a leak might develop has given rise to research, in many parts of the world, into the consequences of a leak of a small amount of water into a large volume of liquid sodium. One technique by which such leaks can be detected is the measurement of the hydrogen produced. Earlier approaches to the problem centred on the measurement of the hydrogen gas released into the argon cover gas; knowing the relation between the hydrogen pressure and the concentration of hydrogen in solution, the magnitude of any leak can then be assessed. However, as discussed above, this relation is dependent on the oxygen concentration in solution in the sodium, so that this aspect has to be studied under reactor conditions. In one such investigation carried out by R. A. Davies and co-workers[193] at the Dounreay Experimental Reactor Establishment in Scotland, a circulating sodium loop containing 70 kg of sodium was used, and oxygen added in the form of sodium monoxide or sodium hydroxide. In each case the Sievert's law constant K (where $K = [H]_{soln}/p_{H_2}^{\frac{1}{2}} \text{ppm torr}^{-\frac{1}{2}}$) was measured. With sodium purified, by gettering, to an oxygen content of 2 ppm, K was about 4.4, and did not change significantly in the temperature range 330–570°C. At an oxygen content of 30 ppm, K increased to about 5.2, indicating that some of the added hydrogen was now associated with dissolved oxygen.

For the practical reasons just mentioned, it is also important to know the total time required for the completion of all the reactions resulting from the introduction of water into a large excess of sodium, in order that impurity monitoring devices can be suitably adjusted. The initial reaction

$$Na + H_2O \rightarrow NaOH + \tfrac{1}{2}H_2 \qquad (9.3)$$

is rapid, and produces a film of sodium hydroxide at the metal surface. We are considering conditions under which the amount of hydroxide produced is relatively small, so that the hydroxide, or its reaction products, will eventually be completely soluble in the large excess of sodium.

Whittingham and co-workers[194,195] have carried out an impressive programme of research at the Central Electricity Generating Board laboratories at Berkeley (UK) on the thermodynamic and kinetic aspects of oxygen–hydrogen interactions, in the course of which the reaction of sodium hydroxide with liquid sodium was shown to proceed as follows:

$$2Na_{(l)} + NaOH \text{ (s or } l) \xrightarrow{\text{fast}} Na_2O_{(s)} + NaH_{(s)} \qquad (9.4)$$

$$Na_2O_{(s)} \xrightarrow{\text{slow}} O^{2-}_{soln} \qquad (9.5)$$

$$NaH_{(s)} \xrightarrow{\text{slow}} H^-_{soln} \rightleftharpoons H_{2(g)} \qquad (9.6)$$

with a contribution from the reaction

$$Na_{(l)} + NaOH \text{ (s or } l) \xrightarrow{\text{fast}} Na_2O_{(s)} + \tfrac{1}{2}H_{2(g)} \qquad (9.7)$$

The importance of reaction 9.7 was assessed by experiments in which

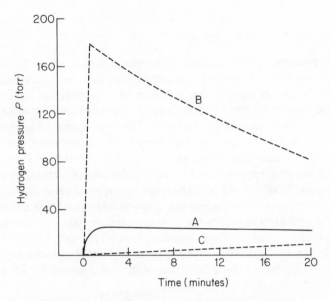

Figure 9.5 Hydrogen gas evolution following the addition of a sodium hydroxide pellet to liquid sodium at $335°C$

pellets of sodium hydroxide were dropped onto the surface of liquid sodium contained in an evacuated steel vessel, and changes in hydrogen pressures were monitored by a pressure transducer. The pressures observed at $335°C$ are shown by the full line A in Figure 9.5. The broken line B represents the pressure changes which would be predicted if initial reaction occurred entirely accordingly to equation 9.7, followed by hydrogen absorption; and line C represents the pressure changes according to equations 9.4–9.6. At $335°C$, the maximum recorded hydrogen pressure is only 11 per cent of the predicted value according to equation 9.7; the contribution from this reaction varies from 3 per cent at $300°C$ to 16 per cent at $500°C$, so that sodium hydroxide decomposes in a large excess of liquid sodium predominantly in accordance with equations 9.4–9.6.

An important advance became possible as a result of the development of electrochemical meters (which will be discussed in a later chapter) which enable direct measurement to be made of the concentration of hydrogen or oxygen in solution. The rate at which sodium hydroxide is converted into its reaction products in liquid sodium is governed by the rate of solution of Na_2O and NaH in the liquid metal (equations 9.5 and 9.6); using the electrochemical meters to follow the accumulation of oxide and hydride ions in solution, it is found that the half-life for the decomposition of sodium hydroxide in sodium is 1–2.3 minutes at $500°C$ (depending on whether the hydrogen or oxygen meter is used), increasing to 16–17 minutes at $300°C$.[195]

The eventual outcome of the sodium hydroxide reaction is, of course, the introduction of H^- and O^2 ions in solution, which then take part in the equilibrium

$$H^- + O^{2-} \xrightleftharpoons{\hspace{1cm}} OH^- + 2e \qquad (9.8)$$

Hydrogen pressure measurements have been useful in showing that such an equilibrium exists in sodium solution, but are not suitable for the determination of an equilibrium constant for the reaction because of the assumptions which have to be made regarding the possible influence of O^{2-} in solution on the Sievert's constant for hydrogen. However, since electrochemical meters can give a direct record of hydrogen and oxygen in solution, equilibrium constants can now be determined and this has been achieved using two approaches, i.e. using solutions of sodium hydroxide, or by the separate addition of hydrogen and oxygen to liquid sodium. In the first case, it was observed[195] that following addition of sodium hydroxide to excess sodium, the equilibrium values for hydrogen and oxygen concentrations were lower in all cases than the total amount of each species added as hydroxide. The deficit can be attributed to the formation of OH^- in solution (equation 9.8), and on this basis values for the equilibrium constant K, where

$$K = \frac{[OH^-]}{[H^-][O^{2-}]}$$

have been determined over a range of temperatures. Results are given in Table 9.2. The consistency is remarkable when we consider that they involve concentrations of only a few parts per million. For the same reason, K is expressed in Table 9.2 in ppm units also, as these seem particularly appropriate for such solutions. There is no obvious variation of K with temperature.

In the alternative approach,[194] sodium was purified by gettering to reduce the oxygen and hydrogen contents to less than 1 ppm, and hydrogen or

Table 9.2 Equilibrium constants K $(= [OH^-]/[H^-][O^{2-}])$ for the reaction $H^- + O^{2-} \rightleftharpoons OH^- + 2e$

Temperature (°C)	Equilibrium constants			
	from OH^- decomposition		from $H_2 + O_2$ addition	
	(ppm^{-1})	(mol fraction$^{-1} \times 10^3$)	(ppm^{-1})	(mol fraction$^{-1} \times 10^3$)
340	—	—	0.19†	7.8†
370	0.19	7.9	0.22*	9.0*
400	0.16	6.5	—	—
435	—	—	0.18*	7.4*
460	0.13	5.3	—	—
470	—	—	0.11†	4.5†
500	0.18	7.4	—	—

*Hydrogen addition to solution of oxygen in sodium.
†Oxygen addition to solution of hydrogen in sodium.

oxygen added as the gases. All experiments were performed with concentrations of hydrogen and oxygen below their saturation value, with an upper limit of about 100 ppm. From the amounts of the gases added, and the final concentrations of hydrogen and oxygen measured by the meters, it was possible to deduce the equilibrium hydroxide concentration and hence the equilibrium constants. These values are also given in Table 9.2; they agree well with values obtained from hydroxide decomposition, and with recent measurements by other workers. The collected results show no very obvious drift of K with temperature, though Ullman, Kozlev and co-workers,[191,192] have proposed an equation

$$\log K \text{ (atom fraction}^{-1}) = 1.53 + 1340/T$$

for this relationship from which the free energy of formation of the hydroxide in sodium is given by

$$\Delta G_{OH^-}(\text{J mol}^{-1}) = -25660 - 29.3T$$

In practical terms, these results mean that if the hydrogen impurity in liquid sodium is reduced to 1 ppm, about 5 to 15 per cent of the total oxygen in solution in the sodium will be present as hydroxide at temperatures between 300°C and 500°C; and at a hydrogen background of 0.1 ppm only 0.5 to 1.4 per cent of the oxygen present will form hydroxide.

9.4 Reactions with ammonia vapour, and related reactions

It is fortunate that studies of gaseous ammonia with liquid alkali metals have dealt specifically with lithium and with caesium. These represent the two extremes of behaviour; reactions of ammonia with sodium, potassium and rubidium will probably resemble reactions with caesium, and a few measurements which have been carried out on liquid sodium would appear to confirm this.[250]

In liquid caesium

Using liquid caesium,[246] a known quantity of ammonia at a known pressure was introduced to the liquid metal contained in a vessel of similar design to that described in Chapter 2 (Figure 2.6) with provision for stirring the liquid metal by an electromagnetic pump, and the decrease in pressure recorded by means of a pressure transducer. The pressure–time curves observed are shown in diagrammatic form in Figure 9.6, where pressure is shown as a fraction of the initial ammonia pressure. The curves most readily interpreted are those obtained at temperatures in the 130–180°C range (Figure 9.6(a)). Absorption occurred in two stages. A rapid fall in pressure from the initial value of about 50 kNm^{-2} (375 torr) to half the initial value, in accordance with the equation

$$Cs + NH_3 \longrightarrow CsNH_2 + \tfrac{1}{2}H_2 \qquad (9.9)$$

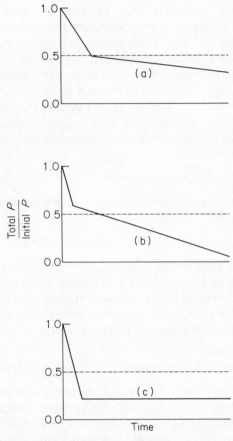

Figure 9.6 Absorption curves for caesium–gaseous ammonia reaction (diagrammatic). (a) 130–180°C; (b) above 180°C; (c) below 130°C

occurred in 25 s at 160°C; the curve then showed a clearly defined break, followed by a slow absorption attributed to the solution of hydrogen produced in the initial reaction. From absorption curves of this type Pulham and Hill[246] were able to draw some interesting conclusions. Assuming the stoichiometry of the reaction to remain unchanged throughout each reaction, separate curves for ammonia absorption and hydrogen evolution can be constructed. From the variation of ammonia pressure with time, it was concluded that the reaction was first-order, giving first-order rate constants $K_p (s^{-1})$ of 0.140 and 0.255 at 133°C and 160°C respectively. Arrhenius plots gave an activation energy of 31.5 kJ mol^{-1} for the direct reaction of gaseous ammonia with liquid caesium to form caesium amide and hydrogen. This reaction was studied using the same reaction vessel as was used for caesium–hydrogen kinetic investigations; and since the surface area of metal and the gas volume were the same in each case, it is possible to compare the rates of absorption of gaseous ammonia and of hydrogen. At 133°C and 160°C, the K_p values for hydrogen are 0.171×10^{-3} and $0.364 \times 10^{-3} \text{s}^{-1}$ respectively, so that the reaction of ammonia with caesium is approximately 820 and 700 times faster than the reaction of hydrogen at these two temperatures; this is no doubt a consequence of the stronger chemisorption of the polar ammonia molecule at the metal surface.

At higher temperatures, represented by experiments in the 180–225°C range, both the initial ammonia reaction and the subsequent hydrogen reaction occurred at a faster rate, as illustrated in Figure 9.6(b). At 203°C, for example, the inflexion point was reached in less than 10 s. Also, the ratio total P/initial P at which the inflexion occurred increased steadily above 0.5 as the temperature increased. This cannot be explained on the basis of decomposition of the amide ion:

$$CsNH_2 \longrightarrow CsH + \tfrac{1}{2}N_2 + \tfrac{1}{2}H_2$$

since the pressure eventually falls to zero, indicating that the gaseous product consists of hydrogen only. Decomposition of amide to imide would produce extra hydrogen, but imides of the heavier alkali metals are not known. This behaviour is suitably explained by assuming that as the temperature increases the reaction of ammonia and hydrogen become more competitive. Some evolved hydrogen may then react with the metal before the ammonia reaction is complete, so that the total pressure at the inflexion includes an increasing contribution from unreacted ammonia.

As the temperature was decreased below 130°C, the pressure fell rapidly to a value less than half the initial pressure, and the subsequent rate of reaction of hydrogen became very slow (Figure 9.6(c)). The initial rate of reaction became steadily faster because less hydrogen was now produced, so that the initial rate moved closer to the rate for absorption of ammonia alone. As the temperature is progressively decreased, we are also moving into a state in which some ammonia molecules dissolve in the liquid caesium without chemical reaction. As long as some caesium amide is still present, it is feasible to represent the reaction as

$$Cs + 2NH_3 \longrightarrow CsNH_2.NH_3 + \tfrac{1}{2}H_2 \qquad (9.10)$$

Among the alkali metal amide ammoniates, the caesium compound has the highest thermal stability (which is the case also with the analogous hydrated hydroxide $CsOH.H_2O$). According to this reaction, the final gas pressure should be only one quarter of the initial value. When the temperature is lowered to 50°C, the molar ratio of hydrogen produced to ammonia added is only 1:8. Eventually, at temperatures around the melting point of caesium (28.5°C) and below, liquid caesium and ammonia have been found, by electrical conductivity[247,249] and knight shift measurements[248] to be miscible in all proportions without reaction, so that gas pressures would fall rapidly to zero.

The fact that pure liquid ammonia and a pure alkali metal can exist together as stable mixtures at low temperatures is, of course, a well-known part of liquid ammonia chemistry. In such mixtures ammonia is in excess. When solutions of sodium and potassium are frozen, the metal is precipitated, but solutions of lithium give a precipitate of the tetrammine $Li(NH_3)_4$, and caesium solutions give a precipitate which is similar in appearance. The formation of ammines will make a contribution to the stability of these solutions, but it is questionable whether ammines have much significance in dilute solutions of ammonia in liquid caesium at 50–100°C. A more attractive concept is based on solvation; the NH_3 molecule is polar, and the solution probably owes its stability to solvation of NH_3 by (say) four or six caesium atoms.

The same investigation[246] gave some basic data on solutions of caesium amide in liquid caesium. Changes in electrical resistivity of caesium on addition of solid caesium amide showed an almost linear relation between resistivity increase $\Delta\rho$ (see Chapter 8) and concentration, and precipitation temperatures on cooling the solutions were used to determine amide solubility. At 82°C, 100°C and 120°C, solubility was found to be 2.03, 2.73 and 3.51 mol % NH_2^- respectively, giving a value of 16.2 kJ mol^{-1} for the partial molar enthalpy of solution. The average resistivity coefficient $d\rho/dx$ is 2.4×10^{-9} Ωm (mol % NH_2^-); this is small by comparison with a number of monatomic species discussed in Chapter 8, but insufficient data are available on resistivity coefficients of other polyatomic solutes to allow further discussion. Addition of ammonia to liquid caesium at 150°C, followed by hydrogen removal, gave resistivity changes consistent with the simple reaction shown in equation 9.9.

In none of the experiments discussed above was there any evidence of the decomposition of the amide ion in solution. This is as expected, since for any of the alkali metals the enthalpy change in the reaction

$$M + MNH_2 \longrightarrow 2MH + \tfrac{1}{2}N_2 \qquad (9.11)$$

is too small to encourage such dissociation.

In liquid lithium

The amide ion does dissociate completely in liquid lithium, because in this particular case an alternative decomposition route becomes possible. Because of the negative enthalpy of formation of the well characterized nitride Li_3N, and the unique ability of lithium to strongly solvate the N^{3-} ion, this ion forms readily in solution, leading to the decomposition reaction

$$4Li + LiNH_2 \longrightarrow Li_3N + 2LiH$$

A guide to the readiness with which this occurs in solution is given by the corresponding solid-state reaction, for which the enthalpy change is a large negative value $(-165.4$ kJ mol$^{-1})$. There is therefore a close analogy between the reactions

$$OH^- \xrightarrow{\quad e \quad} H^- + O^{2-}$$

and

$$NH_2^- \xrightarrow{\quad 4e \quad} 2H^- + N^{3-} \tag{9.12}$$

In liquid caesium neither dissociation takes place, whereas in liquid lithium each dissociation is complete.

Reaction 9.12 has been studied by measuring resistivity changes which occur on addition of lithium nitride, lithium hydride and ammonia to liquid lithium[175] (Figure 9.7). Line OL represents the resistivity of solutions of nitride alone, and line OM that for hydride alone. Addition of hydride, at point A, to a solution already containing nitride increased the resistivity to point B, and the slope AB was found to be that expected for hydride alone. Similarly, addition of nitride to a solution at point C which contained hydride, increased the resistivity to point B, and the slope CB was that expected for nitride alone. Thus, irrespective of the order in which the two solutes were dissolved, they each exerted their characteristic electron scattering; the resistivity is therefore additive, indicating that nitride and hydride ions do not react chemically in liquid lithium.

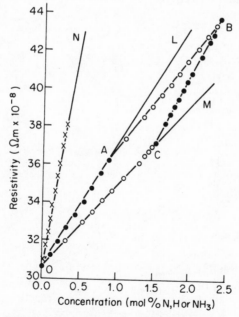

Figure 9.7 Resistivities of solutions in liquid lithium at 420°C. OA and CB denote addition of nitride; OC and AB denote addition of hydride; and ON denotes addition of ammonia

If the NH_2^- ion dissociates completely, then ammonia should react with liquid lithium according to the equation

$$6Li + NH_3 \longrightarrow Li_3N + 3LiH \qquad (9.13)$$

for which the enthalpy change is -392 kJ mol^{-1}. The resistivity change on addition of ammonia vapour is shown by line ON in Figure 9.7. If each molecule of ammonia dissociates to give one nitride ion ($d\rho/dx = 7.1 \times 10^{-8}$ Ωm (mol % N)$^{-1}$) and three hydride ions ($d\rho/dx = 4.9 \times 10^{-8}$ Ωm (mol % H)$^{-1}$) then the slope of line ON should be 21.8×10^{-8} Ωm (mol % NH_3)$^{-1}$. The experimental value is 22×10^{-8} Ωm (mol % NH_3)$^{-1}$, confirming complete dissociation to nitride and hydride ions.

In liquid sodium

The reaction of gaseous ammonia with liquid sodium resembles the reaction with caesium, but is much slower. The absorption curve (Figure 9.8) again falls into two parts, a more rapid reaction with ammonia in which the pressure falls to about half the initial pressure, followed by a slower absorption of hydrogen. However, the initial ammonia reaction now requires about five minutes at 300°C, compared with less than 10 s for caesium at 200°C; and the reaction requires 60 minutes for completion. The curve in

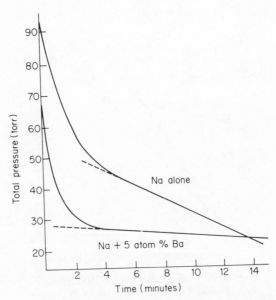

Figure 9.8 Reaction of gaseous ammonia with liquid sodium, and with a sodium–barium alloy, at 300°C

Figure 9.8 is typical of measurements carried out at other temperatures in the range 140–300°C, and with initial pressures between 80 and 160 torr. The sodium amide produced (mp 210°C) is stable at these temperatures but must have a much lower solubility in liquid sodium than has caesium amide in liquid caesium, since it appears as tiny globules of liquid on the surface of the sodium.

In Chapter 11, we shall see that solutions of barium in sodium resemble liquid lithium in that when nitrogen and hydrogen are added, the N^{3-} and H^- ions produced do not interact, and ammonia dissociates to N^{3-} and H^- ions. Measurements of the rate at which ammonia reacts (Figure 9.8) gives some further information on this reaction. Using a 5 atom % barium solution, the first stage is about three times as fast as with sodium alone; but the pressure again falls to about half the original value, indicating that the first step in the reaction is amide formation:

$$Ba + 2NH_3 \longrightarrow Ba(NH_2)_2 + H_2$$

The subsequent rate at which hydrogen pressure decreases is only about one sixth of the pure sodium value, even though it is known[251] that addition of barium increases the rate of absorption of hydrogen into sodium. Since we know that the solution eventually contains only N^{3-} and H^- anions, the decomposition of the amide ion must contribute to the apparent slow hydrogen absorption. The barium amide is unlikely to have any appreciable solubility in sodium (the solvation energy of NH_2^- will be much less than for N^{3-}), so that its decomposition

$$NH_2^- \xrightarrow{\;2e\;} N^{3-} + H_2$$

will contribute hydrogen to the gas phase, and the decrease in pressure due to hydrogen absorption will be partly compensated by the production of hydrogen by amide decomposition. The reaction

$$NH_3 \xrightarrow{\;6e\;} N^{3-} + 3H^-$$

is eventually completed when all the hydrogen has been converted to hydride ion.

Chapter 10

Reactions between liquid alkali metals and water

10.1 Introduction

The vigorous reaction between sodium and water must surely be one of the best known reactions in chemistry. Both scientists and non-scientists remember it from early schooldays as the most impressive demonstration of chemical reactivity, and the expanding usage of sodium as a coolant fluid means that the reaction remains as a potential hazard in the chemical industry. It is not surprising, therefore, that alkali metal–water reactions have been a topic for research for many years. What is surprising is that it cannot yet be claimed that the reactions are fully understood, so that research is still continuing; much of it has not yet appeared in the open literature.

The emphasis is, and has been, on sodium because of its greater use in industry. The complexity arises because there are many pathways by which the reaction can proceed to eventual equilibrium (as was seen in the previous chapter), and many different conditions under which the reaction can be carried out. The reaction conditions with which we are concerned can be broadly classified as follows:

(a) liquid metal + water vapour, metal in excess
(b) liquid metal + water vapour, water in excess
(c) liquid metal + liquid water, metal in excess
(d) liquid metal + liquid water, water in excess

Reactions of the solid metals are not included as a class, since the heat evolved at the beginning of the reaction is usually sufficient to melt the metal. Where this is not the case, surface films on the solid metals are formed, and the kinetics of the reaction are usually concerned with penetration of reactants through films rather than with their direct interaction. It will be seen, from the experiments to be described in this chapter, that not all investigations can be classified rigidly under the above headings; and so far as the study of reaction mechanisms is concerned, a classification is not particularly relevant. The initial reaction of water vapour with liquid metal, for example, obviously involves an excess of liquid metal, and it is the ultimate products of the reactions which depend on whether metal or water is in large excess.

It is relevent here to dispel, in passing, a popular misconception which still persists in the minds of those who are not familiar with research on this topic. If the explosion which results when a pea of sodium is dropped into a sink of water is multiplied by a factor depending only on the quantity of sodium used, then the explosion resulting when kilograms (or tonnes) of the metal are brought into contact would achieve horrific proportions. This is not the case; but the possibility that a leak in the secondary heat exchangers might cause serious damage to a fast nuclear reactor gave rise to extensive research in the period around 1950–1970 when such reactors were first being considered as a commerical proposition. Many of the investigations were engineering in nature, and because of the large scale and expense of this type of work, were undertaken by various atomic energy authorities and agencies. Although we are concerned in this book with the fundamental aspects of alkali metal–water reactions, a few references from this period will illustrate the scope of the research, and the concern that these reactions might indeed represent a major hazard in the operation of such a reactor. The hazards have been outlined by Epstein;[208] his paper is one of a series describing work carried out at the General Electric Company of America. Much of this work was carried out on small-scale reactor equipment,[209] or provided a theoretical interpretation of previous studies.[210] Parallel studies were carried out at Atomic Power Development Associates which involved both small-scale steam incursions into sodium–potassium alloy[211] and simulated studies of the fate of released hydrogen.[212] In the UK, the UKAEA at Dounreay built test rigs which were small-scale models of heat exchangers, and followed closely the events during and after the injection of water into sodium.[213,214] At Harwell, superheated steam was injected into liquid sodium at 450°C, but little change in the temperature or pressure of the system was observed.[196] Interatom in Germany carried out similar tests using the types of tubes employed in heat exchanger construction, and monitored the corrosion which took place;[215] from the results obtained they were able to construct a mathematical model to describe this type of event.[216] Work on heat exchangers in the USSR led to the development of a set of thermodynamic calculations relating to sodium–water leaks.[217] A 'specialists meeting on sodium–water reactions' was held in America in 1968 to correlate this work with the design of reactor equipment,[218] and current trends were reviewed in 1972.[219]

Research in this field is continuing, though the problem has lost much of its urgency. This is due to the realization that the reaction is not so hazardous as was first imagined. Furthermore, the development of hydrogen and oxygen meters makes it possible now to detect small leaks of water into sodium at a very early stage, and there is no longer any doubt that liquid sodium is suitable as a coolant in conjunction with sodium–water heat exchangers.

Nevertheless, our understanding of the fundamental chemistry of water–liquid metal reactions is far from complete.

10.2 Liquid metal–water vapour reactions

Reactions involving surface films

Published research on lithium has been concerned with the solid metal only; this is perhaps owing to its higher melting point, and to the much greater difficulty in manipulating liquid lithium than the other alkali metals. However, some references will be included here since there are likely to be close similarities between the reaction of water vapour with solid lithium, and with the static liquid metal.

Deal and Svec[197] observed that hydrogen evolution was relatively slow, and that the reaction could be conveniently followed by pressure changes in a closed system. Reaction rates were determined at temperatures of 45–75°C, and pressures in the range 20–100 torr. Irving and Lund[198] used X-ray and thermal balance techniques, and water vapour pressures less than 12 torr. In each case, three separate reaction stages could be distinguished. An initial linear increase in rate with pressure was attributed to coverage of the metal by a black film of lithium hydroxide, and this rate changed to logarithmic as the hydroxide film grew in thickness. At somewhat higher pressures the rate was again linearly dependent on pressure, as the composition of the film changed to the hydrate $LiOH.H_2O$. Besson and Pelloux[199] extended the study to higher temperature and pressure ranges, and were able to correlate the change from logarithmic to linear rate law with the dissociation pressure of the hydroxide hydrate; activation energies for the various process were also calculated.

The reaction between liquid sodium and water vapour has been studied quite extensively. When the amount of water added is so small that all reaction products can dissolve in the sodium, H^- O^{2-} and OH^- ions are formed in solution, and this has been treated in detail in Chapter 9. Using larger amounts of water vapour and static liquid sodium, surface films are formed, and the composition of the film depends primarily on temperature and the composition of the gas phase. The general position can be summarized as follows for most experimental conditions.

At the lower temperatures and pressures, the classical reaction to produce sodium hydroxide and hydrogen occurs. At temperatures above the melting point of sodium hydroxide, sodium monoxide also appears in the film; sodium hydride may also be present, depending on whether the hydrogen produced by the reaction is above or below the dissociation pressure of the hydride. Woollen[196] also reported that some sodium peroxide was formed in the reaction of superheated steam with an excess of sodium at 450°C.

The conditions used by various workers have differed widely. Corrsin *et al.*[200] observed the rate at which temperature increased when a sodium droplet was held in a stream of moist argon, and commented on a 'pre-ignition glow' which could be seen on the surface of the metal. Furman[201] used very low water vapour pressures; a stream of helium, at 1 atm pressure, containing only 100–400 ppm of water vapour, was passed over liquid sodium, and the hydrogen content of the effluent gas was measured. No hydrogen was retained in the sodium sample at any temperature between 215°C and 340°C, so that under these extreme conditions the solid reaction product was sodium oxide only. Longton[202] studied pressure changes in a closed system containing static sodium and an initial water vapour pressure of 10 torr, and his results were consistent with the above summary. Addison and Manning[203] also deduced the nature of the film from pressure changes, but extended the initial water vapour pressure used to 650 torr. In their experiments, a phial of sodium was broken inside a glass vessel containing water vapour, held in a thermostat. The sodium splashed, as droplets, on to the sides of the vessel, and subsequent reaction with water vapour was measured by a spiral gauge. At 175°C and 650 torr pressure, there was a momentary flash of flame as sodium was first exposed to the vapour; at 400 torr, a transient glow appeared on the sodium droplets, which is probably the pre-ignition glow observed by Corrsin.[200] Reaction products were again as summarized above. However, at water vapour pressures approaching atmospheric, the sodium hydroxide film can behave as a separate phase and form an effective barrier between water vapour and liquid sodium; water vapour was detected in the gas phase at equilibrium, even though excess sodium was present, and its pressure was equal to the saturation vapour pressure of sodium hydroxide. Time intervals of 1–20 minutes were required to achieve equilibrium, which is consistent with a mechanism involving penetration of reactants through the surface film.

Rate laws have also been determined. Cornec and Sannier[204] and Besson and Pelloux[205] observed a parabolic law followed by a linear rate law, and were able to relate these rates to changes in the nature of the surface film. As a result of these experiments we now have a fairly clear picture of the nature of the films produced by water vapour at a static liquid sodium surface; they represent indirect evidence, since conclusions are based on pressure measurements, but McKnight and Brockway[204] have provided direct confirmation by analysis of the reaction products using reflected X-ray diffraction techniques, when both sodium oxide and sodium hydroxide were identified.

Very little information is available on the reactions of liquid potassium, rubidium or caesium with water vapour. It seems likely that they will resemble, in principle, those of sodium, but there has not been the same obvious industrial application to stimulate research on these heavier alkali metals, and the reactivity is much higher. Some early experiments by Saltsburg[206] in 1956 and 1957 are usually quoted in this connection. A flow method was used in which water vapour and sodium–potassium alloy were mixed by flow at right angles to one another; this greatly reduced the time of contact. The metal flowed through the reaction zone into a quenching tank for analysis of the products, and the gas outflow was also analysed. The technique showed promise, but unfortunately consistent results could not be obtained.

Sodium flames

At temperatures much higher than those employed in the above experiments, i.e. as the sodium approaches its boiling point (883 °C), a vapour phase reaction becomes dominant. The reactions to produce sodium hydroxide and sodium monoxide are strongly exothermic, so that controlled, self-sustaining flames can be produced in which sodium is burning in water vapour, just as it does in oxygen:

$$\text{Na}_{(g)} + \text{H}_2\text{O}_{(g)} \longrightarrow \text{NaOH}_{(l)} + \tfrac{1}{2}\text{H}_2 \qquad \Delta H^\circ = -287 \text{ kJ mol}^{-1}$$

$$2\text{Na}_{(g)} + \text{H}_2\text{O}_{(g)} \longrightarrow \text{Na}_2\text{O}_{(l)} + \text{H}_{2(g)} \qquad \Delta H^\circ = -350 \text{ kJ mol}^{-1}$$

In a preliminary experiment,[207] a sodium disc 22 mm in diameter and 6 mm deep was heated in an insulated nickel container, and water vapour at 1 atm pressure was passed over the sodium at a flow rate of 20 litres/minute. A self-sustaining surface reaction commenced at 100°C, producing a rapid rise in the sodium pool temperature. These reactions heated the pool to incandescence, and at the sodium boiling point a flame region was to be seen above the pool, which superheated by as much as 100°C. The aim of these experiments was to study the corrosive effect of such a flame on reactor containment metals, so that a flame of fixed geometry and steady burning rate was required. This was achieved by boiling the sodium in a metal burner, from which the metal vapour issued through a nozzle 1.5 mm in diameter. The sodium vapour stream was allowed to establish itself for a few seconds; when a water vapour atmosphere was then passed through the surrounding vessel at a rate of 20 litres/minute, a steady yellow–white sodium flame 12 mm long and 7 mm diameter was produced. The apparatus was so arranged that test specimens could be placed in the flame. Lithium metal will also burn brilliantly in a flowing stream of steam mixed with argon.[230]

Rates of reaction with water vapour

In none of the research described above, or in Chapter 9, was it possible to determine at what rate the metal reacts with water vapour, except that the reaction was extremely rapid, the approach to final equilibrium being determined either by the rate of solution of the reaction products, or the rate of penetration of reactants through surface films. To measure the initial reaction rate it is clearly necessary to work with continually renewed metal surfaces at which no film can develop.

A flow technique which satisfies these requirements has been devised at the University of Nottingham,[220] with the possibility in mind also that measured

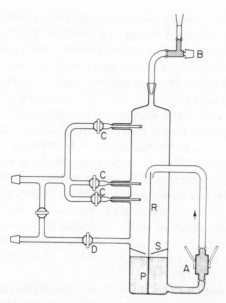

Figure 10.1 Reaction vessel used for
measurements of reaction rates of liquid
NaK alloy with low pressures of water
vapour

reaction rates might reveal some information about the mechanism of the
reaction. Since the rate will be temperature dependent, it is desirable to use
the lowest temperature which is conveniently available, and for this reason
sodium–potassium alloy (50–50 wt %) rather than sodium was used. Very
small vapour pressures were also employed. The reaction vessel is shown in
Figure 10.1. Constructed from Pyrex glass, it consisted essentially of a cylinder 350 mm long and 35 mm diameter operated vertically, down which
passed a stream of water vapour. The dimensions of the vessel were such that the
vapour moved under streamlined flow. A continually fresh metal surface was
provided by circulating the liquid alloy, in the direction shown by the arrow,
by means of the pump A. A glass rod R was suspended from the centre of the
orifice into the metal pool P; the metal flowed down this rod at a rate of about
1 litre/minute, forming a cylindrical, uniform column of liquid, the surface
area of which could be measured. A glass shield S protected the reactor from
any splashing which tended to occur when starting or stopping the flow of
metal. Purified water vapour was admitted from a bulk (containing vapour at
a pressure of 20 torr) by means of the needle valve B, and passed out of the
vessel via tap D into a vacuum frame where the issuing gas was trapped and
analysed. Entry pressures were of the order of 0.1 torr, and pressure
decreased as the vapour passed over the NaK column. At these pressures the
reaction was remarkably controlled, and no visible film was formed on the
metal column.

When the vapour is flowing from a constant entry pressure, a steady state is set up in the reactor whereby the gas composition at any given level remains constant. The change in composition as vapour passed down the reactor was measured by sampling the gas at the three positions marked C in Figure 10.1. The usual chemical methods of gas analysis were not applicable to the minute samples withdrawn, and a quadrupole mass spectrometer was employed. The sampling tubes C were fitted with silicon carbide 'leaks', which reduced the pressure of the samples to the order of 10^{-6} torr, as required by the mass spectrometer. Since the spectrometer can only show the relative proportions of the molecules which enter it, a spiral ionization gauge was also used to measure the total pressure within the mass spectrometer system. By this means it was possible to determine the changes in pressure and composition which occurred as water vapour at a known flow rate and low initial pressure passed over a metal column of known surface area. For reactions above room temperature, the apparatus was essentially the same, but modified in detail to allow the reactor vessel to be wrapped with electrical heating tapes, capable of increasing the temperature to 150°C.

The time of exposure of vapour to the metal was less than 1 s, and a plot of rate against pressure showed the reaction to be first order. From the relationship

$$\mathrm{d}v/\mathrm{d}t = K_{\mathrm{abs}} S\, p$$

where v is the volume change during the reaction, calculated from measured pressure changes, S is the surface area of metal to which the vapour is exposed, and p is the average pressure, the absolute rate constant K_{abs} has been calculated. The rate constant has been defined in these terms by Herold[221] for alkali metal–gas reactions, and used by other workers,[222,223] and is therefore useful for purposes of comparison. Values for K_{abs} are given in Table 10.1 for a range of temperatures. Extrapolation to 250°C allows a comparison to be made with corresponding values for the reactions of hydrogen with sodium and with potassium.[224] It is clear that water reacts at least 10^{-7} times faster than does hydrogen, and has a significantly lower activation

Table 10.1   First-order rate constants for the reaction of sodium–potassium alloy with water vapour

System	Temperature (°C)	K (cm s^{-1} Pa^{-1})	Activation energy E (kJ mol)
NaK$_{(l)}$ + H$_2$O$_{(g)}$	25	1.3×10^{-4}	
	50	8.2×10^{-4}	39
	100	2.4×10^{-3}	
	250*	5×10^{-1}	
Na$_{(l)}$ + H$_{2(g)}$	250	2.5×10^{-8}	72.4
K$_{(l)}$ + H$_{2(g)}$	250	9.0×10^{-8}	66.5

*Extrapolated.

energy. It can be argued that by the very nature of this flow experiment, some molecules of water may be passed through the reactor without having the opportunity to impinge on the metal surface. This, however, must be a limitation in any method, and the linearity of the dv/dt against p results would suggest that this does not occur. In any case, if such a defect is inherent in the experiments, the value of the rate constant (Table 10.1) will be lower than the true value, and it will be interesting to see whether future experiments can produce even higher K_{abs} values.

Mechanism of reaction

There seems to be no reason to doubt that the mechanism of the reaction of water vapour with a liquid alkali metal will be the same as with the solid metal. This has been elucidated by Bremner and Volman,[225] who studied the fractionation of hydrogen isotopes which occurred in the reaction of partially deuterated water vapour with alkali metal mirrors at 23°C. The reaction was followed using both static and flow conditions. In the static method, water vapour was injected into a cylinder 200 mm long and 60 mm diameter, which carried a metal mirror on its inside wall, and the products were analysed by mass spectrometer. In the flow method, a stream of water vapour in argon at atmospheric pressure was passed through a cylindrical reaction tube 16 mm diameter, over a metal mirror 130 mm long, then analysed similarly. In view of the rapidity of the reaction already discussed, it is of particular interest that Bremner and Volman's results point to the fact that every collision of a water molecule with the metal surface results in its removal from the gas phase. The first stage in the reaction is

$$M + H_2O \longrightarrow MOH + H$$

which is the proven gas phase reaction. This involves the rupture of one of the O–H bonds in the H_2O molecule, which is associated with the chemiluminescence observed in the reaction of water vapour with sodium.[200,203,226] The hydrogen atoms are adsorbed at the surface, and interact to form H_2 molecules which desorb into the gas phase. It is the recombination step which can be followed by a study of isotope distribution. As a result of the reaction, the ^1H isotope is enriched; if we write

$$\underset{(a)}{^1H/^2H \text{ (in water vapour)}} + M \longrightarrow \underset{(b)}{^1H/^2H} \text{ (in product gas)}$$

then (a)/(b) is about 2, and this enrichment is independent of the water:metal ratio provided that the metal is in excess. The reaction occurs between adsorbed hydrogen atoms as a result of quantum mechanical tunnelling, for which the relative probabilities are $^1H:^2H = 9.9$. The order of effectiveness of metal films to bring about ^1H enrichment was found to be

$$Na \cong K < Li < Rb \cong Cs$$

although the differences were not large.

10.3 Reaction of alkali metals with liquid water

Some factors governing reaction rates

In the introductory section to this chapter, a number of references were made to experiments in which liquid water was injected into the liquid metals, or vice-versa. These were engineering in character, carried out in steel vessels, and designed largely to study such effects as pressure pulses etc. which might arise in a nuclear reactor cooling system. In 1965 Ford[228] published a comprehensive review of the sodium–water reaction (with 90 references). This, and the 1970 review by Henry,[229] were concerned in particular with the sodium–water steam generator, and the effects of a leak in the system. The significance of the various possible reactions was considered, but in this type of experiment the measured rates depend primarily on the surface areas of the reactants in contact, and hence on the degree of mixing of the reactants. Surface areas in contact are increased at temperatures above the melting point of sodium hydroxide (318°C), and by dissolution of reaction products; they can be decreased by barriers formed by solid or liquid reaction products, or by water vapour.

There has also been a series of small-scale laboratory experiments designed to find out more precisely the factors governing rates of reaction, and some of the physical properties which have been considered are listed in Table 10.2. It is a general observation that the chemical reactivity of the alkali metals increases from lithium to caesium, and this applies to their reactions with water. However, it is not possible to attribute this to any one property. One essential difference between the reactions with vapour and liquid water is that the hydroxide produced can dissolve away from the scene of the reaction into excess liquid water, whereas the solubility of the hydroxide in the liquid metal is very low; the solubility of sodium hydroxide in liquid sodium is only 0.0038 per cent at $100°C$[231] and in consequence the rate at which the hydroxide dissolves will be slow. Lithium hydroxide has a relatively low solubility in water compared with the other hydroxides (Table 10.2) and this may account

Table 10.2 Physical properties of the alkali metals which might influence the rate of their reaction with liquid water[230]

	Li	Na	K	Rb	Cs
(a) $-\Delta H$ (kJ mol^{-1}) for $M_{(s)} + H_2O_{(l)} \rightarrow MOH_{(aq)} + \frac{1}{2}H_{2(g)}$					
	508	469	481	475	478
(b) No. of atoms ($\times 10^{14}$) exposed on immersion of 10 mm cube					
	26.1	36.1	18.2	18.5	15.7
(c) Solubility of metal hydroxides in water (mol % at $T°C$)					
	53.6	105.0	190.7	175.6	263.8
	(20°)	(0°)	(15°)	(15°)	(15°)
(d) Melting points of the metals (°C)					
	180.5	97.8	63.7	38.9	28.6

in part for the relatively slow rate of reaction of lithium with water. This has to be coupled with the higher melting point of lithium. The other alkali metals fuse during reaction with excess water; the molten metals fragment, and this increases substantially the number of metal atoms in contact with water. With lithium, the water behaves as a heat sink, and the temperature of the metal does not rise high enough for fusion to occur. The number of metal atoms exposed to the water by the same area of the various metals[230] was calculated to rise from lithium to sodium. However, the variation down the rest of the group is not in agreement with observed reaction rates. Again, there seems to be no direct connection between energetic factors and reaction rate; lithium reacts slowest, but it has the highest enthalpy of hydrolysis.

In comparing reaction rates, it is important to remember that there are various physical factors which can give a very distorted picture of the true reaction rate. The most obvious instance is the sodium–water explosion, carried out in the presence of air; the violence of the explosion arises from a hydrogen–oxygen chain reaction, triggered no doubt by contact with the hot metal. However, if a small pea of sodium is dropped into water contained in a flask which is continually flushed with argon, the sodium merely reacts with no evident combustion, and is visually akin to the lithium–water reaction in air. Even with potassium, where small pieces of the metal react violently in the presence of air, only a rapid but steady reaction with water is observed under 1 atm of argon.[230] Another feature which gives a false picture of the true reaction rate is the barrier of hydrogen gas which can form between metal and water,[215] and Newlands and Halstead[232] have made a close study of this effect. Measured quantities (~0.02 ml) of cold water were injected from a syringe into sodium in the temperature range 200–450°C in the form of fine droplets about 0.15 mm radius, and their behaviour recorded using a high-speed cine film. On initial contact, a blanket of hydrogen and water vapour was rapidly generated round each droplet, and within 0.02 s. of each droplet entering the sodium, a steady-state situation was created in which further reaction of the water was controlled by a quiet film-boiling process. A mathematical model was developed to estimate the rate of further reaction of the water with sodium, assuming that the water exists as a droplet at the centre of a growing cavity of hydrogen and water vapour within the sodium. The model indicates that for droplets of this size, the time for their total consumption by sodium is 1.3 s (200°C) to 0.7 s (450°C). These results have practical significance, since in a sodium loop in which the sodium is flowing at about 6 m/s, the water (and hence the highly corrosive product sodium hydroxide) could be transported considerable distances from a leak site before reaction was complete.

Thermal explosions

The experiments discussed above have established that explosions related to alkali metal–water reactions which are chemical in origin result from explosive reaction of products (e.g. hydrogen–oxygen explosions), and that if these secondary explosions are prevented, e.g. by the use of inert atmospheres, chemical explosions do not occur. The possibility remains, however, that a 'thermal' explosion might take place. When a hot liquid is brought into contact with a cooler volatile liquid, there is a rapid generation of vapour from the more volatile liquid, giving rise to a thermal explosion, and many large-scale energetic thermal explosions have occurred in the chemical, petroleum, foundry and paper-making industries. Thermal explosions are slower than explosions which are chemical in origin, but the pressure pulses which are set up can be sufficiently powerful to be disruptive. Explosions which are classified as thermal arise purely from the rapid production of vapour of the volatile liquid, and it may be something of a misnomer to apply the term to alkali metal–water interactions, where hydrogen is also produced. Nevertheless, it will be seen from the following discussion that this is justified. We shall therefore first outline some of the general features associated with thermal explosions, and then consider how far the alkali metal–water thermal explosions conform to the general pattern.

The scale of the experiments covers a wide range, from single drops impacting on a liquid surface to large-scale tests involving kilograms of the two liquids. The essential feature is a large temperature difference, and the stimulus for the many experiments which have been carried out arises from the possibility of a melt-down in the core of a fast nuclear reactor, when molten fuel (uranium dioxide, m.p. 2800°C) could come into contact with the liquid sodium coolant. Although no large-scale energetic interactions between the fuel and sodium appear to have been reported, the possibility remains; and as a result thermal explosions have been the topic for extensive investigation in most countries in the world having nuclear facilities or nuclear potential. A report prepared for the Nuclear Installation Inspectorate (UK) in 1978, which reviewed research on the molten fuel-coolant interaction (MFCI) phenomenon, included reference to 296 published papers or research reports. Because the mechanisms involved in thermal explosions are not yet fully understood, it was recognized that it was not necessary to restrict investigation to nuclear fuel–sodium systems, and that a study of any two liquids having widely different temperatures could contribute useful information. Much of the work has therefore been concerned with liquid metal–water systems (particularly aluminium and tin) but has also been extended to the low-temperature range by mixing water with low boiling hydrocarbons or their derivatives (e.g. the freons). The liquids have been mixed in various modes which are related to the ways in which they might come into contact in an industrial situation: (a) the free-contacting mode, in which the hot liquid is merely poured into the cold liquid, producing coarse mixing; (b) the shock-tube mode, in which the hot fluid is contained at the bottom of a shock tube and the cool liquid is allowed to impact upon it; (c) the injection mode, in which one of the liquids is injected as a jet into a volume of the other liquid, usually resulting in the disintegration of the jet into discrete droplets; and (d) the single drop mode, in which drops of the hot liquid fall into a pool of the cool, more volatile liquid.

In many cases a delay is observed before the onset of the explosion, and in the single drop mode the interaction has been found to have an oscillatory nature.[233] One of the most important observations is that explosions only occur within clearly defined temperature ranges, which are defined by the temperatures of the two liquids. For a given temperature of the cool liquid there are upper and lower limits of hot-liquid temperature between which explosions will occur, and there is a characteristic upper limit to the cool-liquid temperature above which explosion does not occur. The cut-off temperatures are sharp.[233] Two main theories have been advanced to explain the explosion mechanism, and they could hardly have less in common. In the simultaneous nucleation theory of Fauske[234] the possibility of an explosion following liquid–liquid contact is said to be determined by the 'instantaneous contact temperature' T_i and the spontaneous nucleation temperature T_N; if T_i is greater than T_N, then a thermal explosion is possible. In the detonation theory of Board and Hall[235, 236] a small initiating disturbance, possibly related to the superheating of the coolant, produces a shock wave which propagates the interactions through the originally course mixture. The nucleation theory is more directly linked to the temperature limits for explosion, but the detonation theory provides a better explanation for the fragmentation of the hot fluid which accompanies a thermal explosion. Explanations are rendered more difficult by the fact that there are pairs of liquids (e.g. liquid aluminium and water) which do not interact violently in single drop experiments, but which are explosive in large-scale experiments. However, both groups of workers emphasize the necessity for fragmentation to occur; only in this way can large surface areas, and the consequent high rates of energy transfer, be achieved.

Comparing these general observations with those made on the interaction of sodium samples submerged in water, Newman et al.[207] postulated that a dispersion (i.e. fragmentation) mechanism was again necessary for an explosive reaction. In experiments carried out at the Berkeley (UK) Nuclear Laboratories, these authors suspended sodium samples, weighing between 2 and 15 g, under cold water in a weighted wire mesh cage, and followed the reaction by high-speed photography and by the insertion of thermocouples into the sodium samples. There was a delay time between immersion of the cage and the explosion, during which time smoke-filled bubbles were produced, with flashes of light. This delay time was about 4 s for samples above 4 g, but below this size either longer delay times up to 20 s were observed, or no explosion occurred. During the last 0.4 s of the delay time, the explosion was preceded by small periodic pressure pulses about 60 ms apart. Four main stages were detected in the explosion sequence. The first small impulse of permanent gas grew in 4 ms, followed by a period of up to 0.2 s during which the gas bubble rose and water again came into contact with the metal. This produced a large white and possibly heterogeneous region which grew and collapsed in 30 ms, followed by the further growth and collapse of a gas bubble within 30 ms. There is the suggestion here of a rhythmic pulsing in the reaction, which appears more clearly in the reaction of sodium–potassium alloy discussed in the next section. It is also to be correlated with the series of escalating growth and collapse of vapour blankets, at intervals of 10 ms, believed to be responsible for dispersion in thermal explosions between liquid tin and water.[235] An unsheathed thermocouple in the sodium sample showed an initial rise in temperature to about 600–700°C, followed by a fall to 250–350°C; final temperature rises up to 1000°C were then recorded, spaced about 30 ms apart.

In a second series of experiments, drops of water (0.05 ml) were allowed to fall on to a small amount of sodium (1 g) at a rate of 2 drops per second, under inert atmosphere. After impingement by 20 drops, a small explosion occurred. The same temperature variation was observed as in the first experiment; the sodium temperature at first rose to 400°C, then fell to below 300°C before rising to (or above) the boiling point of sodium at the time of the explosion. Again, much depends on the conditions of the experiment. It will be recalled that when a small pea of sodium is dropped on to a large excess of water under argon, no explosions have been observed.

In the liquid tin–water experiments, the dispersion mechanism responsible for thermal explosions was believed to be initiated by the violent transition boiling of the water. At temperatures above 95°C, stable film boiling occurs, and no explosions take place. With this in mind, similar sodium immersion experiments to those described above were carried out, but with water at an initial temperature of 95°C. Although the production of hydrogen was faster, no explosion occurred, which would appear to support the same initiation step.

An attractive alternative mechanism for dispersion of the sodium[207] proposes that the rapid chemical reaction rate quickly superheats the sodium, which is then violently dispersed by the rapid growth of metal vapour bubbles. Experiments with water vapour have confirmed that superheating does occur; heat transfer rates are such that only the metal surface may reach superheat temperatures in the very short times applicable in these reactions, but this is no detriment to this theory. However, this mechanism should still operate at water temperatures about 95°C, and it is most likely that each of these mechanisms makes some contribution to the overall reaction.

Injection experiments with sodium—potassium alloy

The studies of sodium–water explosions described above have left no doubt that the overall reaction proceeds in a series of separate stages, but by observation of a single explosive event it is not possible to identify these stages at all clearly, in view of the extremely short time scale involved. In particular, if the reaction involves a series of regular pulses, arising perhaps from rhythmic bubble formation and release, the time scale is only long enough to allow a few such pulses to be recorded. The Nottingham group[227] has undertaken a more detailed study of processes which occur when the metals and water first come into contact, under conditions which avoid explosions, and the following account describes some aspects of their work during the years 1973–79. The aim was to inject a fine jet of the liquid alkali metal under water in glass apparatus, so that both the initial reaction and the subsequent behaviour of the metal and the gas produced could be continually observed and recorded.

The apparatus used is illustrated in Figure 10.2. Sodium–potassium alloy of composition near 50–50 mol % (melting point 6°C) was used in order that the liquid metal could be manipulated at room temperature. Initial experiments showed that it was indeed possible to inject a thread of the metal into water, and that a steady flow could be achieved. The size of the orifice is critical; it has to be small, firstly because large orifices introduce dangerously large quantities of the alloy into water, and secondly because longer unbroken jets are produced from small orifices.[237] An optimum size was found to be 0.1–0.15 mm diameter. This gave an unbroken jet of alloy 20 mm long when

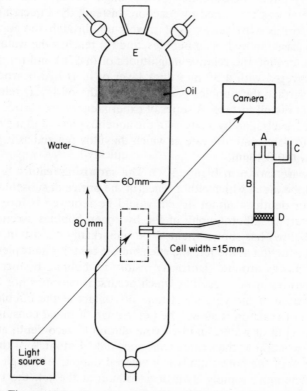

Figure 10.2   Reaction vessel for injection of NaK alloy into water

ejected into argon gas, but this shortened to 2–4 mm when injected into a low-density inert liquid. The alloy was introduced into the barrel B of the injector through the septum cap A, and argon pressure at point C forced the alloy through the grade O sintered glass filter D into the orifice, which was prepared from precision bore, heat resistant glass. Under a driving pressure of argon of 600 torr at C, the metal flowed from a 0.1 mm diameter orifice at a rate of 20 mg/s. The reaction vessel consisted of a vertically mounted glass column 300 mm long and 60 mm diameter containing purified water, and the orifice projected 20 mm into the vessel. A continuous flow of argon was maintained through the space E above the water surface.

The original experiments were carried out in a simplified form of the apparatus in Figure 10.2, with no provision for photography. On injecting the metal, there was the expected reaction around the orifice, with evolution of hydrogen bubbles which rose to the surface. However, the space E almost immediately became filled with copious white hydroxide fumes, and it was evident that some of the alloy was rising through the water column and reaching the water surface. This phenomenon merited more detailed study, though it was now obvious that it would not be possible to determine reaction rates simply by direct measurement of the length of jet which could exist before the metal was consumed by reaction. Instead, the injected metal (a) reacts at the orifice, (b) reacts as it travels up through the water layer, presumably encased in hydrogen bubbles, and (c) reaches the water surface. In order to determine the relative magnitudes of (a), (b) and (c), the water surface was covered with a 30 mm thick layer of inert hydrocarbon oil; on reaching the surface the bubbles passed through the oil layer, releasing the alloy droplet at the oil surface. A series of experiments were carried out using different water levels, and by analysis of the aqueous phase and the oil layer it was possible to determine the rate at which the alloy reacted as it passed up through the water column.

The results are shown in Figure 10.3. The surprising feature is the small proportion of the metal which either reacts at the orifice or subsequently with water; a water depth of about 40 cm would be required before no metal reached the surface. Photography of the hydrogen bubbles ascending from the jet (see below) has shown them to be remarkably consistent in volume and rate of production at the jet, and each bubble contains a droplet of metal. This droplet moves around vigorously inside the bubble, bouncing off the bubble walls owing to the reaction which occurs; it becomes hot enough to glow, but because of the very short time of contact at the bubble walls its apparent rate of reaction is slow. The percentage of metal consumed varies linearly with depth of water, and by extrapolation to zero depth at 20°C we find that the reaction at the orifice consumes only 4 atom % of the sodium, and 10 atom % of the potassium which entered the jet. Although potassium therefore reacts more rapidly than does sodium at the jet, the two metals react at much the same rate as the droplets pass through the water layer.

The products formed when this small proportion of the metal reacts at the

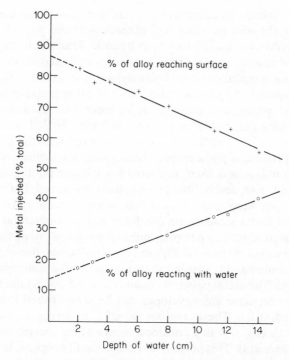

Figure 10.3 Reaction of NaK droplets passing
through water at 20°C

jet have been determined by two different methods. Firstly, knowing the
bubble size and frequency, it is possible to derive the volume of hydrogen
produced at the jet, and to compare this with the quantity of metal known to
have been consumed there. The amount of hydrogen (at 20°C) found was
only about one third of that expected from the reaction

$$M + H_2O \longrightarrow MOH + \tfrac{1}{2}H_2 \qquad (10.1)$$

Clearly, some hydrogen was leaving the jet in a form other than as a gas, and
the only reasonable possibility was that the jet reaction was producing metal
hydride also, and that both hydroxide and hydride were incorporated in the
metal droplets leaving the jet. Secondly, supporting evidence was obtained
from the behaviour of the gas bubbles on leaving the jet. Knowing the original
bubble size, and allowing for (a) hydrostatic pressure and (b) the increase in
volume due to hydrogen produced by reaction of the metal droplet if this gave
hydrogen and hydroxide only, it was possible to calculate the rate of
expansion of the bubbles. The actual rate was found to be greater than that
calculated, because the metal droplets contained hydride, and the hydrogen
produced on the hydrolysis

$$MH + H_2O \longrightarrow MOH + H_2 \qquad (10.2)$$

158

is twice that produced by equation (10.1). On this basis, it was estimated that of that part of the ejected alloy which reacted at the jet, 63 mol % was converted to hydroxide and 37 mol % to hydride. Direct confirmation of the presence of hydride in the metal droplets which collected in the oil layer on the water surface was also obtained by analysis; these results indicated that all hydride was hydrolysed within a water depth of 50 mm above the orifice.

The physical processes occurring at or near the jet were studied by photography, using the apparatus as shown in Figure 10.2. The section of the glass column around the capillary was replaced by optically flat glass plates 15 mm apart, to produce a photographic cell around the orifice. An electronic flash of duration 1 μs was used, and double-exposure photographs enabled the rate of bubble rise, and bubble growth, to be measured. For the metal jet itself, photographs were taken with a high-speed camera at 1000 frames/s. These indicated that a steady-state condition did not operate at the jet, but that the gas was produced by a regular pulsing movement along the jet. This is illustrated by the sketch shown as Figure 10.4. In all cases, there is a section of metal jet AB immediately outside the orifice, about 0.4 mm long, which has no gas envelope. The metal passes through AB in 0.5 ms. At the beginning of each cycle, two separate gas envelopes can be seen around the jet beyond point B (sketch (a)). These regions expand to become one elongated envelope (sketch (b)). As this envelope moves along the jet, another small envelope develops at B. This process is repeated until bubble C becomes large enough to detach. There is a remarkable regularity in the bubble release; under the conditions described, bubbles of closely similar volume were

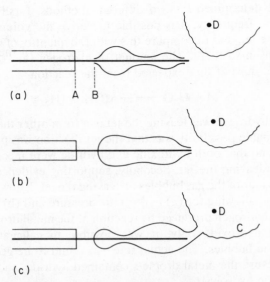

Figure 10.4 Formation of hydrogen bubbles at a jet of NaK alloy in water

released every 25 ms, and it may be relevant to compare this with the bubble growth and collapse at 30 ms intervals which was noted in the course of the sodium–water explosion studies. The jet eventually breaks up, and each bubble normally carries one metal droplet D. Photographs of these droplets also indicate that they are very similar in size.

The relative importance of the various factors responsible for this pattern of behaviour cannot yet be assessed, but experiments described earlier in this chapter have pointed to a number of the factors which must be responsible for the 'clean' jet AB, and for the profile of the gas envelope around the jet. They are (1) the heat evolved when sodium (and presumably NaK) reacts with water can raise the temperature at the surface of the metal to around 600–700°C in milliseconds; (2) sodium and potassium react much more rapidly with water vapour than with hydrogen; and (3) hydride, as well as hydroxide, is formed by reaction at the jet. These factors can be put together to provide the following possible interpretation.

The simplest explanation of the fact that no gas is produced as the metal passes through the initial clean section AB of the jet is that the reaction

$$M + H_2O \longrightarrow MH + MOH \qquad (10.3)$$

is occurring at the metal surface. This is a simple process which involves only the breaking of one O–H bond, and no solution rates are operative since turbulence of the metal in the jet will readily incorporate both hydroxide and hydride into the metal. At the same time the metal is raised to a high temperature, and at B an envelope of water vapour is formed. Thereafter, two reactions can occur:

$$\text{(a)} \quad M + H_2O_{(g)} \longrightarrow MOH + \tfrac{1}{2}H_2 \qquad (10.4)$$
$$\text{(b)} \quad 2M + H_2 \longrightarrow 2MH \qquad (10.5)$$

(a) is rapid and can account for the apparent collapse of the first envelope, and the continued production of hydrogen along the jet provides the second envelope. This hydrogen accumulates and eventually forms bubble C. (b) is much slower than (a), but it may make some contribution to the appreciable amount of hydride present in the metal as it leaves the jet.

In view of the very high temperatures to which the jet is raised by reaction with water, it is perhaps surprising that relatively small increases in the temperature of the water into which the jet is injected have a considerable effect on the rate of reaction. At a water temperature of 65°C a flame appeared at the orifice, and in subdued light the rising bubbles were illuminated from within. The reaction is more controllable at 50°C and below, and measurements at 40°C and 50°C are compared with those for 20°C in Figure 10.5. The relative rates for sodium and potassium are not appreciably influenced, but extrapolation to zero water depth shows that four times as much metal is consumed at the orifice at 50°C as at 20°C. The rate at which hydrogen is released at the jet increases by a much smaller factor, suggesting that the products of reaction at the jet contain a higher proportion of hydride as the

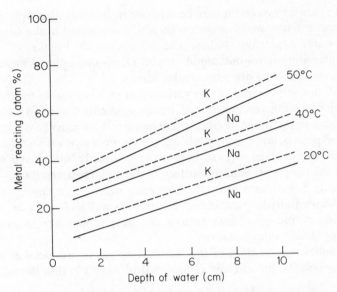

Figure 10.5 Influence of temperature on reaction of NaK
droplets passing through water

temperature increases. Consistent with this, the increase in volume of the
bubbles which occurs as they rise through the aqueous phase, and which can
be attributed to hydrolysis of hydride, also increases with increasing temperature. The slopes of the graphs shown in Figure 10.5 for all three temperatures
are roughly equal, indicating quite clearly that the rate at which the entrapped
droplets react is independent of temperature. This may be partly coinciden-
tal. The amount of metal available for this stage is smaller at higher temperatures, but the main reason is probably a physical one. As the droplets strike
the bubble walls at the higher temperature, the more vigorous reaction will
produce a more vigorous pulse of hydrogen and water vapour; the droplet will
therefore 'bounce' from the wall more rapidly, and in consequence its reaction time will be reduced.

Sodium–potassium alloy was used in the above experiments for experimental convenience, but various aspects of the results would indicate that the
behaviour of the alloy is also broadly characteristic of that of the single
metals. This has been confirmed by experiments using liquid sodium alone.
The metal and injector were heated by oil jackets, so that the liquid metal
issued from the orifice at 120–130°C, into water at 30°C. The sodium jet
resembled that formed by the alloy, and again broke up into droplets which
rose to the water surface, each one trapped in its own protective bubble of
gas. A linear relation was again observed between the quantity of metal
dissolved and its distance of travel through water; the slope of the graph (0.42
atom % per mm travel) was rather larger than that found for sodium in the
alloy (from Figure 10.5, 0.31 atom % per mm), but this can be accounted for

by the longer times of contact available, at the bubble walls, to the rather less reactive metal. Extrapolation to zero water depth indicated that 14.2 mol % of the injected metal reacted at the jet, and this value is very close to that obtained by interpolation of the data in Figure 10.5 to 30°C.

Mechanisms

The mechanism of reaction of water vapour with an alkali metal surface (which involves adsorption, polarization of the H_2O molecule, bond cleavage, electron transfer and combination of H atoms) has already been discussed. These processes may also be operative with liquid water, but the use of liquid water introduces a new dimension in that electrons released from the metal surface can now be transferred to the aqueous medium in the form of solvated electrons $e^-_{(aq)}$. This species has very limited stability where the solvent is water, but there is little doubt that it plays a part in the mechanism of reaction; conditions can easily be arranged whereby the blue colour characteristic of the solvated electron can be observed visually when drops of water and sodium–potassium alloy are brought into contact. A discussion of the extensive chemistry of the solvated electron is not appropriate here, and it has been fully described in a number of reviews.[238–242] The formation of the solvated electron can be represented by

$$M + H_2O_{(l)} \longrightarrow M^+_{(aq)} + e^-_{(aq)} \qquad (10.6)$$

A guide to the readiness with which this will occur with the different alkali metals is given by their ionization potentials and work functions, which decrease down the group, in agreement with the known increase in reactivity with water. Kinetic evidence has been put forward[241,243] which indicates that molecular hydrogen arises from a combination of two solvated electrons:

$$e^-_{(aq)} + e^-_{(aq)} \longrightarrow H_2 + 2OH^-_{(aq)} \qquad (10.7)$$

rather than from a combination of individual H atoms produced by the much slower reaction:

$$e^-_{(aq)} + H_2O_{(l)} \longrightarrow H + OH^-_{(aq)} \qquad (10.8)$$

This mechanism would treat the formation of hydride at a metal jet as a secondary reaction between hydrogen gas and the metal, but the large amount of hydride which features in the reaction product would suggest that H atoms are also involved in the mechanism, as in the reaction with water vapour.

Chapter 11

Reactions of nitrogen with lithium and the group II metals in liquid sodium

11.1 Introduction

Lithium is the only alkali metal which reacts with nitrogen (see Chapters 4 and 5). Since solution of nitrogen in the alkali metal, as well as the formation of the nitride, involves transfer of electrons from metal to nitrogen, it follows that lithium is the only liquid alkali metal in which nitrogen will dissolve, and solubility values have been given in Chapter 5. We have also already discussed Born–Haber cycles for the formation of nitrides of sodium and the heavier alkali metals, which show that nitride formation is not thermodynamically favourable, so that no solution of nitrogen occurs.

As well as dissolving in pure liquid lithium, nitrogen will also dissolve in a liquid lithium–sodium mixture; sodium acts essentially as an inert diluent for the lithium, and the extent of nitrogen solubility is a function of the lithium content of the liquid mixture. The same situation arises with solutions of the group II metals in liquid sodium. Reactions of nitrogen with solutions of barium in liquid sodium have proved to be particularly amenable to study; most of the available techniques have been applied to the kinetics and stoichiometry of this reaction, and most of the factors governing such reactions have been defined using these solutions. The various aspects of the reaction of nitrogen with solutions of barium in liquid sodium will therefore be discussed first.

11.2 The barium–nitrogen reaction

Precipitation studies

These reactions[244] were carried out in the type of steel vessel shown in Figure 2.6, in which the liquid metal was circulated by means of an electromagnetic pump at 300°C. The vessel was modified to permit the insertion of a sampling device; this consisted of a glass tube, closed at the end by a sintered glass filter pad (porosity 2×10^{-3} mm) through which samples of the liquid metal could be drawn. Solid barium was added to liquid sodium in the vessel, and the metal circulated through the pump until all barium dissolved. Concentrations near 4.4 atom % Ba were used. A predetermined quantity of nitrogen contained in a vacuum frame was then exposed to the liquid metal. When nitrogen absorption was complete, the liquid metal was sampled through the filter. Analysis of the sample gave the proportion of nitrogen in solution, the remaining nitrogen being present in the form of barium nitride precipitate.

In Figure 11.1 the amount of nitrogen found in solution is plotted against the total amount of nitrogen added (in arbitrary units). The ultimate product, barium nitride, is insoluble in liquid sodium, but the steps through which the reaction passes before the final condition is reached are remarkable. In fact, the behaviour illustrated in Figure 11.1 is typical of many metal–nitrogen reactions carried out in a metallic medium, but quite unlike precipitation reactions in aqueous solutions, where reagents which produce an insoluble product (such as in the addition of sodium chloride to aqueous silver nitrate solution) usually give a precipitate as soon as the reagents are mixed. In the $Ba-N_2$ reaction, the nitrogen in solution increased until the reaction was half completed, then fell to zero. In the stage AB of the reaction, all the nitrogen absorbed into the metal was found in solution, presumably as the nitride ion N^{3-}. In terms of simple solubility, if point B represented saturation of the liquid with nitrogen, then further additions of nitrogen should not influence

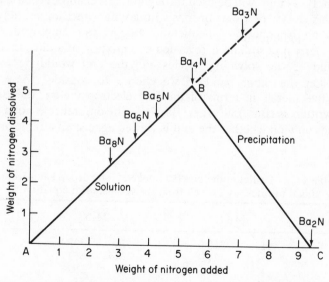

Figure 11.1   Reaction of nitrogen with barium in liquid sodium,
at 300°C

the amount in solution. Instead, precipitation of a barium nitride began at point B. This diminished the barium content of the solution; with further additions of nitrogen beyond point B precipitation continued, and the amount of dissolved nitrogen decreased, until at C all barium was precipitated as nitride and no nitrogen was dissolved in the remaining liquid sodium.

The behaviour illustrated by Figure 11.1 provides one of the best examples of the application of the solvation concept in liquid metal chemistry. For a given quantity of barium in solution, nitrogen dissolves until the ratio represented by Ba_4N is reached and thereafter precipitation commences immediately. For dilute solutions of barium, this general behaviour is independent of the concentration of barium used in the reaction. In the range AB, the N^{3-} ion is believed to be held in solution as a result of the solvation energy arising from the solvation of N^{3-} by four Ba atoms, to give soluble Ba_4N units. Insufficient enthalpy values are available for the construction of the appropriate cycle, and hence for the calculation of the actual solvation energy of N^{3-} by barium atoms. However, lattice enthalpy and solvation enthalpy are of the same order of magnitude, and some clue to the magnitude of the solvation energy can be derived from comparison of the lattice enthalpies of group I and group II nitrides (Table 11.1). Owing to the greater charge and smaller size of the group II metal ions, the lattice enthalpies of their nitrides are appreciably greater. The greater attraction between the N^{3-} and M^{2+} ions will also be reflected in the solvation enthalpies, so that in a liquid consisting of Na^+, Ba^{2+} and free electrons, the N^{3-} will be solvated preferentially by Ba^{2+}. The experiments indicate that this preferential solvation is strong enough to set up Ba_4N units in solution.

The term 'Ba_4N unit' will be used throughout this chapter because attempts to define this cluster of atoms or ions with greater precision are unjustified, and depend on the concept of the metallic state which is employed. Thus, if a solution of barium in sodium is regarded as a mixture of atoms, then barium atoms would be the solvating species and the unit would be written as $[Ba_4N]^{3-}$. On the other hand, if we choose to consider these strongly electropositive metals in terms of the free electron theory, the Ba_4N unit would be written as $(Ba^{2+})_4N^{3-}$, i.e. $[Ba_4N]^{5+}$. The difference is not a real one, and depends only on whether the unit is viewed against an atomic or an ionic

Table 11.1   Lattice enthalpies $(kJ (mol N)^{-1})$ for group I and group II metal nitrides, as derived from the Kapustinskii equation[245]

M_3N		M_3N_2	
Li	5104	Be	7394
Na	4711	Mg	6517
K	4201	Ca	5823
Rb	4030	Sr	5572
Cs	3808	Ba	5221

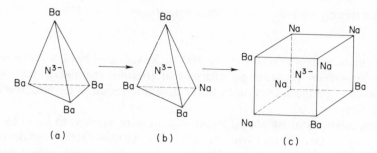

Figure 11.2 Solvation of N^{3-} in solutions of barium in liquid sodium

background. Purely as a matter of convenience, we shall regard N^{3-} as solvated by metal atoms.

When the nitrogen added to the solution increases beyond the Ba_4N ratio (i.e. beyond B in Figure 11.1), barium atoms must be replaced by sodium atoms in the solvation shell of some of the N^{3-} ions, and the first steps in this process are illustrated in Figure 11.2. Because of the strong solvation of N^{3-} by barium, the $Ba-N^{3-}$ distance will be short, and the Ba_4N unit is likely to have the tetrahedral structure shown at (a). Replacement of one Ba atom by an Na atom gives structure (b). However, reasonable extrapolations from the solvation energy data given in Chapter 7 suggest that the solvation energies of N^{3-} by barium and by sodium are about 4000 and 3000 kJ (mol N^{-1}) respectively, so that the solvation energy in unit (b) is weaker than in (a). In consequence, the average metal–N^{3-} distance will be greater, which will encourage the coordination shell to convert to one having a greater coordination number, as shown at (c). This involves the introduction of more Na atoms into the coordination shell, with a consequent further reduction in solvation energy. Since the Ba_4N unit is responsible for nitrogen solubility, precipitation will therefore occur whenever this unit can no longer be maintained. During the precipitation stage (BC in Figure 11.1) the quantity of barium nitride which separates is such as to maintain the Ba_4N ratio in the liquid until precipitation is complete.

The solid product

The quantities of barium and nitrogen which have reacted when precipitation is complete correspond to a nitride of composition Ba_2N, and analyses carried out in the range BC indicate that this is the nitride which separates throughout the precipitation process. The identity of the product has also been confirmed by X-ray powder diffraction. This solid must contain free electrons, and it is of interest that Ba_2N, rather than the ionic solid Ba_3N_2, separates from the metallic environment. Dibarium nitride is one of the family of group II metal nitrides M_2N which have the anti-$CdCl_2$ structure, and is normally prepared by the decomposition

$$2Ba_3N_{2(s)} \longrightarrow \tfrac{1}{2}N_2 + 3Ba_2N_{(s)}$$

which requires temperatures above 500°C. Reaction of barium and nitrogen in liquid sodium provides an efficient alternative method which can be conducted at temperatures much lower than that required to decompose the solid Ba_3N_2.

When mixtures of Ba_2N and liquid sodium were allowed to stand for long periods in contact with nitrogen at 300°C, a very slow further uptake of gas occurred. Even allowing for the fact that the nitride as produced is wetted by sodium, which cannot act as a carrier for nitrogen, there is no doubt that Ba_2N is indeed the stable nitride in a liquid metal environment.

Variations in electrical resistivity

These experiments[177] were carried out using the apparatus described in Chapter 3 (Figures 3.5 and 3.6). The barium solution was circulated continuously through a steel capillary loop, and the resistance of the liquid metal thread was recorded as nitrogen was metered into the reaction vessel. The results shown in Figure 11.3 are for a 2 mol % barium solution, but the

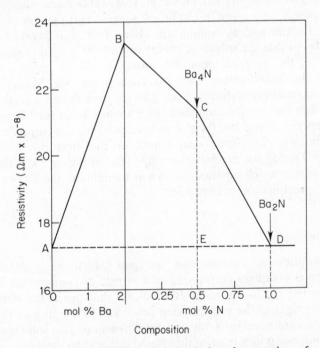

Figure 11.3 Resistivity changes during the reaction of nitrogen with a 2 mol % solution of barium in liquid sodium, at 300°C

same behaviour was observed using barium concentrations in the range 0.3–7.0 mol %. The resistivity of pure sodium is represented by the broken line AD. During solution of 2 mol % barium, the resistivity increased rapidly to point B, and at this point nitrogen was added to the liquid. The N^{3-} ion has a high resistivity coefficient, and if it remained uncombined in the solution, resistivity should continue to rise. In the event, resistivity falls on nitrogen addition to a point C which corresponds to the composition Ba_4N, and the line BC is linear. No precipitation occurs during this stage, so that the electron scattering ability of dissolved barium is greatly reduced when it is incorporated with N^{3-} in solution as the strongly bonded Ba_4N unit. Since the region BC is linear, each N^{3-} ion takes up the same number of barium atoms in its solvation shell throughout this stage. In effect, therefore, when point C is reached, five scattering centres ($4Ba + N^{3-}$) have been replaced by one. The dimension CE gives the value 8.8×10^{-8} Ωm (mol % N)$^{-1}$ for the resistivity coefficient of Ba_4N in sodium which has been quoted in Table 8.2.

Precipitation begins as nitrogen is added beyond point C. The barium and nitrogen contents of the solution are progressively depleted, and resistivity falls regularly to the pure sodium value. At point D, only a mixture of sodium and solid Ba_2N remains. With dilute barium solutions, CD is linear. At the higher concentrations, the resistivity changes in this region are often irregular. This has no theoretical significance, and is due to occasional accumulation of nitride precipitate in the capillary tube. This tends to set an upper limit to the barium concentration which can be used in this type of equipment, but the general behaviour outlined in Figure 11.3 will be typical of solutions much more concentrated than 7 mol % barium, which was the maximum used in this study.

Rates of nitrogen absorption

The experiments described above have been concerned with solutions in which the nitrogen–barium reactions have been completed, and equilibrium has been achieved. Further information on the reaction can be obtained from a study of the rates at which nitrogen is absorbed into a solution of barium in sodium. These rates have been measured, from time to time, over a number of years. Preliminary experiments were carried out by Wallace[252] and by Davies[253] during the years 1963–67, and more extensive studies were conducted by Lindley in 1969.[250] The experimental work did not present any unusual difficulties, but the interpretation of reaction rates did not become evident until the precipitation and resistivity studies described above were carried out in 1975–76. Reaction rates were measured using the simple steel vessel shown in Figure 2.3(b), and stirring was effected using a rotating horseshoe magnet. Absorption rates were monitored by following the decrease in pressure of a known volume of nitrogen introduced to the metal. Barium was added to about 20 g of liquid sodium to give solutions in the concentration range 1–10 atom % barium, and during a reaction the nitrogen

168

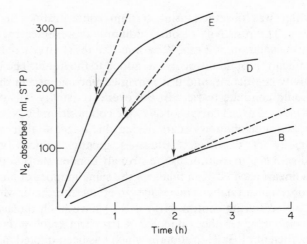

Figure 11.4   Nitrogen absorption curves at 300°C.
Atom % Ba: E, 8.95; D, 7.91; B, 4.23

pressure decreased from an initial 450 torr to around 300–350 torr. Three typical absorption curves are shown in Figure 11.4.

In each experiment, the first notable observation was that for an initial period the rate at which nitrogen was absorbed remained constant. This occurred in spite of the fact that nitrogen pressure was falling, so that the absorption rate was independent of pressure, at least over this narrow pressure range. Presumably, then, the N_2 molecule plays no part in the mechanism of reaction, and we can consider that a chemisorbed layer of N atoms is formed at the surface, and that this forms a constant reservoir from which nitrogen atoms can be withdrawn for reaction. The rate of reaction, as determined from the linear section of the absorption curves, varies smoothly with barium (bulk) concentration. Unfortunately it is not possible to express this process in quantitative terms since the barium concentration at the surface is not known. Other researches have shown that in solutions of group II metals in liquid sodium the solute tends to concentrate at the surface, but the extent to which this occurs in barium solutions has not yet been determined.

In any given experiment (Figure 11.4) the linear section of the curve extends to a point at which the ratio of barium present to nitrogen absorbed reaches 4:1, and the extent of this first stage in the reaction is therefore dependent on the quantity, rather than the concentration, of barium in the solution. During this stage, the barium in solution is changing from a state in which it is 'free' to a state in which each barium atom is employed in solvating an N^{3-} ion. However, since the absorption rate remains constant, the change in the function which the barium atoms are performing has no influence on their ability to convert adsorbed N atoms to dissolved N^{3-} ions. This may seem surprising at first sight, but it may be no more than a reflection of our limited understanding of the fundamental nature of liquid metal mixtures.

Beyond the points represented by the Ba$_4$N arrows in Figure 11.4, precipitation begins. The barium concentration (and quantity) in the solution becomes depleted, and absorption rates are rapidly diminished. These results therefore correlate closely with the precipitation studies discussed above, though this is not a good method for determining the Ba:N ratio at which precipitation commences because of the difficulty in defining the 'break point' in the absorption curves with sufficient precision. It will now be clear that while the break point depends only on the quantity of barium present, the reaction rate is a function of barium concentration, and this applies to the second (precipitation) stage of the reaction as well as to the linear stage. The slope of the curve at any point on the second stage is determined by the barium remaining in solution, so that each absorption curve represents a part of a single composite curve. This is illustrated in Figure 11.5. (For purposes of correlation, the letters used to denote separate experiments are the same in both Figures 11.4 and 11.5.) In experiment D, the reaction rate at first remains constant, then diminishes along DC. At C the reaction rate is the

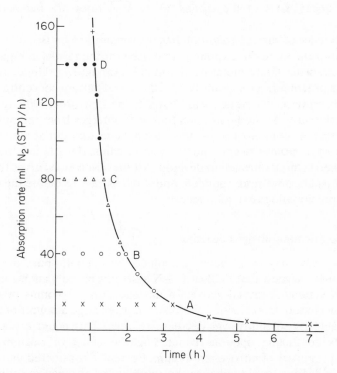

Figure 11.5   Rates of absorption of nitrogen by solutions of barium in liquid sodium, at 300°C. Atom % Ba: D, 7.91; C, 6.35; B, 4.23; A, 2.50

same whether this condition has been reached by precipitation from a more concentrated solution, or whether it represents a freshly prepared solution of that concentration. The same argument applies for other solutions B and A, so that given standard experimental conditions there is a single curve DCBA to which all reactions will conform, and which will be joined, in any individual experiment, at a point which depends only on the barium concentration.

Where the property being measured (e.g. resistivity) is related to changes in bulk composition, the choice of method to be used for stirring the liquid metal during a gas–metal reaction is largely a matter of experimental convenience. Methods already described involve stirring by an external magnet (Figure 2.3(b)) or by means of a liquid metal loop, in which the liquid is drawn from the reaction vessel by an electromagnetic pump and returned as a jet ejected from a nozzle into the gas phase (Figure 2.6). Where rates of gas–metal reactions are concerned, however, there can be a significant difference in the results obtained. In the 'stirred pot' method, the liquid is stirred, by a circular movement, but a condition is set up in which the surface, though irregular in profile, is steady and largely unbroken. At such a surface the excess barium concentration, and the chemisorbed layer of nitrogen atoms, have time to develop, and this is believed to account for the observation in Figure 11.4 that over the linear parts of the curves, absorption rates are independent of pressure.

The influence of surface condition has been examined by Bee,[257] who used a steel pot of the same dimensions as that used for the results in Figure 11.4, but circulated the liquid metal by the metal loop and jet. Under the same conditions of temperature, quantity of sodium and barium concentration, the rate of absorption of nitrogen was found to be first order with respect to nitrogen pressure. At the liquid metal jet as it emerges from the orifice, the metal surface is completely fresh; the barium concentration at the surface is the same as in bulk solution, and the age of the surface is too brief for a chemisorbed layer of nitrogen to develop. All the absorption curves (compare Figure 11.4) are now more rounded, and it is less easy to define the position at which precipitation first commences.

11.3 The calcium–nitrogen reaction

This system has not been examined in detail, but preliminary experiments[253] suggest that the Na–Ca–N system is broadly analagous to the Na–Ba–N system described above. In this case, small amounts of nitrogen will dissolve in sodium–calcium solutions, but when large amounts of nitrogen are available its solubility will decrease to zero. This could explain some apparently conflicting opinions concerning the effect of calcium on the corrosive properties of nitrogen in sodium. Epstein[254] considered that calcium impurity in sodium would enhance the solubility of nitrogen in sodium, and promote the nitriding of steel. On the other hand, experiments by Cafasso,[255] using an excess of nitrogen, showed diminished corrosion, and these findings form the basis of a patent on corrosion prevention.[256] These observations could be in complete accord if the solubility of nitrogen in dilute solutions of calcium in sodium follows the general pattern shown in Figure 11.1 for barium solutions. Calcium is the main impurity introduced in the electrolytic production of sodium from $NaCl-CaCl_2$ melts, and reference has already been made in Chapter 3 to the use of excess of oxygen to remove calcium impurity by precipitation as calcium oxide. Nitrogen could be used for the same purpose, with the added advantage that nitrogen in excess of that required to precipitate the calcium as nitride is quite insoluble in the purified sodium.

11.4 The strontium–nitrogen reaction

Though similar in principle, the reactions of strontium solutions show some interesting differences from those of barium solutions. Reactions of strontium solutions are much more difficult to carry out (and to interpret) because of the remarkable ability of strontium solutions to creep over the internal surface of the steel containing vessels. When a barium solution was allowed to cool and solidify at the end of a reaction and the pot cut open, the solution was seen to have retained its normal position, so that surface area had remained constant during a reaction. In contrast, the strontium solution had crept over the entire internal surface of the container, with a consequent increase in the surface area exposed to reaction. This creep occurs more rapidly with increased temperature and strontium concentration, and is probably complete within a few minutes at 400°C. The most significant observation, however, is that the spreading only occurs when nitrogen is added. The spreading of a liquid metal over a solid metal is normally attributed to a chemical reaction at the point where three phases are in contact; for instance, the spreading of sodium over iron, cobalt or nickel is enhanced by addition of calcium or barium, because the added group II metal can reduce the oxide film on the solid metal. This cannot be the case here, since spreading depends on the addition of nitrogen. It may be that a ternary nitride of nitrogen with strontium and one of the steel component metals is formed as a thin film, though no ternary nitrides of this type appear to have been described in the literature. Once formed, this film seems able to undergo further reaction with the solution, since it is thicker than can be accounted for by a simple wetting process. This film formation on the metal surface interferes to the greatest extent in attempts to study the strontium–nitrogen reaction by measurement of resistivity changes. A study by Bussey[258] gave clear evidence that the capillary method (Figures 3.5 and 3.6), which proved so satisfactory for barium solutions, was not suitable for strontium solutions; during the reaction the capillary tube became restricted, and a different technique will be necessary for resistivity measurements on these solutions.

Rates of reaction of nitrogen with strontium solutions were measured with success using pots stirred by magnet.[257] The problem discussed above was

172

obviated by adding a small amount of nitrogen prior to commencement of the experiment, and allowing time for complete wetting of the internal surfaces of the vessel to occur. Thereafter, the surface area of metal could be regarded as constant during an experiment. Altogether, 35 absorption curves of the type shown in Figure 11.4 were obtained, at temperatures in the range 300–500°C and with strontium concentrations up to 20 atom %, but the general pattern differed from that of the barium solutions. Most of the curves showed no linear section, and only at the highest concentrations and lowest temperatures did a linear section appear. As a result, the curves could not be used, with any confidence, to determine the Sr:N ratio at which precipitation commenced. This behaviour can be explained on the assumption that strontium concentrates at the liquid metal–gas surface to a much smaller extent than does barium. In consequence, chemisorption of nitrogen is reduced, and the absorption rate becomes pressure dependent. Again, surface excess of strontium will decrease at higher temperatures; in the concentration range 6–10 atom % strontium and at 300°C, absorption curves showed some initial linearity, but this disappeared at 400°C.

A corresponding study using the loop and jet apparatus confirmed that nitrogen absorption rates were first order with respect to both nitrogen and strontium concentrations. In order to remove the pressure variable, results can be expressed in terms of variation in reaction rates at constant pressure, and these are shown in Figure 11.6. To achieve this, the nitrogen was added in a series of small aliquots at the same starting pressure and the initial absorption rates measured as additions of nitrogen progressed. (Each rate is given as a point in Figure 11.6.) This technique has the added advantage that

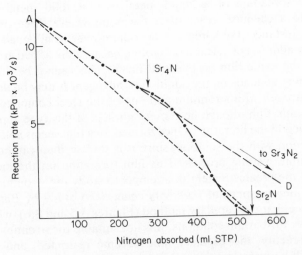

Figure 11.6   Reaction rates of nitrogen with a 5.5 atom % solution of strontium in liquid sodium, at 330°C. Nit-Nitrogen pressure at 44 kPa throughout

any insoluble nitride is formed in a series of minute quantities which then have the opportunity to disperse evenly through the system. Initially, these points lie precisely on the line AD representing the ratio Sr_3N_2, indicating that as nitrogen dissolves the electrons required to convert 2N to $2N^{3-}$ are provided by three Sr atoms. The formation of N^{3-} ions in these solutions has always been assumed, but experimental support of this sort is welcome. One feasible interpretation of these results is that the ratio Sr_4N indicates the maximum quantity of nitrogen which can be absorbed if each N^{3-} ion in solution is to be solvated by four Sr atoms. Thereafter, precipitation of solid Sr_2N begins, and absorption rates fall to zero at the composition Sr_2N, which corresponds closely with the behaviour of the barium solutions. This interpretation is based on the assumption that nitrogen does, in fact, dissolve in solutions of strontium in sodium until the stage represented by Sr_4N is reached. This must remain an assumption until adequate analyses have been performed, and at the time of writing such analyses are not available.

Unfortunately analysis, like all other techniques, seems fraught with difficulties so far as these solutions are concerned, and this may be due to the physical nature of the Sr_2N precipitate. In aqueous solution, for example, some precipitates form slowly, some quickly; some separate in a crystalline form which is readily filtered, and some separate slowly as colloids. It is likely that a similar range of behaviour exists as precipitates separate from liquid metals. With the barium solutions, Ba_2N separated sharply and filtration appeared to present no difficulties, but analysis of the strontium solutions is a different and difficult problem.

As produced in this way, the product is wetted by sodium. When the sodium is removed by distillation under vacuum, the Sr_2N remains as a dark purple crystalline mass, which ignites spontaneously when exposed to air or oxygen.[257] Its identity has been confirmed by X-ray diffraction and chemical analysis. The product can take up further nitrogen slowly, but complete conversion to Sr_3N_2 requires 24 h at 400°C under 1 atm pressure of nitrogen. The conversion is much slower than this when the Sr_2N is wetted by liquid sodium.

11.5 The lithium–nitrogen reaction

Solutions of lithium in sodium resemble those of strontium in many respects. In the reaction of lithium solutions with nitrogen there is again a solution stage followed by precipitation, but there is some uncertainty about the solubility of nitrogen in lithium–sodium mixtures. The problem is simplified a little by the fact that lithium forms only one nitride so that the same product Li_3N is produced by precipitation from the liquid metal as is formed by reaction between the elements themselves.

The problems associated with resistivity measurements, which have been referred to in connection with strontium solutions, arise also with lithium solutions, and it does appear that the physical nature of the precipitate of

Li_3N must be different depending on whether it separates from pure lithium, or from a lithium–sodium mixture. Resistivity measurements on solutions of nitrogen is pure lithium using the capillary tube method (discussed in Chapter 8) were accurate and reproducible, and precipitation of the nitride caused no irregularities. In contrast, similar measurements on solutions of nitrogen in lithium–sodium mixtures were erratic, owing to the accumulation of precipitate in the capillary tube.[260]

The similarity between lithium and strontium solutions is emphasized by the observation that lithium solutions also creep over the surface of the steel containing vessel on introduction of nitrogen.[261] A steel vessel containing some pure sodium was heated at 400°C for 18 h in order to facilitate surface nitriding of the steel; lithium (to give a 1 g atom % solution) was then added, but after a further 7 h at 400°C no creep occurred. When nitrogen was then admitted, the solution showed complete creep over the entire internal surface of the vessel within 1 h. This behaviour is attributed to the formation of ternary nitrides such as Li_3FeN_2[259] at the point where the three phases meet; ternary nitrides of chromium and nickel are also known.

Some information on the Li–Na–N system has been obtained from a study of nitrogen absorption rates. Absorption curves of the type shown in Figure 11.4 have been obtained by the stirred pot method for concentrations up to 30 atom % lithium. At all lower concentrations (up to 13 atom % Li) the curves resemble those for strontium solutions; the curves are rounded, with no initial linear part, for reasons which probably resemble those advanced above for the strontium solutions. In the range 13–30 atom % lithium, the curves do show an initial linear section. If we assume that the break point in these absorption curves indicate the onset of precipitation, then the curves do show one significant feature, i.e. that the number of lithium atoms which solvate the N^{3-} ion in solution is greater than the number of barium atoms required. In an earlier investigation Addison and Davies[262] suggested that Li_6N represented the solvated unit, but a closer re-examination of this work[261] indicated that the unit was represented better by the formula $Li_{8.5}N$.

The reaction has also been studied by the loop and jet method, using the technique described for strontium solutions to obtain absorption rates at constant pressure.[261] A small amount of nitrogen was again added before the experiment began, to facilitate the complete creep of the lithium–sodium mixture over the internal walls of the reactor. The results are shown in Figure 11.7, and they bear a close resemblance to the corresponding results for strontium solutions shown in Figure 11.6. The behaviour represented by Figure 11.7 is typical of lithium concentrations up to about 8 atom %; at higher concentrations (up to 18 atom % Li) the results deviate somewhat, but the general pattern remains the same. Because the reaction is occurring at a fresh surface, the absorption rate is first order with respect to lithium concentration. As successive aliquots of nitrogen are absorbed, absorption rates decrease along the line AC, which represents the conversion of lithium metal in solution into Li_3N. When the ratio of lithium metal in solution to

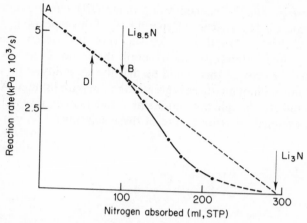

Figure 11.7 Reaction rates of nitrogen with a 4.2 atom %
solution of lithium in liquid sodium, at 400°C. Nitrogen
pressure at 42 kPa throughout

nitrogen added reaches about $Li_{8.5}N$ (at point B), there is a sudden change in
behaviour; if we regard nitrogen as being in solution up to point B, then on
addition of further nitrogen, insufficient lithium metal remains to form the
solvated unit $Li_{8.5}N$, and precipitation commences.

Although this interpretation is an attractive one, the results in Figure 11.7
do not provide a proof that the nitrogen added over the range AB does in fact
remain in solution; the nitride Li_3N produced on absorption may be in
solution, or may be continually precipitated. This is a good illustration of one
of the major difficulties inherent in the study of chemical reactions in liquid
metal media. In molecular liquids, it can be seen immediately whether a
reaction product is soluble or insoluble, and if insoluble whether it separates
in crystalline, amorphous or colloidal form. Because liquid metals are opaque
this facility is not available, and whether or not a product is in solution has
often to be deduced from indirect evidence. Filtration should provide direct
evidence, but this is not always as easy as with molecular solvents. The barium
solutions discussed above filtered readily, but both strontium and lithium
solutions presented great difficulty; this may be related to the creep
properties of these solutions, which may lead to the build up of surface layers
within the filter itself.

One example of the type of indirect approach to this problem of solubility
which can be used involves the addition of solid lithium nitride to the lithium
solution before reaction with nitrogen commences. It is argued that if
precipitation commences at the beginning of nitrogen absorption, any added
solid Li_3N will be ignored, and absorption rates will commence at point A
(Figure 11.7). If on the other hand the added nitride dissolves, bringing the
dissolved nitrogen content to a point D, then a smaller amount of added

nitrogen (corresponding to concentration range DB) will be required before the composition $Li_{8.5}N$ is reached. Experiments of this type are in accord with the second of these alternatives which supports the concept that nitrogen dissolves in dilute lithium solutions until the ratio $Li_{8.5}N$ is achieved. However, one corollary of this would be that the solubility of lithium nitride in some lithium–sodium mixtures is greater than in pure lithium. At first sight this seems unlikely though not impossible, and further research is clearly necessary to establish the true nature of these solutions.

Chapter 12

Formation, dissociation and stability of heteronuclear polyatomic anions

12.1 Introduction

Earlier chapters have shown that, despite the electronic environment within a liquid metal, it is possible for some dissolved monatomic anions to associate with one another given favourable circumstances. The products of such association may be stable polyatomic ions capable of being isolated from the liquid metal (e.g. OH^- in liquid caesium), or they may form as a result of solvation (e.g. Ba_4N in liquid sodium), in which case they have no existence outside the liquid metal medium. Whether or not any given polyatomic anion will dissociate on contact with a liquid metal depends on a number of factors, among which the solvation energy of the dissociation products, and the free energy change on dissociation into the separate monatomic anions, are important. Both terms vary with the metal concerned, so that whether or not dissociation occurs also depends on the metal solvent employed; the behaviour of the OH^- ion in liquid lithium and liquid caesium is a good illustration of this.

In recent years a number of investigations have been carried out with solutions containing two of the elements hydrogen, oxygen, nitrogen, carbon, silicon and germanium, and this chapter will describe experiments carried out on several pairs of these elements to determine the stability of possible polyatomic anions. The solvents used are restricted almost entirely to sodium (because of its use as a coolant in the fast breeder reactor) and lithium (which is a candidate for the breeding medium or primary coolant in future fusion reactors). Most non-metals have higher solubilities in liquid lithium than in liquid sodium, so that unless the research is directed towards some particular technical objective, lithium is the preferred solvent; it will be seen that most interactions have indeed been studied in liquid lithium. The stimulus for this work arises from the need to know, in studies of corrosion by the liquid metal and in the development of purification methods, what are the actual species present in solution.

177

12.2 Examples of dissociation

We have already seen (Chapter 9) that lithium hydroxide undergoes complete dissociation to oxide and hydride in liquid lithium, and that lithium amide dissociates to nitride and hydride. In considering the stability of other polyatomic ions, we may use as a guide the enthalpy changes involved in the corresponding solid state reactions. For example, the two reactions

$$LiNH_{2(s)} + 4Li_{(l)} \longrightarrow 5Li^+_{soln} + N^{3-}_{soln} + 2H^-_{soln} \tag{12.1}$$

and

$$LiNH_{2(s)} + 4Li_{(s)} \longrightarrow Li_3N_{(s)} + 2LiH_{(s)} \tag{12.2}$$

differ only in that the simple salts are produced either as solutes dissolved in liquid lithium, or as solids. Since lattice and solvation energies for the simple salts are remarkably similar we may use the enthalpy changes (in the absence of adequate free energy values) as a guide to the stability of polyatomic ions in solution. Enthalpy values for some such anions are given in Table 12.1; all enthalpy values for reaction with lithium are negative, suggesting that nitrate, carbonate and cyanide ions should also dissociate fully in liquid lithium. The most negative values occur with oxygen-containing salts, owing to the driving force of the very large negative enthalpy of formation of Li_2O.

Experimental work supports these predictions. When carbonate or nitrate were encapsulated with liquid lithium at 600°C for up to 65 h followed by hydrolysis, carbon and nitrogen were recovered as acetylene and ammonia respectively, consistent with dissociation of carbonate into acetylide and nitrate into nitride.[263] Cyanide also dissociated into acetylide and nitride, and it is interesting to recall that, in contrast, sodium cyanide can be dissolved in liquid sodium, then crystallized from the liquid unchanged. These dissociation processes can also be followed by change in resistivity of the liquid metal. In the case of lithium nitrate, for instance, the expected dissociation reaction is

$$LiNO_{3(s)} + 8Li_{(l)} \longrightarrow 3Li_2O_{(s)} + 3Li^+_{soln} + N^{3-}_{soln} \tag{12.3}$$

The oxide is represented as solid since its solubility at 400°C (0.05 mol %) is small in comparison with that of the nitride (1.45 mol %), and it will make very little contribution to the resistivity change during the dissociation. Lithium nitrate was added to liquid lithium at 400°C up to a concentration of

Table 12.1 Enthalpies (kJ mol^{-1}) of formation (a) and enthalpies of reaction with lithium (b) for lithium salts at 25°C

(a)				(b)	
LiCN	−127	Li_2C_2	−59	LiCN	−68
$LiNH_2$	−182	LiH	−90	$LiNH_2$	−164
$LiNO_3$	−482	Li_3N	−165	LiOH	−183
LiOH	−487	Li_2O	−580	Li_2CO_3	−554
Li_2CO_3	−1216			$LiNO_3$	−1422

0.6 mol %; the increase in resistivity corresponded almost exactly with the increase to be expected on addition of the corresponding amount of lithium nitride.[264]

The dissociation of acetylene in contact with liquid lithium was illustrated in a similar manner. Small quantities of acetylene were added successively to liquid lithium at 442°C, and the equilibrium resistivity measured after each addition; the acetylene is completely absorbed, but leaves a small quantity (8 per cent of the original gas volume) of the hydrogenation products C_2H_4, C_2H_6 and $n-C_4H_{10}$. The plot of resistivity increase against acetylene concentration gave a straight line, with a slope which coincided precisely with that which would be given by dissolved hydrogen if dissociation occurred according to

$$C_2H_2 + 4Li \longrightarrow Li_2C_2 + 2Li^+ + 2H^- \qquad (12.4)$$

The solubility, and electron scattering, of lithium acetylide is too small at this temperature to affect the resistivity significantly. These results also mean, of course, that no compounds (e.g. $LiHC_2$), soluble or otherwise, will be formed if lithium acetylide and lithium hydride are added to liquid lithium.

The tendency of these heteronuclear compounds to dissociate in liquid lithium is emphasized by the decomposition of the ceramics BN and Si_3N_4 which under most environments are highly stable refractory compounds. Boron nitride requires high temperatures (>600°C) for complete decomposition, although some dissociation was observed at 320°C; unidentified products of reaction hydrolysed to borates on immersion in water. Decomposition of silicon nitride was evident at temperatures as low as 250°C. The reaction product formed under experimental conditions in which a limited amount of additional nitrogen was available was identified as Li_2SiN_2.[265-267]

The study of the reactions of nitrogen with solutions of the group 4 elements in liquid lithium and liquid sodium has provided results of particular interest. In lithium, the nitride ion does not react with dissolved germanium, tin or lead. In the case of germanium, the resistivity increased linearly with concentration on addition of $Li_{22}Ge_5$ to liquid lithium at 450°C; subsequent addition of Li_3N gave a further linear increase, with a gradient equal to that for the addition of nitrogen to pure lithium, so that no interaction occurs between germanium and nitrogen in this medium.[167] The results are reminiscent of those obtained for dissolved hydrogen and nitrogen which were illustrated in Figure 9.7. With carbon and silicon in liquid lithium, however, dissolved nitride does react to give polyatomic products of remarkable stability, and these reactions will be discussed in the following sections.

12.3 Reactions between Li_3N and Li_2C_2 in liquid lithium

However this reaction is carried out, the final product is the dilithium salt of cyanamide, Li_2NCN. Prior to the work on the silicon–nitrogen reaction (see below) this was the only known example of a salt containing a heteroatomic anion which was stable towards liquid lithium, and it still appears to be the only such compound which can be dissolved in, and crystallized from, the liquid metal. For this reason the compound itself has been studied in some detail, and the intermediate stages in the reaction have been followed by resistivity measurements.

Nitride in excess

Typical resistivity changes are shown in Figure 12.1.[268] The line OP represents the linear increase in resistivity as nitrogen is added to pure lithium in increments up to a concentration of 0.82 atom % nitrogen. To this unsaturated solution was added successive amounts of elemental carbon (which is more easily handled than is Li_2C_2). If carbon dissolved, as Li_2C_2, independently of the nitride present (as would be the case on addition of LiH), the resistivity would continue to increase along some line such as that shown as a broken line in Figure 12.1. Instead, addition of carbon at point P caused an immediate large, linear decrease in resistivity (along PQ). At Q there is a sharp break in the resistivity curve, and resistivity falls much more slowly (along QR) as carbon is added beyond point Q. At a carbon content beyond R, lithium was distilled off, to leave as residue the compound Li_2NCN mixed with excess lithium acetylide.

The final products can be identified readily by removal of the metal medium, but unfortunately the intermediate stages cannot be identified directly in the absence of techniques which allow direct recognition of species in solution. This is particularly to be regretted in the Li–C–N system, as it

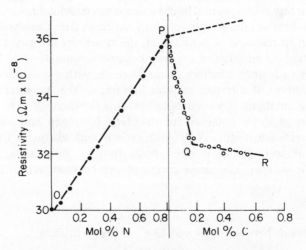

Figure 12.1 Resistivity changes during the reaction between Li_3N and Li_2C_2 in lithium, at 400°C

seems that a number of species are probably produced in the liquid metal which are not accessible in molecular liquids. Under the circumstances, the nature of the intermediate stages is subject to a good deal of speculation and several interpretations of these results have been advanced.[72,268]

One interpretation which is of particular interest and is consistent with experiment is as follows. Carbon added at point P dissolves initially as C_2^{2-} units, but in the relatively high concentration of N^{3-} ions is split into monatomic C units immediately. Two observations give clues to the fate of these C units: (a) the CN^- ion is not formed, as this is known to dissociate in liquid lithium, and (b) the C:N ratio at point Q is 1:4. The species present at Q is not the dilithium salt of the cyclic compound tetrazole, Li_2CN_4, since separate experiments have verified that salts of tetrazole dissociate to the more stable cyanamide salts in liquid lithium. The dilithium compound is not available, but the monolithium compound $LiHCN_4$ dissociates according to the equation[72]

$$LiHCN_4 + 8Li \longrightarrow Li_2NCN + 7Li^+ + 2N^{3-} + H^- \qquad (12.5)$$

We may consider, however, that in the region PQ each C atom is solvated by four nitride ions, giving the solvated complex CN_4, and the solvation energy is sufficient to keep the C atom in solution so long as sufficient N^{3-} ions are available. The situation is similar to that in which nitride ions were stabilized in solution in sodium so long as sufficient Ba atoms were present to form the solvated complex Ba_4N. (For the same reasons as were advanced in Chapter 11, it is not feasible to assign specific electronic charges to the CN_4 unit.) On formation of the CN_4 unit, five potential electron scattering centres are reduced to one; hence the rapid fall in resistivity along PQ.

On addition of carbon beyond Q, the quantity of nitride available is no longer sufficient to maintain carbon as the CN_4 unit in solution, and the reaction

$$CN_4 + C \longrightarrow 2CN_2 \qquad (12.6)$$

occurs (ignoring electric charges). The cyanamide has relatively low solubility, and continues to precipitate until all nitrogen originally present in solution at P is converted to the cyanamide salt Li_2NCN.

Carbon in excess

The final product is again Li_2NCN. Any intermediate products would be expected to differ from those formed when nitrogen is in initial excess, and to have N:C ratios less than two. We may again consider that the first nitride ions to enter the system will attack C_2^{2-} ions to give monatomic C units. These C atoms will form whichever C–N compound is most stable in liquid lithium, which is the NCN^{2-} rather than the CN^- ion; no excess nitride is available to form CN_4 units. In a typical experiment,[72] 0.32 atom % carbon in the form of graphite was added to lithium at 475°C. Nitrogen was then introduced in successive volumes, allowing 24 h between each addition for equilibration. The resistivity varied only slightly until all carbon was converted to cyanamide, after which the resistivity increased rapidly, at a rate characteristic of the solution of nitrogen in lithium. These experiments confirm that cyanamide is the final product in the presence of excess carbon too. However, both Li_2C_2 and Li_2NCN have low solubilities and low $d\rho/dx$ values in lithium, and this accounts for the fact that the resistivity changes very little during the course of the reaction; the resistivity method is therefore unsuitable for the detection of intermediate stages, but experiments using a different technique have shown that products having a N:C less than two can be formed.

Whether such products would be stable in a pure liquid lithium medium seems doubtful, but carbon–nitrogen reactions carried out in dilute solutions of lithium in liquid sodium give a product having empirical formula LiC_2N.[269] Results obtained using an 8 atom % solution of lithium in sodium at 400°C are shown in Figure 12.2. In each experiment a known quantity of carbon was added to the liquid, and the total amount of nitrogen absorbed was then measured. Nitrogen absorption increased as the carbon content increased. Reaction depends entirely on the presence of lithium, since there is no reaction between carbon and nitrogen in liquid sodium at this temperature, and both carbon added and nitrogen absorbed are therefore plotted in Figure 12.2 in terms of atoms of non-metal per atom of lithium present in the liquid metal. In the absence of carbon (or if carbon did not react) the nitrogen absorption would correspond to the formation of Li_3N. If the reaction product was Li_2NCN, $LiCN$ or Li_2C_2, nitrogen absorption should follow one of the corresponding broken lines in Figure 12.2. In fact experiments show that the

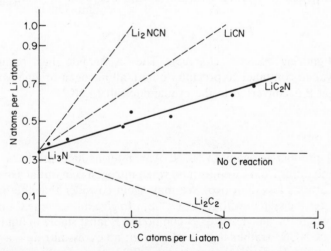

Figure 12.2   Carbon–nitrogen interactions in 8 atom % lithium in sodium, at 400°C

nitrogen absorption values lie close to a line which represents the formation of a product having empirical formula LiC_2N. Attempts have been made to give a chemical identity to this product by distilling off the sodium, but the product then remains as a polymeric mass, and polymerization has no doubt been brought about by the distillation conditions. Isolation by removing the sodium using liquid ammonia is also unsatisfactory, as some decomposition appears to take place.

It is therefore not possible, at this time, to deduce a reaction mechanism, but the following mechanism, although speculative, is in accord with experiment. We may consider that the first nitride ions which enter the liquid will break the $C\equiv C^{2-}$ bond; the subsequent reaction of the monatomic C unit can be represented as

$$C^{n-} \longrightarrow CN^- \longrightarrow NCN^{2-} \tag{12.7}$$

In a pure lithium medium the reaction proceeds to the cyanamide stage, since this is more stable than the cyanide. In the liquid sodium solution, however, cyanide is the more stable anion (reasons for this will be discussed later), and this is therefore regarded as being formed as a first step. The only available species with which the CN^- ion can then react is the acetylide ion, and this can give rise to the following known reaction:

$$2CN^- + C\equiv C^{2-} \longrightarrow N\equiv C-C\equiv C-C\equiv N + 4e \tag{12.8}$$

Dicyanoacetylene is known to polymerize readily, and can also take up two electrons to give the dilithium salt $Li_2C_4N_2$; this simplifies to the empirical formula LiC_2N observed by experiment.

12.4 Reaction of carbon with nitrogen in liquid sodium

The reaction only takes place at high temperatures. In a laboratory experiment, graphite (0.19 mol) was added to stirred sodium (2.18 mol) in a stainless steel vessel, and the absorption of nitrogen with time was followed by pressure changes. There was no measurable absorption at $400°C$ or $600°C$, but reaction could be detected at $700°C$ and was more rapid at $800°C$. At $700°C$ the reaction required 120 h for completion and, after allowance was made for absorption of nitrogen by the steel, the final stoichiometry ($N/C = 1.05$) corresponded to the formation of sodium cyanide, even though excess of nitrogen was available. The sodium was distilled off under vacuum at $600°C$, leaving a residue of sodium cyanide which was identified by chemical analysis and X-ray powder diffraction.[271] These observations are consistent with the increase in carbon solubility in sodium, owing to the formation of CN^- species, which occurs under a pressure of nitrogen,[272] and with the recovery of unchanged sodium cyanide from sodium.[273] It is also interesting to note that the Castner process for the manufacture of sodium cyanide involved reaction of sodium with carbon and nitrogen at $650°C$.[270]

In view of the negligible solubility of nitrogen in liquid sodium, it seems likely that the reaction occurs at the metal surface, which would also account for the slowness of the reaction. It is also probable that, by virtue of its immersion in sodium, carbon may be in a more activated form because of entry of electrons from the metallic medium, with consequent weakening of the graphite structure.

The carbon–nitrogen reaction in sodium is therefore quite different from that in lithium. Under the conditions described, the equilibrium

$$CN^- + N^{3-} \; \underset{Na}{\overset{Li}{\rightleftharpoons}} \; NCN^{2-} + 2e \qquad (12.9)$$

lies entirely on the right-hand side in liquid lithium, and on the left-hand side in sodium. The remarkable stability of Li_2NCN in lithium is well illustrated by the reactions discussed above. Additional evidence is provided by the direct addition of the NCN^{2-} ion to liquid lithium in the form of the salt NaHNCN.[264] When known quantities of this salt were added successively to lithium, the change in electrical resistivity was precisely that expected from the reaction

$$NaHNCN_{(s)} + 2Li_{(l)} \longrightarrow Na^+ + Li_2NCN_{(s)} + H^-_{soln} \qquad (12.10)$$

rather than the much greater change which would have resulted from the dissociation of the NCN^{2-} ion. The remarkable stability of Li_2NCN towards lithium can be rationalized in terms of its high enthalpy of formation. Using the values for the enthalpy of formation of Li_3N and Li_2C_2 given in Table 12.1, we see that the enthalpy of formation of Li_2NCN will be at least -360 kJ mol^{-1}, and greatly in excess of the value for LiCN. The high value is attributed to the high lattice enthalpy of its tetragonally distorted body-centred cubic structure, and to the strength of the covalent bonding with the $N{=}C{=}N^{2-}$ complex ion. Appropriate physical data for the corresponding sodium compounds is not available. However, remembering that Na_2C_2 is much less stable than Li_2C_2, and that Na_3N does not exist, we might expect that Na_2NCN would also have a lower stability. Trends shown by the groups I and II metal cyanides and cyanamides, if extrapolated, also suggest that the difference between enthalpies of formation of NaCN and Na_2NCN is much smaller than in the case of lithium. Again, at the higher temperatures required to bring about the carbon–nitrogen nitrogen in liquid sodium, Na_2NCN may be thermally unstable. All these points are consistent with the experimental observation that CN^-, rather than NCN^{2-}, is the product of the carbon–nitrogen reaction in liquid sodium. No corresponding studies appear to have been carried out using the heavier alkali metals as reaction media.

12.5 Preparation of Li_2NCN

Nitrogen and carbon are among the commonest impurities in liquid lithium, and the formation of this very stable compound is of interest wherever lithium is employed as coolant on an industrial scale. Its preparation will be described here mainly as an illustration of the way in which liquid lithium can be used as a preparative medium.

Until recently the compound was poorly characterized, but it has now been prepared by heating lithium carbonate with urea at temperatures up to 800°C,[275] and in a purer form (96–98 per cent) by the thermal decomposition of LiHNCN at 900°C.[276] In the liquid metal method,[274] pure starting materials were obtained from the elements. Lithium nitride was made by reaction of nitrogen with a solution of lithium (7 g) in liquid sodium (23 g) at 400°C in a steel beaker contained in a distillation vessel. The reaction temperature chosen is sufficiently above the consolute temperature (303°C) of the metal mixture that metallic phase separation is avoided. Nitrogen reacts with the lithium present to form the nitride, giving eventually a suspension of Li_3N in pure sodium. The sodium was distilled off at 300–400°C under 10^{-5} torr pressure, leaving a residue of ruby-red hexagonal plates of Li_3N. This method of preparation ensures complete reaction of the lithium, and avoids the high temperature needed for distillation of unreacted lithium, with consequent dissociation of Li_3N. Dilithium acetylide can be prepared by heating lithium (6 g) with carbon (Johnson Matthey Specpure, 6 g) at 650°C for 100 h under argon. In this case excess lithium can be distilled off at 600°C under 10^{-5} torr pressure, leaving a white residue of Li_2C_2. Li_2NCN may then be prepared by heating Li_3N with Li_2C_2 in 4:1 mol ratio at 600°C for 10 h, under argon. Liberated lithium is distilled off as before, leaving a residue of off-white Li_2NCN.

In an alternative method Li_2NCN has been prepared by crystallization directly from liquid lithium. The apparatus was essentially that shown in Figure 2.6, in which a steel reservoir was equipped with an electromagnetic pump and capillary loop. The reservoir was charged with lithium (30 g) under argon, and the apparatus heated to 530°C in an air oven. In order to encourage homogeneity, Li_2C_2 was added in a series of small blocks (20 × 0.07 g). Nitrogen was then introduced, and the reaction followed by resistivity changes. When the C:N ratio reached about 4:1, the solution was allowed 24 h for equilibration, and the lithium again distilled off, as before. When the reservoir was opened under argon, the salt Li_2NCN was seen to be present as hygroscopic, colourless, single plates which were mixed with, but easily separated from, unchanged Li_2C_2. These plates were suitable for use in the determination of the crystal structure of Li_2NCN.[274]

12.6 The carbon–nitrogen reaction in a solution of barium in sodium

Evidence was given in section 12.3 that carbon and nitrogen do not interact in liquid sodium alone at 400°C, but that they do react when lithium is added to the sodium. This is one example of what is probably a much wider generalization, i.e. that if any metal which (a) reacts readily with nitrogen and (b) dissolves in liquid sodium is mixed with sodium, it will then cause a

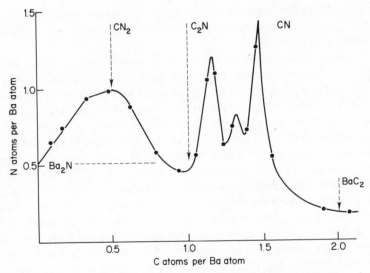

Figure 12.3 Carbon–nitrogen interactions in 4 atom % barium in sodium, at 400°C

carbon–nitrogen reaction to take place by introducing N^{3-} ions into the solution.

Barium has been found to be effective in this respect, and some results are shown in Figure 12.3. Because barium is responsible for the reaction, the quantities of carbon and nitrogen are again (as in Figure 12.2) expressed in terms of atoms reacting per barium atom present. In each experiment (shown by separate points) carbon was first added to the solution (4 atom %) of barium in sodium, stirred to equilibrate, and the maximum amount of nitrogen which the solution would take up was then measured. The two limiting conditions are quite evident. In the absence of carbon, Ba_2N is precipitated; when the Ba:C ratio reaches 1:2, BaC_2 is formed. This is insoluble, so that all barium is removed from solution. The BaC_2 is wetted by sodium, and since nitrogen is insoluble in sodium, little nitrogen is absorbed at this stage. With intermediate quantities of carbon the C:N ratios vary widely, but each of the ratios representing products already discussed in this chapter (CN_2, C_2N and CN) feature as obvious maxima or minima in the curve. Various interpretations of this behaviour have been advanced[129, 271] but they are still to be regarded as tentative since the solubility, and hence the reactivity, of some of the intermediate products are unknown. The presence of cyanamide has been recognized by chemical tests, but these tests may also include other closely related products. What is apparent, however, is that this technique could be remarkably versatile. Future work could with advantage use the graph in Figure 12.3 as a guide, select conditions for the formation of each single product, and by careful separation and analysis attempt their identification.

12.7 Reaction between silicon and nitrogen in liquid lithium

As with the carbon–nitrogen reactions, the product can vary depending on the relative amounts of reactants available in the solution. In a survey of silicon–nitrogen reactions,[280] liquid lithium was contained in stainless steel crucibles which were housed in a steel distillation vessel. Silicon was added as powder and nitrogen as its compound Li_3N, and the reaction mixture equilibrated for 168 h at 600°C. Reaction products were identified by X-ray diffraction of the powder residue after distillation. Additionally, X-ray diffraction patterns of the product were taken within the lithium matrix before distillation to ensure that no decomposition of the original product had taken place during distillation. Silicon:nitrogen ratios from 1:1 to 1:4 were used, and the reactions are represented by the following equations:

$$2Si + 2Li_3N + 0.4\,Li \longrightarrow Li_2SiN_2 + 0.2\,Li_{22}Si_5 \quad (12.11)$$

$$Si + 2Li_3N \longrightarrow Li_2SiN_2 + 4Li \quad (12.12)$$

$$Si + 3Li_3N \longrightarrow Li_5SiN_3 + 4Li \quad (12.13)$$

$$Si + 4Li_3N \longrightarrow Li_5SiN_3 + 4Li + Li_3N \quad (12.14)$$

The compound Li_2SiN_2 is therefore formed under conditions in which no additional nitrogen is available. Given an excess of nitrogen, Li_5SiN_3 is formed. It is interesting to note that at low nitrogen levels a mixture of Li_2SiN_2 and $Li_{22}Si_5$ is produced rather than the compound $LiSi_2N_3$, and that at high nitrogen levels excess Li_3N and Li_5SiN_3 are formed rather than Li_8SiN_4.

The course of the reaction can be followed by the resistivity method. Silicon is very sparingly soluble in liquid sodium, but has an appreciable solubility in liquid lithium (2.2 atom % Si at 450°C), which is similar to the solubility of nitrogen. Dissolved silicon (as the element or its compound $Li_{22}Si_5$) has a high resistivity coefficient ($d\rho/dx = 10.4 \times 10^{-8}\,\Omega m\,(mol\,\%\,Si)^{-1}$), so that changes in resistivity can be used to follow interactions between silicon and nitrogen in solution.[129,277] Resistivity changes may also be used to determine the times required for establishment of equilibrium.[277] Thus, on addition of 0.2 atom % nitrogen to a solution of silicon in lithium at 475°C the resistivity decreased, indicating that reaction was occurring; 12 h were required for equilibrium at this temperature, compared with a few minutes for solution of nitrogen in lithium alone. Similarly, 24 h were required for equilibrium on adding 0.2 atom % silicon to a solution of nitrogen in lithium, compared with 1 h for the solution of this quantity of silicon in lithium alone. Equilibrium values for the resistivity changes observed on adding silicon, followed by nitrogen, to liquid lithium at 475°C are shown in Figure 12.4. There is a linear increase in resistivity as silicon is added (OA) to give an unsaturated solution at A. Nitrogen was then added in increments, causing a progressive decrease in resistivity (AB) which fell linearly to a value at B near the value for pure lithium. Along AB, nitrogen is reacting with dissolved silicon to give a virtually insoluble product, and the linearity of AB indicates that only one product

188

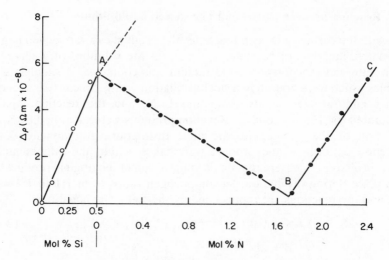

Figure 12.4   Changes in the resistivity of liquid lithium at 475°C on successive additions of silicon and nitrogen

is being formed during the reaction. Addition of nitrogen beyond B merely leads to a solution of lithium nitride, and the increase in resistivity along BC corresponds to that expected for such a solution.

Because both nitrogen and silicon have high resistivity coefficients, a set of experiments similar to those illustrated to Figure 12.4 can be carried out, adding nitrogen first, then silicon. The results can be treated in the same way, and the same reaction product is formed irrespective of the order in which the elements are added.

The Si:N ratio in the reaction product can be derived from the gradient of AB, assuming the resistivity change to be due solely to the loss of solute from the solution, or from the ratios given by point B. From a series of such experiments the ratio was found to be between 1:3.1 and 1:3.6, implying that the precipitating phase is either Li_5SiN_3 or Li_8SiN_4. The reaction product was isolated by vacuum distillation of excess lithium, and X-ray powder diffraction showed the product to be entirely Li_5SiN_3. This was confirmed by X-ray diffraction analysis of the product prior to distillation, i.e. while still embedded in a solid lithium matrix, so that no decomposition occurred during the isolation stage. The compound Li_5SiN_3, formed by reaction (12.13), is therefore the most stable ternary nitride in this system. Thermochemical arguments, based on the reaction between solid nitride and silicide, predict that the standard enthalpy of formation of Li_5SiN_3 is at least as negative as -642 kJ mol^{-1}.

To discount further the possibility that Li_8SiN_4 was originally produced in the liquid metal environment, and subsequently converted to Li_5SiN_3 during distillation, Li_8SiN_4 was subjected to distillation conditions. A mixture of

Li_8SiN_4 and Li_2SiN_2 was produced, with no evidence for Li_5SiN_3. The compound Li_2SiN_2 was also prepared, and an intimate 1:1 mixture with Li_8SiN_4 was heated under distillation conditions, but no interaction was observed. This gave further confirmation that the characteristic product of the liquid metal reaction is indeed Li_5SiN_3, provided that sufficient nitrogen is available for the formation of this compound. The compound is a member of the $Li_{2n-3}M^{n+}N_{n-1}$ series, like Li_9CrN_5 which is also produced in liquid lithium.[278,279] The stability of these compounds towards liquid lithium is thought to be associated with their cubic anti-fluorite $(Li, M)_2N$ type lattice structure, in which the metal atoms occupy all the tetrahedral sites in a face-centred cubic array of nitrogen atoms.

12.8 Carbon dioxide and carbon monoxide

Carbon dioxide is mentioned here because reference to the literature on sodium–carbon dioxide reactions might suggest that some association of carbon dioxide occurs in the formation of sodium oxalate. Finar[304] states that sodium oxalate can be synthesized by passage of carbon dioxide over sodium heated to 360°C, and this is also reported in Gmelin's *Handbuch*.[305] According to Gilbert,[306] when a spoon containing burning sodium is plunged into an atmosphere of carbon dioxide, a cloud of finely divided carbon is formed, and the sodium oxide which is also formed reacts with further carbon dioxide to give sodium carbonate; some carbon monoxide was also produced, but no sodium oxalate was mentioned. The product of reaction of carbon dioxide with sodium in the form of its amalgam at 25°C and 180°C contained a high proportion of sodium oxalate and no carbon monoxide was detected, whereas analogous reactions with potassium, rubidium and caesium amalgams did give some carbon monoxide as an additional product.[307] A close examination of the experimental conditions under which these reactions were carried out suggests strongly that oxalate is produced by processes which occur outside the metal itself, i.e. by subsequent interaction between initial products, and that the oxalate ion is not formed from carbon dioxide within the liquid metal medium. The sodium–carbon dioxide reaction has been studied in the temperature range 110–200°C by following the pressure changes which occur when carbon dioxide at an initial pressure of about 20 torr is exposed to a large excess of liquid sodium. Stirring by magnet was gentle in order that surface films could be observed.[308] After an induction period of a few minutes, the pressure fell to zero at an almost constant rate, though a small increase in rate could be observed as the reaction proceeded. On introduction of carbon dioxide, a white film was formed immediately; this persisted throughout the reaction, but became dark in colour as the reaction proceeded. Spectroscopic examination of the gas phase confirmed the absence of carbon monoxide. No unusual features were observed which could be attributed to the formation of oxalate, and the reaction, under these conditions, is adequately described by the equation

$$4Na + CO_2 \longrightarrow 2Na_2O + C$$

followed by

$$Na_2O + CO_2 \longrightarrow Na_2CO_3$$

Carbon monoxide, however, provides quite a different story. The reaction is a complicated one which is not yet fully understood, but there seems little doubt that polymerization of carbon monoxide occurs within a liquid metal medium to give the largest molecule known to be synthesized in this medium. Studies of the reactions between alkali metals and carbon monoxide have mainly used the metal in solution in liquid ammonia, when a reactive product of empirical composition $[M(CO)]_x$ is formed. This is often referred to loosely as an alkali metal carbonyl, and it has been suggested[309] that it is composed of an acetylene diolate $M^+ \, {}^-OC{\equiv}CO^- \, M^+$ together with organometallic compounds of the same empirical composition. With liquid potassium, rubidium and caesium, carbon monoxide again undergoes direct reaction at temperatures only slightly above the melting points of the metals, and with liquid potassium at higher temperatures the formation of the potassium salt of hexahydroxybenzene was reported. There is conflicting evidence in the literature about the corresponding reaction of carbon monoxide with liquid sodium. At the low temperatures and pressure used in the potassium, rubidium and caesium reactions, Buchner[310] was not able to observe any absorption of carbon monoxide by liquid sodium, but at higher temperatures and pressures the formation of sodium carbonyl was reported,[311] and Miller[312] has claimed that at 300°C, with 1 atm pressure of carbon monoxide, hexasodium hexacarbonyl is one of the products. It transpires (see below) that the reason for the apparent conflict in results arises from the fact that the sodium reaction is characterized by an induction period followed by quite rapid absorption of carbon monoxide; under some conditions the induction period can extend to many days, giving the impression that no interaction will occur.

Since carbon and oxygen are two of the most common impurities in liquid sodium, the behaviour of carbon monoxide in sodium has obvious significance in the operation of sodium-cooled fast nuclear reactors. The reaction was investigated by Sinclair et al.[313] at the UKAEA Dounreay Experimental Reactor Establishment in the temperature range 200–500°C with pressure of carbon monoxide less than 1 atm, and by Addison et al. at the University of Nottingham[308] using temperatures in the range 110–200°C. This, and other attempts at Nottingham,[314] have led to the general conclusion that the reaction is unpredictable, and no conditions have been found under which the reaction can be made to proceed in a reproducible fashion; it is not possible, therefore, in the present context to provide absorption rate curves which can be regarded as typical of the behaviour of carbon monoxide at the sodium surface.

Nevertheless some important observations regarding reaction products have been made. The course of the reaction falls broadly into three stages, but the readiness with which the stages can be identified separately varies with temperature, pressure, and particularly with the relative amounts of carbon monoxide and sodium available for reaction. The observations made during stage 1 are those which would be expected if this stage represented solution of carbon monoxide to saturation. It covers only a small part of the total reaction, and is fairly rapid, requiring only a few hours at 200°C. The quantity of carbon monoxide taken from the gas phase increases as the quantity of sodium is increased, and the rate of absorption is a function of the carbon monoxide already taken up. If sodium is removed by vacuum distillation at 300°C at the end of this stage, no solid remains, so that the carbon monoxide must be present as such, or in the form of a very weak polymer which dissociates during the distillation. Towards the end of this stage (at which time, perhaps, the metal is becoming saturated with carbon monoxide) the sodium begins to creep up the sides of the glass containing vessel, and within about an hour has covered the entire internal surface of the vessel. This effect is unique and is the nearest approach to surface activity which has been encountered with the liquid alkali metals.

During stage 2, carbon monoxide absorption is relatively slow; it is this stage which can very widely in its duration, though the duration does seem to be shorter at the higher temperatures. This stage can be interpreted as a period during which dissolved carbon monoxide is undergoing polymerization. When sodium was distilled off at 350°C at the end of this stage, a solid product remained, about half of which was identified[313] as a mixture of sodium salts with the anions of rhodizonic acid (2), tetrahydroxybenzoquinone (3) and hexahydroxybenzene (4). We can envisage the polymerization of the dipolar carbon monoxide molecules as occurring in the first instance to give cyclohexanehexone (1):

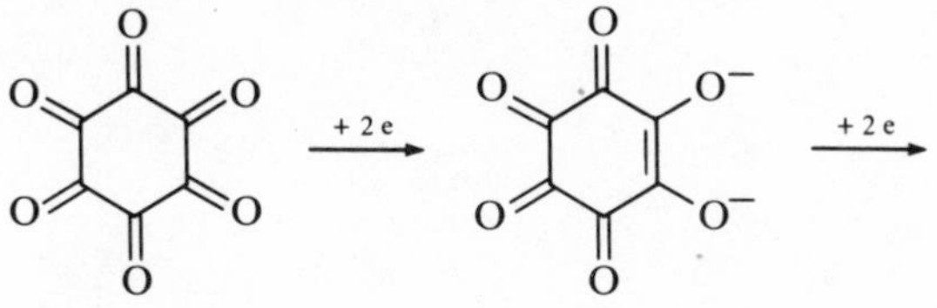

(1)　　　　　(2)

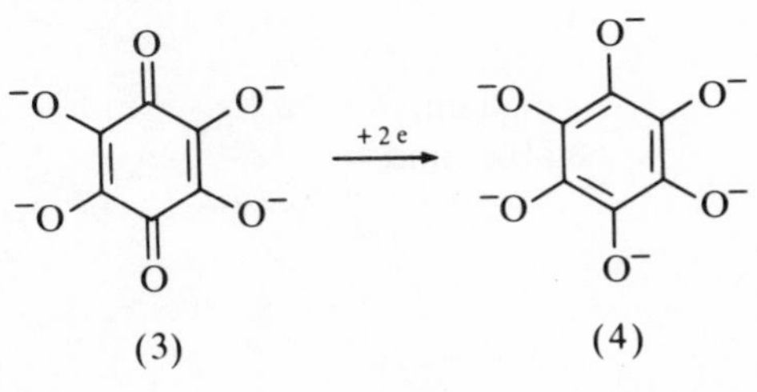

(3)　　　　　(4)

This is feasible in energy terms. The carbon–oxygen bond in carbon monoxide is strong (1073 kJ mol^{-1}, compared with 866 and 813 kJ mol^{-1} for C≡N and C≡C respectively) and the energy required to dissociate the bond in the reaction

$$2CO + 6Na \longrightarrow 2Na_2O + Na_2C_2$$

is not sufficiently offset by the energy of formation of sodium oxide and acetylide. This will also be the case with liquid potassium, where the formation of the potassium salt of hexahydroxybenzene is reported, and with rubidium and caesium. On this basis it can be argued that in liquid lithium, dissociation of carbon monoxide to the oxide and acetylide will occur, without the formation of polymer, but this has yet to be confirmed experimentally.

When polymerization to (1) has occurred, its conversion to the various anions mentioned above involves merely the addition of electrons to the molecule, and internal rearrangement of bonds. The product isolated by Sinclair et al.[313] had a C:Na ratio of 2.9, which is near the value of 3 required by compound (2). As the reaction proceeded into stage 3, the absorption rate increased and other (as yet unidentified) products were formed. However, the quantity of products (2)–(4) was little changed, except that the C:Na ratio changed from 3:1 towards 1:1, as required by compound (4). The limited amount of available experimental evidence is nevertheless sufficient to indicate that compounds (1)–(4) are formed by the addition of carbon monoxide to sodium and the heavier alkali metals, and that it might be possible to so arrange conditions as to produce the individual compounds. This topic is well worthy of further research. The first requirement is to devise conditions in which the reaction is reproducible, and this might be possible by the use of the heavier alkali metals or their mixtures.

Chapter 13

Reactions of the liquid alkali metals and their alloys with simple aliphatic hydrocarbons

13.1 Introduction

Very few of the researches published to date on this topic have been concerned with the bulk alkali metals in liquid form. Most investigations have involved the use of alkali metals dissolved in liquid ammonia, or solid alkali metals; alkali metals supported on carbon or anhydrous oxides have also been studies for their possible catalytic properties. Much of the work is not relevant to our present interests and will not be reviewed, but brief reference should be made to research which is related, either experimentally or theoretically, to the experiments to be described in this chapter.

The reaction of acetylene with potassium, rubidium and caesium has been studied using the solid metal and an acetylene pressure of less than 1 atm.[291] Hydrogen acetylides HMC_2 were produced as a film, which was responsible for the diffusion controlled nature of the reactions, and little or no ethylene was produced. Clusius and Mollet[292] showed that ethylene would react with caesium at room temperature (i.e. as the solid). Reaction was only complete after several months, and produced ethane, butane and a caesium–ethylene adduct. On increasing the pressure, polymerization became dominant. Rubidium was found to be less reactive than caesium towards ethylene. Hackspill and Rohmer[293] concentrated attention on the caesium–ethylene adduct, which was formed after several days' contact with the liquid metal at 45°C, and hydrolysis indicated its formula to be $C_2H_4Cs_2$. Kistiakowski[294] was concerned with the hydrogenation of ethylene on a caesium surface; ethylene–hydrogen mixtures were found to react to give ethane even at room temperature. An attempt to repeat the ethylene–hydrogen reaction over sodium produced only hydride below 300°C. The efficiency of lithium and sodium in the hydrogenation of ethylene to ethane has been compared, using the metals supported on alumina.[295] The metals were supported by dispersing the molten metal over alumina at 250°C and

150°C respectively, and the extent to which an equimolar mixture of ethylene and hydrogen reacted was then compared. Pretreatment with hydrogen increased the efficiency, and sodium was more efficient than lithium at 200°C under these conditions. A more recent study[296] compared all the alkali metals, supported on activated carbon. The carbon was first impregnated with the metal hydroxide, then heated to 600–800°C, when water vapour and carbon monoxide were evolved leaving the metal supported on carbon. Hydrogenation experiments were then performed using an equimolar ethylene–hydrogen mixture, and the ability to cause hydrogenation was found to increase in the order

$$Li < Na < Rb < Cs < K \qquad (at\ 20°C)$$

and $\qquad Li < Rb < Cs \qquad\qquad (at\ 200°C)$

The activity of supported alkali metals has also been used in the hydrogenation of hex-3-ene, hex-1-ene and pent-2-yne.[297] Regarding the mechanism of hydrogenation, there have been many proposals which involve alkali metal–organic compound adducts as intermediates (see, for example, References 298 and 299), and as early as 1931 it was suggested that only those organic compounds which could form an adduct with an alkali metal could be reduced by the metal.[300] This is perhaps too rigid a definition, but the ability to form adducts, and their stability, is a useful indication of the extent to which electrons can be transferred from metal to hydrocarbon, and thus of the strength of the chemisorption at the metal surface.

It is difficult to correlate these observations or to obtain a true overall picture because of the diverse experimental conditions, and sometimes the unknown purity of the alkali metals used. Against this background, an organized investigation has been carried out since 1965 at the University of Nottingham which has employed fresh metal surfaces, similar techniques throughout, and modern analytical techniques, so that the results for various hydrocarbons and various alkali metals and their alloys are strictly comparable. This work forms the subject of this chapter. In many cases the experiments have been carried out using very simple apparatus, by following pressure changes as a function of time. These changes, in themselves, can give enough information to determine reaction rates and often final products of reaction. In addition, chromatographic analysis and mass spectrometry have also been applied to the gas phase. Where more than one gaseous species is involved (e.g. a reactant gas and a product gas) pressure changes help to define their relative reactivity at the metal surface.

One of the main problems arises as a result of the formation of surface films which occurs in almost all cases when a static surface is used. In this volume we are less interested in film properties than in the reactivity of gases at a fresh metal surface, and reactions have therefore been studied using liquid metal stirred by magnet (Chapter 2, Figure 2.3) or at a liquid metal jet (Figure 2.6). Even so, some reactions are so rapid that it is not possible to avoid conditions under which reaction rates are controlled by surface films.

In general, sodium does not react with saturated straight chain aliphatic hydrocarbons, and Ogini[301] reports lack of success in attempts to dehydrogenate such compounds with any liquid metal. However, if there is a conjugated system elsewhere in the molecule, dehydrogenation can occur. For example, dehydrogenation of ethylbenzene to styrene:

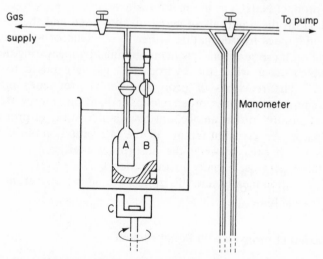

was accomplished using alloys of sodium or potassium with lead, thallium, indium, tin or bismuth, since the removal of two hydrogen atoms extends the conjugation in the molecule. Many aromatic compounds are also inert, and Sittig[289] has described the preparation of stable, finely divided dispersions of sodium in toluene and in xylene. With unsaturated aliphatic hydrocarbons, however, sodium does react.

The apparatus used to study the reactions of propyne and acetylene, is shown in Figure 13.1. The reaction vessel consisted of two concentric cylindrical vessels A and B, which could be independently isolated, and was constructed from Pyrex glass so that the metal surface could be observed continually. B contained about 40 g of liquid sodium, which could be stirred vigorously over an internal baffle by means of the external rotating magnet C. The reaction unit was connected to a simple mercury manometer (of negligible volume compared with B) and a vacuum line, and was heated by immersion in a silicone oil bath at 160°C. The reactant gas was originally contained in A; when temperature equilibrium had been established, the gas was expanded into the evacuated volume (about 100 ml) in B, and A then

Figure 13.1 Apparatus for reaction of gases with excess liquid sodium isolated. By recharging A with the reactant gas, and evacuating B after a reaction had been completed, it was possible to add a series of separate aliquots of gas to the same sample of liquid sodium without disturbing the apparatus. Apart from the propyne and acetylene experiments, the apparatus involving electromagnetic pump circulation (Figure 2.6) was used, in which reaction occurred at the surface of a liquid metal jet, and was constructed of glass or of steel, depending on the temperature.

The reaction with acetylene was first studied because it provides a means of adding very small quantities of carbon to liquid sodium. A major problem in carbon solubility determinations relates to the form in which the carbon is added to liquid sodium. If added as graphite, the stability of the graphite lattice may hinder equilibrium; alternatively, fine particles may break away, and it is not easy in liquid metals to distinguish between a solution and a fine suspension of carbon, especially since the solubility of carbon in sodium is so very small. However, the process $H_2C_2 \rightarrow NaHC_2 \rightarrow Na_2C_2$ occurs in the presence of an excess of liquid sodium, and carbon is therefore presented to the liquid metal in the form of C^{2-} units, thus providing optimum conditions for the solution of carbon. As the work developed, it became clear that hydrogenation reactions were taken place at the alkali metal surface which bore some analogy with corresponding reactions at the surface of transition metals. Again, the reactions which were possible depended on the choice of alkali metal or alloy used, and for these reasons the research was extended beyond the original narrow scope of solubility studies. Previous work on the sodium–acetylene reaction had been mainly concerned with the preparation of the hydrogen acetylide $NaHC_2$, which involved low temperatures and a liquid ammonia medium, but the reactions of liquid sodium, at temperatures above 98°C, were expected to differ because of the likely decomposition of the initial product $NaHC_2$ at the metal surface.

Depending on the liquid alkali metal or alloy used and the temperature, acetylene undergoes hydrogenation either to ethylene or to ethane. Useful information on these reactions has been obtained by measuring the pressure changes which occur when the hydrocarbon gas is exposed to the metal surface.[282-285] The reactions of propyne (methyl acetylene) prove to be somewhat simpler than those of acetylene, and aid the interpretation of the acetylene reactions, so that an account of a limited study of propyne is also included below. An essential feature of the interpretation is the very great differences which exist between the adsorption coefficients of the hydrocarbons; these decrease rapidly in the order $C_2H_2 > C_2H_4 > C_2H_6$ at the surface of transition metals, and this appears to be the case at the surface of the alkali metals too.

13.2 Reaction of propyne with liquid sodium

Propyne gives rise to a simpler reaction than does acetylene because it contains only one hydrogen atom replaceable by sodium. Pressure changes

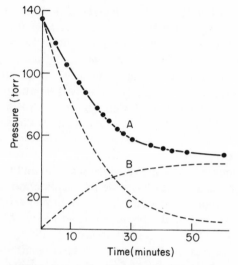

Figure 13.2 (A) Variation of pressure with time in the sodium–propyne reaction at 110°C. (B) Propene gain. (C) Propyne loss

during the reaction at 110°C are shown as curve A in Figure 13.2. The curve shows no discontinuity and reaction is virtually complete within 1 h. The infrared spectrum of the gaseous product showed it to be propene, and there was no evidence for hydrogen or any other hydrocarbon in the gas phase. The final pressure at equilibrium was precisely one third of the initial pressure, so that on reaction with sodium, three molecules of propyne yield one molecule of propene and no other gaseous product.

The results can be interpreted suitably by reference first to the corresponding reaction in liquid ammonia. The solution of sodium provides solvated electrons which convert propyne to the anion:

$$MeC\equiv CH + e \rightarrow MeC\equiv C^- + H \qquad (13.1)$$

with release of hydrogen atoms. The latter are surrounded by dissolved molecules of propyne, which are hydrogenated:

$$MeC\equiv CH + 2H \rightarrow MeCH=CH_2 \qquad (13.2)$$

If all hydrogen atoms released are employed in this hydrogenation, then the overall reaction is

$$3MeC\equiv CH + 2e \rightarrow 2MeC\equiv C^- + MeCH=CH_2 \qquad (13.3)$$

This represents the same 3:1 stoichiometry, so far as gaseous products are concerned, as is found for the reaction with liquid sodium at 110°C. At the surface of the liquid metal we may therefore consider that electrons are provided from the conduction band of the metal for transfer to adsorbed propyne; the product $Na^+C\equiv CMe^-$ is formed as a separate solid phase, and hydrogen atoms are chemisorbed at the metal surface. Since propyne molecules and hydrogen atoms are present together on the sodium surface, and since the 3:1 stoichiometry (equation 13.3) also operates for liquid sodium, it follows that all the H atoms produced at the surface are used in hydrogenating propyne; none forms sodium hydride and none escapes from the surface as H_2 gas. If we assume, by analogy with transition metal surfaces, that alkenes are less strongly adsorbed than are alkynes at the sodium surface, we may consider that propene loses in the competition with propyne for occupancy of the surface; it therefore escapes as soon as it is formed, and is not further hydrogenated to propane. Since reaction at 110°C is represented completely by equation 13.3, it is possible to calculate the composition of the gas phase at any stage of the reaction. The variation in propyne and propene pressures are represented by the broken lines in Figure 13.2.

A temperature of 110°C is near the lower limit available with liquid sodium. At higher temperatures, reaction rates increase rapidly and other effects are introduced. The main effect is the appearance of a break in the pressure curve, which was smooth at 110°C (Figure 13.2, curve A). At 200°C, for instance, reactions 13.1 and 13.2 are complete within 15 minutes. This is followed by a slow fall in pressure, attributed to a polymerization process involving reaction between propene and the sodium salt of propyne with which it is in contact.[283]

13.3 Reaction of acetylene with liquid sodium

The pressure–time curves for acetylene are quite different from those of propyne. Figure 13.3 shows a typical curve. This was obtained at 200°C, but curves of the same shape were obtained within the temperature range 110–200°C, so that the total pressure–time curve in Figure 13.3 may be compared directly with that for propyne in Figure 13.2. With acetylene, the curves reach a plateau P_2 within 10–20 minutes, when the pressure has fallen to about 70 per cent of the initial value P_1. After a period (which is not in itself closely reproducible) during which the pressure is almost constant, a further decrease sets in quite suddenly, and final equilibrium is reached after several hours. The curve is reminiscent of the hydrogenation of acetylene over palladium–alumina,[286] where the two distinguishable stages represent successive hydrogenation to ethylene and ethane; however, over sodium at 200°C different processes are involved. The composition of the gas phase was determined from its infrared spectrum, and the presence of hydrogen could be detected by condensing the gas phase at liquid nitrogen temperature, $-196°C$; any residual pressure was then due to hydrogen.

At final equilibrium, the gas consisted entirely of ethylene. At the plateau pressure P_2, the spectrum was that of ethylene alone, and no acetylene remained. However, some of the gas was not condensable at $-196°C$, and these experiments indicated that the ratio of hydrogen to ethylene at P_2 was

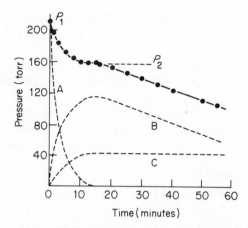

Figure 13.3 Pressure–time curves for the reaction of acetylene with liquid sodium at 200°C. Full line, total pressure; A, acetylene; B, hydrogen; C, ethylene

about 1.4:1. The first stage P_1 to P_2 therefore represents the conversion of acetylene to ethylene and hydrogen, and the subsequent reaction is the solution of hydrogen in liquid sodium. Ethane is not a product of the reaction at this temperature. Some additional experiments support these conclusions:

(a) Pure ethylene was used instead of acetylene, with experimental conditions otherwise unchanged. No reaction was observed at 110–200°C; the sodium surface remained clean after 20 h, and neither the pressure nor the infrared spectrum of the gas showed any change.

(b) Experiments with hydrogen alone showed that the rate of hydrogen absorption corresponded closely with that for the post-plateau region, for the same temperatures and pressures.

(c) The reaction vessel was evacuated on completion of a reaction, and a further quantity of acetylene added. The behaviour shown in Figure 13.3 was repeated.

Further clarification of the reaction is obtained from a consideration of the hydrogen balance throughout reaction. The first stage of the reaction between adsorbed acetylene molecules and electrons from the sodium conduction band will be

$$HC \equiv CH + e \longrightarrow HC \equiv C^- + H \qquad (13.4)$$

and some hydrogen could be retained in the solid product in the form of the hydrogen acetylide $NaC \equiv CH$. However, measurement of the relative quantities of acetylene, ethylene and hydrogen during the stage P_1 to P_2

preclude this possibility; reaction 13.4 is followed immediately by

$$HC\equiv C^- + e \longrightarrow C\equiv C^{2-} + H \tag{13.5}$$

so that both atoms of hydrogen are immediately available at the sodium surface, and the compound Na_2C_2 is the only solid phase retained in the liquid metal.

In the range P_1 to P_2 (Figure 13.3) hydrogen and ethylene are both being evolved from the metal surface, so that the overall reaction when the plateau P_2 is reached lies between the limits represented by the equations

$$2C_2H_2 + 2Na \longrightarrow Na_2C_2 + C_2H_4 \tag{13.6}$$

$$C_2H_2 + 2Na \longrightarrow Na_2C_2 + H_2 \tag{13.7}$$

Analysis of the gases at P_2 shows that less than half the hydrogen produced in equation 13.7 is employed in hydrogenating the adsorbed acetylene (equation 13.6). This is in marked contrast to the behaviour of propyne, where all hydrogen was taken up in hydrogenation. The reason for this difference is to be found in the relative magnitude of the adsorption coefficients for the various gases. Propyne is adsorbed rather less strongly than acetylene on transition metals,[288] and it seems likely that this difference is greater at the sodium surface. Again, twice as much hydrogen is produced from acetylene as from propyne, and hydrogen atoms may be produced at the surface more quickly than they can be employed in hydrogenation. The slower production of hydrogen atoms in the case of propyne, coupled with the fact that propyne is less able than acetylene to compete with hydrogen for occupancy of the surface, can explain why hydrogen is not evolved in the propyne reaction.

In the acetylene reaction, the feature which has greatest importance is the strong adsorption of acetylene at the sodium surface. The adsorption coefficient for acetylene is much larger than that for hydrogen (by a factor of 100 on transition metals[289]) which encourages displacement of hydrogen from the surface. Adsorption of ethylene is relatively weak, so that ethylene is also readily displaced from the surface. We may therefore consider that so long as any acetylene remains in the gas phase, the surface will be covered by a a strongly bonded monolayer, and not until this can no longer be maintained can the sodium–hydrogen reaction commence. (It is of interest that the same general principle operates in the hydrogen chloride–sodium reaction discussed in Chapter 14.) The experimental total pressure–time curve (Figure 13.3) therefore represents the sum of the pressure–time variations for the three separate components of the gas phase, and this is also illustrated in Figure 13.3 by broken lines.

13.4 Reaction of acetylene with a solution of barium in sodium

Several references have been made above to the analogies which appear to exist between reactions of simple alkynes and alkenes at the surface of

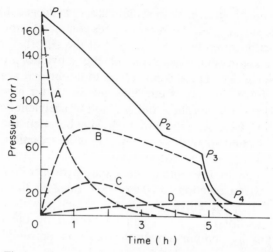

Figure 13.4 Components in the gas phase during the reaction of acetylene with a 4 atom % solution of barium in sodium, at 150°C. A, acetylene; B, hydrogen; C, ethylene; D, ethane

liquid sodium, and at the surface of solid transition metals. With the latter, much use is made of alloys, since d-orbital vacancies can be controlled in this way. With liquid sodium, bonding at the surface presumably involves s- and p- rather than d-orbitals, but since sodium is a good solvent for many other metals, variation in the energy levels of the electrons is readily achieved by the use of solutions of other metals in liquid sodium; in earlier chapters we have discussed the use of barium for this purpose. In reactions with hydrocarbons, the effect of added barium is quite profound.[283] Hydrogenation can now extend as far as ethane even at these low temperatures, but the rate of reaction with acetylene is much less. (This again resembles observations made with hydrogen chloride gas, and for much the same reasons.)

The pressure–time curves for the reaction of a 4 atom % solution of barium in sodium with acetylene at 150°C are illustrated diagrammatically in Figure 13.4. The full line represents total pressure, and reflects the continually changing composition of the gas phase during reaction; the ways in which the pressures of the separate components change are shown as broken lines. There are obviously three stages in the reaction (P_1 to P_2, P_2 to P_3 and P_3 to P_4). During the first stage, acetylene disappears from the gas phase, and at P_2 the gas consists essentially of hydrogen, together with about 10 per cent ethylene. The rate at which the pressure decreases is much smaller (by a factor of about 50) than the corresponding rate with sodium alone. Since ethylene reacts with a sodium–barium solution at this temperature (though not with sodium alone), we may assume that ethylene can adsorb at the sodium–barium surface, and the slower reaction of acetylene with the metal solution can be attributed to the retention of ethylene at the metal surface. With sodium alone, hydrogen and ethylene are produced at the surface, but the latter escapes completely into the gas phase, so that the reaction rate observed was that of the adsorbed acetylene only. With barium–sodium solutions, some of the ethylene produced is now retained on the surface. There is no reason to doubt that acetylene is again more strongly adsorbed than is ethylene, but the latter is produced at the surface, hindering the adsorption and reaction of acetylene.

Calculation of the distribution of hydrogen between the various gases shows that the second stage $P_2 \rightarrow P_3$ represents the reaction of the residual ethylene, and also of some hydrogen. Adsorbed ethylene is not as efficient as adsorbed acetylene in preventing the metal–hydrogen reaction, but can reduce the hydrogen reaction to a very slow rate. At the end of the first two stages (at P_3), all ethylene has now reacted with the metal. The hydrogen–carbon bond $H-CH\equiv$ can be broken by the barium–sodium solution but not by sodium alone; the reaction presumably follows the sequence

$$C_2H_4 \xrightarrow{+e} HC=CH_2^- \xrightarrow{+e} HC=CH^{2-} \xrightarrow{-2e}$$

$$HC\equiv CH \xrightarrow{+2e} HC\equiv C^- \xrightarrow{+e} C\equiv C^{2-} \quad (13.8)$$

each stage involving transfer of electrons to or from the metal surface. A small amount of the hydrogen produced is used in further hydrogenation of ethylene to ethane (Figure 13.4). The solid product will probably be BaC_2 rather than Na_2C_2, in view of the small but significant difference in the ΔH_f values for the two compounds (-81 and -17 kJ mol^{-1} respectively). The possibility that barium hydrogen acetylide might be produced can be dismissed because of its very low thermal stability.

At P_3, all hydrocarbon has disappeared from the gas phase, and pressure changes thereafter (P_3 to P_4) reflect the unhindered metal–hydrogen reaction. The small amount of residual gas was found to be ethane, and barium in solution can therefore bring about a hydrogenation step which is not perceptible with liquid sodium alone at this temperature. The efficiency of this hydrogenation to ethane might well vary widely with the metal dissolved in liquid sodium, and this would be a worthy topic for further study.

13.5 Ethylene reactions

Ethylene is formed as an initial gaseous product when acetylene reacts with any of the alkali metals or their mixtures, and we need not therefore continue the discussion of acetylene reactions. It is the hydrogenation of ethylene to ethane which is now of major interest, and there are some general points which help to rationalize all the ethylene reactions.

The rates of reaction of ethylene are very much slower than those of acetylene, which is a reflection of its weaker adsorption at the metal surface.

In the acetylene–sodium reaction discussed above, ethylene is displaced by acetylene, and when all acetylene has disappeared the reaction of ethylene is too slow to be observable at 100–200°C. Nevertheless, in an environment where no stronger adsorber (such as acetylene) is present, ethylene does have some adsorptive ability, and we can again discuss its reactions in terms of the behaviour of adsorbed ethylene molecules. The rates of reaction are increased by raising the temperature; furthermore, since hydrogenation involves transfer of electrons from the metal surface, ethylene reactions occur more readily with the heavier alkali metals. Hydrogenation of ethylene has been found to occur at rates which are experimentally convenient by using liquid potassium at 230–400°C.[285] These results will be discussed first, and then compared with corresponding but more limited results for liquid sodium.

Liquid potassium

The reaction was studied using the apparatus (Figure 2.6) in which the liquid metal is injected as a jet into the gas by means of an electromagnetic pump; the reaction was followed by pressure changes and analysis of the gas by gas–liquid chromatography. At the lower end of the 230–400°C temperature range, self-hydrogenation was found to occur precisely according to

$$3C_2H_4 \longrightarrow 2C_2H_6 + 2C \quad \text{(or } K_2C_2\text{)} \tag{13.9}$$

A typical pressure–time curve (at 295°C) is shown in Figure 13.5. The pressure decreases smoothly to a constant finite value, and the reaction

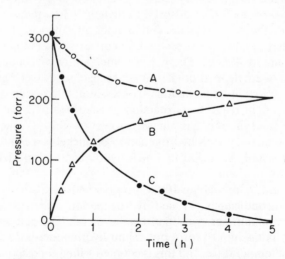

Figure 13.5 Partial pressures of ethylene and ethane in the reaction of ethylene with liquid potassium at 295°C. A, total pressure; B, ethane gain; C, ethylene loss

required about 5 h at 295°C for completion; this should be compared with 1 h at 110°C for propyne (Figure 13.2) and 15 minutes for acetylene at 200°C (Figure 13.3). At all intermediate stages in the reaction, the gaseous mixture was condensable at liquid nitrogen temperature (−196°C), which excluded the presence of hydrogen, and gas analysis showed the mixture to consist only of ethane and unchanged ethylene. The sole gaseous product on completion of the reaction was ethane, and the ratio of initial to final pressures was 3:2, in agreement with equation 13.9. This reaction therefore shows a distinct similarity to the reaction of propyne with sodium, in that under appropriate conditions maximum possible efficiency in self-hydrogen-ation at the metal surface is achieved. The solid product varied with the temperature of the reaction. At 350°C the product was carbon, but at lower temperatures some potassium acetylide, K_2C_2, was found in the product; this was indicated by the production of acetylene on hydrolysis by water vapour. This compound does not appear to have been characterized since it was first reported by Moissan in 1903; but by analogy with the sodium compound Na_2C_2 it would be expected to have low thermal stability.

The reaction is again suitably interpreted in terms of a surface reaction. The ethylene molecule would appear to be more strongly adsorbed on liquid potassium than on liquid sodium, and the more electropositive nature of potassium means that electrons are more readily available for transfer from the metal to the adsorbed ethylene molecules. The latter will dissociate according to the process expressed in equation 13.8, and in the course of forming C_2^{2-} ions, each dissociated ethylene molecule provides four adsorbed hydrogen atoms. We may consider these to be retained on the liquid potassium surface (rather than escape as H_2 molecules), where they are all used in the hydrogenation of adsorbed ethylene. The ethane produced has little affinity for the metal surface, and escapes into the gas phase; there was no evidence that ethane was subsequently converted to methane, at least at temperatures up to 400°C. There are some points of similarity between hydrogenation of ethylene at potassium and nickel surfaces,[285] and in view of the time taken for complete reaction on potassium, it has been suggested that hydrogenation might result from reaction of adsorbed hydrogen atoms with ethylene molecules in the gas phase. However, this is a less attractive alternative; the analogy with hydrogenation on nickel is a limited one, since conditions are readily arranged in which hydrogenation on nickel results in the formation of methane.

At temperatures approaching 400°C progressively less ethane is produced. This decreasing efficiency in self-hydrogenation can be attributed to increasing competition for hydrogen by the metal. The rate of reaction of hydrogen with potassium to form potassium hydride increases exponentially with increasing temperature, and this is coupled with a corresponding increase in the solubility of hydride in the metal. With increasing temperature, therefore, progressively more adsorbed hydrogen forms hydride and dissolves, rather than taking part in hydrogenation; ethylene then encounters a hydrogen-deficient surface, and carbon (or K_2C_2) is formed rather than ethane.

The above experiments suggest that provision of adsorbed hydrogen in excess of that available from ethylene should enhance hydrogenation. In further experiments[285] carried out between 265°C and 330°C, hydrogen in equimolecular proportions with ethylene was treated with potassium as before. The mixture is stable, of course, in the absence of potassium. The extent of hydrogenation was now almost 100 per cent, i.e. much greater than that indicated by equation 13.9, and is attributed to the copious formation of adsorbed hydrogen atoms at the metal surface.

Liquid sodium

At temperatures of 250–350°C, the reaction of sodium with ethylene resembles closely that of potassium, except that the reaction is much slower. Ethylene is consumed, and ethane is the sole gaseous product.[290] The reaction can be represented by a graph which is very similar to that shown for potassium in Figure 13.5, except that at 295°C the time scale now extends beyond 60 h, compared with 5 h for potassium. For purposes of comparison, the content of the gas phase was determined after 24 h for both sodium and potassium over a series of temperatures, and results are shown in Figure 13.6; it is clear that in each case reaction occurs much more rapidly with increasing

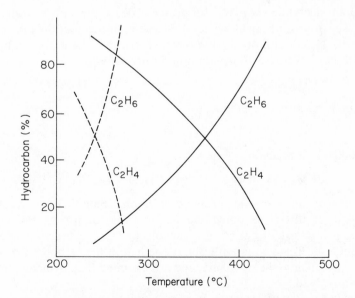

Figure 13.6 Reaction of ethylene with liquid sodium and liquid potassium; composition of gas phase after 24 h at various temperatures. Full lines, sodium; broken lines, potassium

temperature, but that sodium reacts more slowly with ethylene than does potassium under identical conditions.

At temperatures above 350°C, small amounts of other hydrocarbons appear in the gas phase. At 370°C, traces of methane, propene, propane, but-1-ene, n-butane and cis-2-butene were detected, and at 430°C the total concentration of these hydrocarbons reached 3 per cent of the gas phase after 24 h exposure. The appearance of hydrocarbons containing odd numbers of carbon atoms at these higher temperatures indicates that the reaction must also involve rupture of the C–C bond in ethylene to give methyl and ethyl radicals, which will undergo further reaction with adsorbed ethylene, and cracking and polymerization of ethylene at an alkali metal surface at temperatures above 500°C may become significant. These aspects have been extended by a study of the reactions of propene and isobutene (2-methyl-propene).[290] With propene, at pressures around 0.5 atm and temperatures in the 350–450°C range, the total pressure over liquid sodium fell smoothly with time to a constant final and finite pressure; at 368°C the reaction was complete in about 80 h. The gas phase then consisted of propane and methane in almost equimolar amounts, with about 2 per cent of ethane. The results can be explained on the basis of a surface reaction which is analogous to that of ethylene, but also involves the rupture of C–C bonds at these higher temperatures. We can consider that both C–H and C–CH_3 bonds undergo dissociation following adsorption, i.e.

$$CH_3.CH{=}CH_2 \longrightarrow CH^{\cdot}_3 + 3H^{\cdot} + Na_2C_2 \qquad (13.10)$$

The sodium acetylide may well suffer some decomposition to sodium and carbon. The almost negligible yield of ethane indicates that the dimerization $2CH^{\cdot}_3 \rightarrow C_2H_6$ is not a favourable reaction. Instead, the adsorbed hydrogen atoms are used in the hydrogenation of adsorbed propene:

$$CH_3.CH{=}CH_2 + 2H^{\cdot} \longrightarrow CH_3.CH_2.CH_3 \qquad (13.11)$$

and in the formation of methane:

$$CH^{\cdot}_3 + H^{\cdot} \longrightarrow CH_4 \qquad (13.12)$$

giving the overall reaction

$$2CH_3.CH{=}CH_2 \longrightarrow CH_4 + CH_3.CH_2.CH_3 \qquad (13.13)$$

which yields methane and propane in equal quantities, according to experiment. If this represented the only process occurring, the pressure should remain constant during the reaction. In fact, experiments at 368°C, 386°C and 400°C show that only about one half of the carbon introduced as propene remained in the gas phase, and it is assumed that involatile polymers are also produced.

Because of polymer formation it is not possible to compare reaction rates in quantitative terms, but some qualitative comparisons can be made. Thus, the rate of reaction of propene with potassium is faster than with sodium.

With potassium, the reaction with propene is complete at 295°C in 14 h, whereas at the higher temperature of 400°C reaction with sodium requires 60 h for completion. This is analagous to the behaviour of ethylene. Again, the rate of reaction decreases as ethylene is progressively substituted by methyl groups. This is evident from the slower rate of reaction of propene compared with ethylene at the sodium surface. With potassium also, the effect is obvious; with liquid potassium at 295°C, the reactions of ethylene and propene require 5 h and 14 h respectively for completion. This general effect has been illustrated further by the reaction of isobutene with liquid sodium. Following the same mechanism as outlined above for propene, the dissociation following adsorption occurs according to:

$$(CH_3)_2C{=}CH_2 \longrightarrow 2CH\,{}^{\bullet}_3 + 2H^{\bullet} + Na_2C_2 \qquad (13.14)$$

followed by

$$(CH_3)_2C{=}CH_2 + 2H^{\bullet} \longrightarrow (CH_3)_2C.CH_3 \qquad (13.15)$$

and

$$CH\,{}^{\bullet}_3 + H^{\bullet} \longrightarrow CH_4 \qquad (13.15)$$

However, insufficient hydrogen is now produced by dissociation of one isobutene molecule to complete reaction 13.15 and also to form methane; as a result, isobutane is the main gaseous product, and a larger proportion of ethane is formed by dimerization of $CH\,{}^{\bullet}_3$ radicals than occurred with propene. The reactions 13.14 and 13.15 represent only part of the total process, and polymerization to polymers of low volatility also takes place. The sodium–isobutene reaction is very slow. Below 380°C no reaction could be detected after 50 h exposure, and at 426°C only 20 per cent of the isobutene had reacted after 25 h.

The rate of reaction of these alkenes at a sodium or a potassium surface therefore decreases in the order

ethylene > propene > isobutene,

i.e. with the increasing degree of substitution of the ethylene molecule. This could be a steric effect, in that fewer hydrocarbon molecules can be accommodated on the surface as substitution is increased, but must also be due in part to a changing ability to adsorb at the surface, and perhaps also to an increasing reluctance to undergo dissociation at the metal surface.

Sodium–potassium alloy

The reactions of alloys with ethylene, covering a range of alloy compositions, have been studied at 295°C and under conditions identical with those employed for the two pure metals. Both sodium and potassium react at 295°C to produce ethane as the sole gaseous product in a stoichiometric self-hydrogenation process, and all alloys studied conformed to this

208

Table 13.1   First order rate constants for the reaction of sodium, potassium and their alloys with ethylene at 295°C

Atom % K	K (h^{-1})
Pure sodium	0.0033
11.2	0.025
19.1	0.056
34.0	0.089
75.2	0.31
Pure potassium	0.35
[6.3 atom % Cs in Na	2.24]

behaviour. Reactions at all compositions were first order with respect to ethylene, and rate constants derived from graphs of $\ln p$ against time are given in Table 13.1. It is evident that the addition of potassium to sodium has a considerable effect on the rate constant, whereas the addition of sodium to potassium has a relatively small effect. This is entirely consistent with the widely different reaction rates observed for the two pure metals.

Sodium–caesium alloy

Caesium has the lowest work function value of any alkali metal, so that electrons will be most readily available for transfer to an adsorbed hydrocarbon. Addition of caesium to liquid sodium should therefore have an even greater influence on reaction rate than has potassium. A further important aspect is that the hydrogenation of ethylene at the surface of this alloy can be studied at the lower temperatures at which reactions on sodium alone are too slow to be measured. Reaction at the surface of pure caesium has not yet been studied, but addition of only 6.3 atom % caesium to sodium does in fact make a profound difference.[290] In the temperature range 215–350°C ethane is the sole gaseous reaction product, and reaction is complete within 100 h. The reaction is identical to that described for sodium and potassium, except for the reaction rate. The process is again first order with respect to ethylene pressure, and a rate constant for 6.3 atom % solution of caesium in sodium at 295°C is included in Table 13.1. This constant is greater, by a factor of 7×10^2, than for sodium alone, and it is a logical conclusion that the rate would increase still further at higher caesium concentrations.

In the temperature range 100–160°C, appreciable quantities of n-butane appear in the gas phase (together with a trace of buta-1-3-diene) and the proportion of n-butane increases as the temperature is lowered, eventually reaching 9 per cent of the total residual hydrocarbon after exposure at 104°C. This is in direct contrast to the behaviour on pure sodium, where small quantities of n-butane were formed only above 350°C, the quantity increasing as the temperature increased. The sodium–caesium mixtures at low temperature provides a unique set of conditions in which ethylene can undergo dissociation at a measurable rate, yet the temperature is sufficiently low to offer a degree of stability to intermediate species which have only fleeting existence at higher temperatures. Thus, the ethyl radicals which represent a step in the predominant hydrogenation process, i.e.

$$C_2H_4 \xrightarrow{\ H\ } C_2H{_5}^{\cdot} \xrightarrow{\ H\ } C_2H_6$$

are either sufficiently long-lived, or sufficiently strongly adsorbed under these conditions, that dimerization to n-butane becomes a possible alternative to further hydrogenation. A more detailed study of these reactions at the lower temperatures available with alloys of higher caesium content, or with liquid caesium itself, coupled with examination of the solid products of reaction, would appear to be well worth while.

Liquid lithium

The relatively small size of the lithium atom gives rise to many notable differences between its chemistry and that of the other alkali metals. To the various examples of this which have arisen in previous chapters we can now add its reaction with ethylene. In preliminary experiments employing reaction at a jet of the metal, it was immediately obvious that a different reaction was involved. A film formed at the metal surface which was cohesive enough to block the circulation and free flow of the metal, and to obtain acceptable results it was necessary to modify the apparatus.[290] Two liquid metal loops of larger diameter were incorporated into the reaction vessel, each circulated by independent electromagnetic pumps, to give two jets which were diametrically opposed within the reaction vessel. These jets impinged directly on one another, so that a constant and larger fresh surface was available for reaction with ethylene. The pressure fell smoothly, and almost linearly with time, for most of the reaction, with a small increase in reaction rate towards the end of the reaction. The rate was almost independent of pressure, indicating a zero order of reaction which is consistent with strong adsorption of ethylene at the metal surface. The pressure eventually decreased to near-zero, and in the temperature range 228–306°C the reaction was complete within 70 and 10 minutes respectively. Gas analysis during the course of the reaction showed the presence of ethylene, with a small amount of hydrogen.

The solid product consisted of a mixture of lithium acetylide and lithium hydride; the presence of the hydrogen acetylide $LiHC_2$ is unlikely as this is the least stable of the hydrogen acetylides of the alkali metals.[303] The reaction can be represented as

$$6Li_{(l)} + C_2H_{4(ads)} \longrightarrow Li_2C_{2(s)} + 4LiH_{soln} \qquad (13.16)$$

Regarding the mechanism, we may consider the C_2H_4 molecules to be adsorbed as a first step, then to dissociate into C_2^{2-} and $4H^{\cdot}$ at the surface,

Table 13.2   Data related to alkali metal hydrides

	Li	Na	K	Rb	Cs
Dissociation pressures of metal hydrides MH at 295°C (kNm^{-2}) (Reference 302)	2.8×10^{-6}	1.23	1.09	2.40	1.87
Relative rates of reaction of metals with hydrogen at 250°C (referred to Na rate as unity)	76	1	6.7	—	59

as with sodium and potassium. With the latter, the adsorbed H atoms then hydrogenate other adsorbed C_2H_4 molecules. With lithium, however, the metal itself is able to enter into strong competition with ethylene for reaction with hydrogen, for reasons which are illustrated in Table 13.2. Dissociation pressures show that lithium hydride is much more stable than the hydrides of the other alkali metals, and (what is probably even more relevant in this case) the rate of reaction of lithium with hydrogen is very much faster. The hydrogen atoms produced at the surface by dissociation of ethylene are therefore taken up by reaction with the metal, rather than being used in hydrogenation. For most of the reaction, the solution of hydrogen in lithium is hindered by the presence of ethylene on the surface; when all the ethylene has reacted, solution of hydrogen can occur unhindered, which accounts for the small increase in rate towards the end of the reaction.

Lithium–sodium alloy

Reaction of ethylene with this alloy is not so predictable as was the case with the sodium–potassium alloy, since each component metal reacts differently with ethylene. The experiments were carried out at 295°C to permit direct comparison with results for pure lithium and sodium.[290] However, there is then a restriction on the concentration range available since there is a two-phase immiscibility region between 45 and 75 atom % lithium. Results have been obtained over the 0–45 atom % lithium concentration range, and it will become evident from Figure 13.7 that measurements over the 75–100 per cent lithium range are unnecessary. As the lithium concentration was increased, the pressure–time curves changed in shape from those characteristic of sodium to those characteristic of lithium. Ethane was the only major gaseous product in all cases, but the ratio C_2H_6 produced:C_2H_4 consumed changed from 0.66 (for sodium) to near-zero (for lithium), and the rate of reaction increased with increasing lithium concentration. The remarkable features (illustrated in Figure 13.7) are that no more than a few per cent of lithium added to sodium is sufficient to almost completely suppress the formation of ethane, and that as the concentration approaches 45 atom % lithium, ethane is no longer produced. If we accept that in these

211

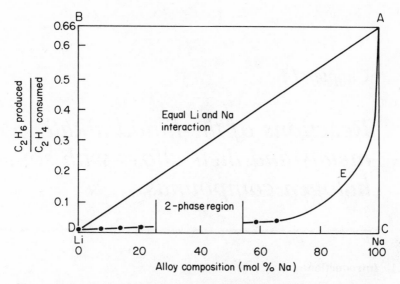

Figure 13.7 Reaction of ethylene with liquid lithium–sodium alloys at 295°C. If only sodium reacted, results would lie along AB. If only lithium reacted, results would lie along CD. Curve E shows experimental results

alloys the composition of the gas–metal surface differs from the bulk concentration, then these results can only mean that as the lithium concentration approaches the immiscibility limit, the surface condition approaches that for pure lithium. The reaction rate with ethylene is increased disproportionally by small additions of lithium to sodium, and we may recall from Chapter 4 that the addition of lithium to sodium increases the rate of reaction with hydrogen, the effect being appreciable for the addition of as little as 1.1 atom % lithium.

Chapter 14

Reactions of the liquid alkali metals and their alloys with some halogen compounds

14.1 Introduction

The major emphasis in this chapter is on the chemistry of reactions of sodium with various halides. As a preamble, the behaviour of hydrogen chloride at a clean sodium metal surface is discussed, and it will be seen that in some respects this behaviour resembles the behaviour of unsaturated hydrocarbons discussed in Chapter 13. The introduction of a halogen atom into an inert hydrocarbon usually creates a molecule which is no longer inert; for instance ethane is inert to liquid sodium, whereas ethyl chloride reacts readily, and with some other organic halides reaction can be explosive. These aspects are discussed in section 14.3 below, followed by a more detailed account of the reactions of ethyl chloride; the sodium–ethyl chloride reaction is important because of its extensive use in the industrial preparation of tetra-ethyl lead. The reactions with boron trichloride are included because of their intrinsic interest, the possibilities for boron production, and for comparison with carbon tetrachloride.

The alkali metals continue to be important reagents in both organic and inorganic chemistry, and a wide range of experimental conditions are employed. The metal may be in the form of gaseous, liquid or solid metal, as a despersion in an inert solvent, or as a solution in liquid ammonia. One consequence of this, however, is that it is difficult to make comparisons between reactions carried out under widely different conditions, or to decide whether any given reaction is truly characteristic of the reactant and the alkali metal alone. The results described in this chapter for hydrogen chloride, methyl chloride, ethyl chloride and boron trichloride have been obtained using the gaseous reactant in contact with a liquid metal surface which is maintained free from surface films either by stirring the metal (as described for Figure 13.1) or, more usually, by carrying out the reaction at the surface of a flowing jet of metal. The results are therefore directly comparable with those presented in the previous chapter.

212

14.2 Hydrogen chloride

In this case, satisfactory results were obtained using magnetic stirring of the metal (Figure 13.1). The pressure changes observed during the reaction with pure sodium[281] are shown as curve A in Figure 14.1. The pressure falls rapidly, and at a constant rate, until a pressure P' is reached which is precisely half the original pressure. This stage requires only 3–4 minutes at 160°C. Thereafter the pressure falls more slowly, eventually reaching zero. Analysis of the gas phase by mass spectrometry showed the HCl was consumed during the first stage, and when P' was reached the gas phase consisted entirely of hydrogen. The overall reaction

$$2Na + HCl \longrightarrow NaH + NaCl \qquad (14.1)$$

therefore proceeds in two distinct steps:

$$2Na + 2HCl \longrightarrow H_2 + 2NaCl \qquad (14.2)$$

followed by

$$2Na + H_2 \longrightarrow 2NaH \qquad (14.3)$$

The rate of reaction 14.2 was measured over a temperature range of 106–160°C; a straight-line relation between $\log(d\Delta P/dt)$ and $1/_T$ was found, giving an apparent activation energy of 53 kJ mol^{-1}.

The reaction can be interpreted satisfactorily in terms of the following mechanism:

(a) HCl, being a polar molecule, is strongly adsorbed at the metal surface.
(b) Electron transfer from sodium to the adsorbed HCl molecule occurs:

$$Na + HCl_{(ads)} \longrightarrow Na^+Cl^-_{(s)} + H^{\bullet}_{(ads)}$$

The NaCl solid will collect as a separate phase (see below), and H$^{\bullet}$ atoms are adsorbed at the surface.

(c) As this continues, the surface is preferentially occupied by HCl molecules. Electron transfer occurs more readily to HCl than to H$^{\bullet}$, so that H$^{\bullet}$ leaves the surface as hydrogen gas, i.e.

$$2H^{\bullet}_{(ads)} \longrightarrow H_{2(g)}$$

(d) This continues so long as there is sufficient HCl in the gas phase to form a unimolecular layer on the sodium surface.

(e) When all HCl has been exhausted, the metal–hydrogen reaction takes place, i.e.

$$H_{2(g)} \longrightarrow H_{2(ads)} \longrightarrow 2H^{\bullet}_{(ads)} \xrightarrow{2e} 2H^-_{surface} \longrightarrow 2H^-_{soln}$$

to complete the overall reaction 14.1.

In these experiments, more sodium chloride is produced than can be dissolved in the sodium at these temperatures, so that a separate solid phase of sodium chloride is always present. However, its presence influences

214

reaction rates at the beginning of the experiment only. Immediately on addition of the first aliquot of HCl, a golden film formed on the metal surface. This was cohesive, so that little reaction occurred over the first 10–15 minutes. The film darkened with time, which can be attributed to penetration by sodium. This contributed to a change in the nature of the film from cohesive (amorphous) to crystalline. This change, in turn, exposed fresh metal surface, and the reaction rate increased. This effect was observed, but to a lesser extent, following addition of the second HCl aliquot. With all subsequent aliquots, the pressure decreased linearly with time as shown in Figure 14.1. Crystalline sodium chloride collected behind the baffle (Figure 13.1), without any influence on reaction rates.

The addition of barium to the sodium (4 atom %) does not alter the reaction with hydrogen chloride in principle, but has a profound effect on reaction rates (Figure 14.1). The pressure–time curve again falls into two clear stages, and the break in the curve occurs at half the original pressure. The mass spectrum of the gas confirms that all HCl is again consumed when the breakpoint P' is reached. Although the bulk concentration of barium is only 4 atom %, the concentration of barium in the surface may well be greater than this; separation and analysis of the solid product at stage P shows this to be largely $BaCl_2$, so that reaction of HCl is now taking place with barium rather than with sodium. The pronounced difference from the pure sodium reaction lies in the fact that the HCl reaction is much slower (the P^0 to P'

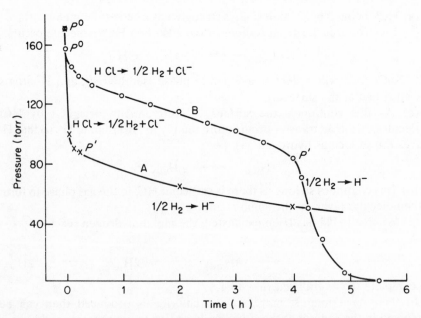

Figure 14.1 Hydrogen chloride reaction with sodium and sodium–barium alloy, at 160°C

stage requiring 4 h, as against 3–4 minutes with sodium alone) and the subsequent reaction with hydrogen much more rapid. It has been suggested[281] that the relative slowness of the HCl reaction might be due to the persistence, and the greater cohesive properties, of the $BaCl_2$ film, which restricts reaction. However, if this should be the case, a degree of irregularity might be expected, whereas the pressure-time curves are regular, smooth and reproducible, and it is more attractive to regard this behaviour as indicating a shift in the relative abilities of HCl and H⁺ to adsorb at the metal surface. Separate experiments have shown that the addition of barium to sodium can enhance the rate of reaction with hydrogen by two orders of magnitude, and this is shown also by comparing the rates of reaction in the second stage of the curves in Figure 14.1. Hydrogen is therefore more strongly adsorbed at a sodium surface which contains barium, and will compete more effectively with HCl molecules for occupancy of the metal surface. Access of HCl to the surface is therefore restricted, and its rate of reaction is reduced accordingly.

14.3 Polyhalogenated carbon compounds

The relative reactivity of an alkane and an alkyl halide towards sodium was established as early as 1855 in one of the simplest and best known reactions in organic chemistry, the Wurtz reaction:

$$2RX + 2Na \longrightarrow R.R + 2NaX$$

in which an alkyl halide RX (usually the iodide) reacts with sodium to produce an alkane R.R which is no longer reactive towards the metal. The method was extended soon afterwards (in 1864) by Fittig to the synthesis of hydrocarbons containing an alkyl and an aryl group (Ar):

$$ArX + RX + 2Na \longrightarrow Ar.R + 2NaX$$

Since those early days, organic chemistry has provided a multitude of examples of reactions between the alkali metals and halogenated carbon compounds. Taking the simple equation

$$CX_4 + 4Na \longrightarrow C + 4NaX$$

where X is a halogen or hydrogen, it is clear that an important driving force encouraging the forward reaction will be the free energy of formation of NaX.

For purposes of comparison, values of the enthalpies of formation are adequate, and these are listed in Table 14.1. There is a very large difference between the values for the halides and the hydrides, which in itself must go a long way towards explaining the stability of CH_4 as against the reactivity of its halogen substituted compounds. If this was the only factor, we might conclude that reactivity decreased in the order fluoro $\rightarrow$ chloro $\rightarrow$ bromo $\rightarrow$ iodo compound, other things being equal. It will be noted also that the $-\Delta H$ values for fluoride decrease in the order Li to Cs, but increase from Li to Cs in the case of the chloride, bromide and iodide.

Table 14.1   Enthalpies of formation $(-\Delta H_{298}$ kJ mol$^{-1})$
for alkali metal halides and hydrides

	Li	Na	K	Rb	Cs
F$^-$	612	571	563	557	531
Cl$^-$	409	413	436	440	445
Br$^-$	350	363	394	401	409
I$^-$	272	334	330	339	351
H$^-$	90.6	56.4	57.8	54.3	56.4

There are other factors, however, which render Table 14.1 no more than a general guide. For example, it is difficult to ensure that all other conditions are identical when comparing the halides in view of their often widely different physical properties. Again, the values in Table 14.1 provide a thermodynamic justification for reactivity, but give no clue to the different kinetics which might be involved. Another generalization which might be drawn from Table 14.1 is that the reactivity of halogenated carbon compound with an alkali metal will increase with the number of halogen atoms involved. In this connection it is interesting to pull together some experimental observations on the reactions of sodium with compounds in the series

$$CH_4, \ CH_3Cl, \ CH_2Cl_2, \ CHCl_3, \ CCl_4$$

where there is a change from inert to explosive behaviour. The behaviour of methyl chloride vapour has been followed[315] using the same technique as was used in the hydrocarbon reactions discussed in the previous chapter, i.e. by following the pressure changes which are caused by reaction at a jet of liquid sodium, with simultaneous analysis of the vapour phase using the mass spectrometer. The vapour reacted immediately, and reaction was complete in times varying from a few minutes to a few hours, depending on the temperature. The major gaseous products were methane and ethane, with a little propane, and the pronounced influence of temperature on their relative proportions is shown in Table 14.2. The initial reaction

$$CH_3Cl + Na \longrightarrow CH_3^{\cdot} + NaCl$$

produces methyl radicals, and at low temperatures these combine:

$$CH_3^{\cdot} + CH_3^{\cdot} \longrightarrow C_2H_6$$

Table 14.2   Major gaseous products of the methyl chloride–sodium reaction

Gaseous product (mol %)	Temperature (°C)				
	100	150	200	250	300
CH$_4$	12	29	44	63	80
C$_2$H$_6$	78	63	47	32	18
C$_3$H$_8$	6	5	4	2	1

to produce ethane as the main gaseous product. This radical combination process requires a low activation energy, but as the temperature is increased other reactions of the methyl radicals requiring higher activation energies become possible. The hydrogen abstraction reaction:

$$CH_3^{\cdot} + CH_3^{\cdot} \longrightarrow CH_4 + CH_2^{\cdot}$$

has been proposed as one step in the reaction of methyl iodide with sodium vapour;[136] this, and subsequent reactions of the $CH_2^{\cdot}$ radical, such as

$$CH_2^{\cdot} + CH_2^{\cdot} \longrightarrow CH_4 + C \tag{14.4}$$

can account for the considerable increase in methane produced at the higher temperatures, at the expense of ethane.

Methylene chloride and chloroform also react readily with sodium.[317] These experiments were performed by bubbling the organic halide vapour through an excess of liquid sodium at temperatures in the range 260–370°C for a predetermined period of 1–2 h, and then analysing both vapour and liquid metal phases. The reactions were therefore carried out at a reasonably fresh metal surface, but since they were not allowed to go to completion the results cannot be compared too closely with those for methyl chloride discussed above. Both reactions occurred readily; with methylene chloride the initial reaction can be written

$$CH_2Cl_2 + 2Na \longrightarrow 2NaCl + CH_2^{\cdot}$$

The gaseous products were mainly hydrogen and methane in roughly equal amounts, representing over 90 per cent of the total gaseous products, and the carbon residue contained some sodium acetylide. These arise from subsequent reactions of the $CH_2^{\cdot}$ radicals produced at the surface; reaction 14.4 above is no doubt one such reaction, together with

$$CH_2^{\cdot} + CH_2^{\cdot} \xrightarrow{2e} C\equiv C^{2-} + 2H_2 \tag{14.5}$$

It is also interesting to note that ethylene, which might have been expected to form by combination of CH_2 radicals, was not observed in the gas phase. This corresponds to the disappearance of ethane from methyl chloride reaction products at the higher temperatures. With chloroform vapour, hydrogen and methane are again the main gaseous products of reaction with sodium, and similar processes initiated by the CH radical are no doubt involved.

Reactions carried out in the controlled manner described above indicate the reactivity of the polyhalogenated compounds, but do not reveal the vigour with which the more highly halogenated compounds can react with sodium if the mixture is subject to shock. As early as 1925, Staudinger[319] first demonstrated the explosion which can occur when chloroform and sodium in contact with one another are subjected to shock, and reported sodium chloride and carbon as the products. Extending these experiments, Davis and McLean[318] prepared glass phials of sodium, potassium or the liquid alloy containing about 0.3 g of the metal, and sealed these within an outer phial containing 1–2 ml of chloroform. When one of these phials was dropped on to a concrete pavement from a height of 2 m, a loud explosion occurred. The same explosion was made to occur by breaking a phial within a steel bomb, when some of the reaction products could be identified. These consisted of carbon, hydrogen and hydrogen chloride, together with some hexachlorobenzene; the latter may have been produced by pyrolysis of chloroform brought about by the high temperature of the explosion, and not be reaction of sodium itself.

Sodium or potassium with carbon tetrachloride also provide potentially explosive mixtures, and it is reported[320] that the sensitivity to shock of mixtures of carbon tetrachloride with sodium–potassium alloy is 150–200 times greater than that of mercury fulminate. Fortunately, the use of carbon tetrachloride in fire extinguishers has now been largely abandoned. But in laboratories using the alkali metals, small fires are always possible, and it is wise to keep constantly in mind that carbon tetrachloride has no place in the liquid alkali metals laboratory!

The reactivity of the polyhalogenated plastics with alkali metals is also important when apparatus for use with these metals is being designed. In the handling of most corrosive liquids, polyhalogenated plastics offer considerable advantages. Products such as Kel-F (a polymer of chlorotrifluoroethylene) and Teflon (a polymer of tetrafluoroethylene) are finding extensive use for the construction of reaction vessels, and particularly as gaskets in taps and valves, and are generally much more resistant than the polypropylene-type rubbers previously used. However, the reverse is the case with liquid alkali metals. The polypropylene rubbers are reasonably stable, even in contact with caesium, which is consistent with the general principles outlined in Chapter 13. As the halogen content of the polymer increases, so does its reactivity towards the alkali metals, so that polymers such as Teflon are quite unsatisfactory in taps and gaskets. The reactivity is greater with the heavier metals; with caesium, reaction occurs immediately, and the plastic turns black owing to the production of carbon. Solutions of sodium in liquid ammonia are also used to etch Teflon surfaces prior to bonding with other materials. The etching occurs rapidly even at room temperature.

Similar observations have been made with respect to the series of fully saturated fluorocarbon liquids, such as the FLUTEC series[340,341] which are compatible with most of the common metals and have extensive electrical, heat transfer, medical and other applications. The producers warn that the alkali metals (and perhaps calcium strontium and barium) may provide exceptions to the general compatibility of these liquids with metals, and this has been tested[342] in the case of two liquids having molecular formulae

$$F_3C.CF_2.CF_2.CF_2.CF_2.CF_3 \qquad (\text{b.p. } 57°C)$$

and

(b.p. 140°C)

Vapours of these liquids were exposed to the surface of a jet of sodium–potassium alloy. In each case, even at 25°C and pressures as low as a few torr, reaction occurred immediately. A skin of reaction product formed which inhibited reaction; if this was broken by stopping the jet intermittently, the pressure fell to zero so that no gaseous product was formed. With the high-boiling fluorocarbon, the solid product appeared to be a mixture of sodium fluoride and carbon. The low-boiling fluorocarbon gave a golden solid product, so that some polymer may have been formed too.

14.4 Ethyl chloride

The reactions between alkyl halides and the alkali metals have been studied extensively. One reason, mentioned above, is the continued use of the alkali metals in synthetic organic chemistry. Additionally, however, these reactions have provided an excellent vehicle for the study of reaction mechanisms, and much of the research was carried out at a time when the relative significance of free radicals and organometallic compounds as intermediates was being argued. As a result, the reactions have been carried out using the metals as vapours, liquids and solids, and in 1972 Garrett[315] listed 27 pieces of research which had as their main objective the elucidation of reaction mechanisms.

Ethyl chloride is referred to here as being of special importance because of its reaction with the intermetallic compound NaPb. The equation

$$4NaPb + 4C_2H_5Cl \longrightarrow (C_2H_5)_4Pb + 3Pb + 4NaCl$$

was established by Groggins[321] and is based on the pioneer work of Löwig as early as 1852.[322] Tetraethyl lead holds a unique position among organometallic compounds because of its pronounced antiknock properties, i.e. an exceptional ability to prevent detonation of fuel in internal combustion engines. The present-day commercial process uses this equation; the reaction constitutes the principal industrial consumption of sodium metal in the UK and, since the 1920s, in the USA.[323] The mechanisms of the various stages of this reaction are still not entirely clear, and some research work at the University of Nottingham[315,324] has been carried out as a contribution towards the elucidation of these mechanisms. The reactions with liquid sodium alone have been carried out under the same experimental conditions (i.e. employing a jet of liquid metal) as was used in the gaseous hydrocarbon reactions, so that direct comparison is possible, and some reference has been made to corresponding reactions with sodium–potassium alloy. A study has also been carried out on the behaviour of ethyl chloride at the surface of a solution of lead in liquid sodium, to ascertain whether there is any analogy with the reaction at the surface of the solid alloy which is used in the commercial process.

With liquid sodium

The first and most obvious difference between this reaction and that of the hydrocarbons is the formation of a solid film at the surface of the jet. In a typical experiment at 200°C, the film forms immediately and is copper-coloured. The film is sufficiently cohesive to form a solid tube through which the liquid metal passes, and is reminiscent of the solid formed in the liquid lithium–ethylene reaction. After several minutes the film changes in colour to grey, and appears to crystallize, probably owing to penetration by sodium metal. It then falls away from the jet and collects as a grey powder in the reaction vessel. This behaviour is characteristic of the higher initial starting pressures of ethyl chloride; as the pressure falls, the quantity of solid product formed is not sufficient to form a tube around the jet. Analysis showed this product to consist largely of sodium chloride, with small amounts of carbon and sodium acetylide, the latter being recognized by the production of acetylene on hydrolysis. During this hydrolysis, a brown oil also separated which was immiscible with water; this consisted of a chloro-substituted polymer, probably based on dichloroethane.

During the course of these reactions, in which 50–80 ml of ethyl chloride vapour at initial pressures of about 200 torr were exposed to sodium, the rate at which ethyl chloride disappeared from the gas phase was monitored by mass spectrometry.[315] At first, this rate increased with time as the surface film formed, crystallized and separated from the jet. Thereafter, a state of dynamic equilibrium was achieved; the logarithm of ethyl chloride pressure changed linearly with time, so that the reaction was first order with respect to ethyl chloride. Some values for the first order rate constants using pure sodium are given in Table 14.3. The values change very little with temperature, which indicates that the activation energy for the rate-determining step is very low. These results yield an activation energy of the order of 1–2 kJ mol^{-1}; this suggests that the adsorption of ethyl chloride molecules at the surface of the metal may well be the rate-determining step in

Table 14.3 First order rate constants for the reaction of ethyl chloride with sodium and its alloys

Temperature (°C)	$K_p(min^{-1})$		
	Pure Na	Na/3.2 atom % Pb	Na/21 atom % K
50	—	—	0.047
100	—	—	0.258
126	—	—	0.295
155	0.145	—	0.332
203	—	—	0.339
212	0.151	—	—
252	0.154	—	0.444
260	—	0.076	—
281	—	0.082	—
296	0.184	—	—
320	—	0.095	—
350	—	0.111	—

Table 14.4   Composition of the gaseous phase after reaction of ethyl chloride with some liquid alkali metals and alloys*

Liquid metal	Temperature (°C)	Composition (mol %)		
		C_2H_6	C_2H_4	C_4H_{10}
Pure Na	150	48	47	5
	200	62	32	6
	250	70	24	6
Na/21 atom % K	100	58	32	10
	200	87	8	5
Na/80 atom % K	50	67	17	15
	100	84	11	5
	150	90	7	3
Pure K	100	89	5	6
	150	93	6	1
Na/3.2 atom % Pb	250	65	25	9
	300	89	2	9
	350	93	0	7

*Composition determined after 1 h; in all cases, reaction of ethyl chloride was complete within this time.

the reaction, since the rates of adsorption processes are usually found to be only marginally temperature dependent, with low activation energies.[325]

The main gaseous products of reaction are ethylene, ethane and butane. One hour was allowed for complete reaction of the ethyl chloride, and the composition of the gas phase at this point in time, as determined by mass spectrometry and gas–liquid chromatography, is shown in Table 14.4. In some cases, and especially at the higher temperatures, reaction of ethyl chloride is completed well within this time; the values in Table 14.4 are therefore not to be regarded as highly precise, but they do illustrate the principles involved. Following adsorption at the sodium surface, we may regard the first reaction as producing ethyl radicals at the surface:

$$C_2H_5Cl + Na \longrightarrow NaCl + C_2H_5^{\bullet}$$

Two main routes are then available to the ethyl radicals, i.e. radical combination:

$$C_2H_5^{\bullet} + C_2H_5^{\bullet} \longrightarrow C_4H_{10}$$

or hydrogen abstraction:

$$C_2H_5^{\bullet} + C_2H_5^{\bullet} \longrightarrow C_2H_6 + C_2H_4$$

Radical combination to give butane occurs to a much smaller extent than does the combination of methyl radicals (Table 14.2), and at the lower temperatures the hydrogen abstraction reaction which yields ethylene and ethane in equal amounts would seem to represent the major process. As the temperature is increased, the ethane content also increases at the expense of ethylene. This can be accounted for in part by the reaction

$$3C_2H_4 \xrightarrow{\ 2e\ } 2C_2H_6 + C_2^{2-}$$

which has been discussed earlier and is also responsible for the presence of sodium acetylide in the solid product. Other processes are no doubt also operating, such as the reaction of C_2H_5 radicals with adsorbed C_2H_5Cl molecules. If the experiment is continued for a prolonged period after all ethyl chloride has reacted, the ethylene content of the gas continues to decrease, eventually leaving only ethane and butane as the gaseous phase. Calculations of the mass balance of carbon and hydrogen show that not all the ethyl chloride which has reacted is accounted for in the gaseous products, the difference being due to the formation of the polymer which is present together with the sodium chloride, in the solid product.

With liquid potassium, and NaK alloys

With these liquid metals, the solid product was rather more cohesive than with sodium, but similar in composition. When dynamic equilibrium was established, the reaction rate was again first order with respect to ethyl chloride. The reaction appears to follow the same mechanism as outlined above for sodium, the main difference being the more rapid rate with potassium or the NaK alloys; Table 14.3 shows the increase in the rate constant brought about by the introduction of 21 atom % potassium into liquid sodium. This increased reaction rate also influences the composition of the gas phase. Whereas with pure sodium the gas phase after 1 h at 150°C consists essentially of an equimolar mixture of ethylene and ethane, reaction with pure potassium will take place in a much shorter time. The original gas composition will again be the equimolar ethylene–ethane mixture, but potassium also reacts rapidly with ethylene (see Chapter 13). As a result of this subsequent reaction, almost all the ethylene is converted to ethane at 150°C within the hour (Table 14.4), leaving a gas phase consisting almost entirely of ethane. With potassium, too, the radical recombination process to give butane is relatively unimportant. The composition of the gaseous phase after reaction with NaK alloys (Table 14.4) can be interpreted quite satisfactorily on the basis of the increase in reaction rate with potassium content of the alloy.

With a solution of lead in liquid sodium

These reactions were carried out in steel vessels, using a 3.2 atom % solution of lead. The liquidus temperature for this solution is 240°C, so that in order to retain lead in solution it was essential to employ temperatures above 240°C. Tetraethyl lead begins to decompose at 100°C, and since all solutions of lead in liquid sodium must necessarily be at temperatures above 100°C, tetraethyl lead is not formed in these reactions. The solid product consists essentially of sodium chloride, rather than lead chloride. The kinetics of the reaction resemble those of sodium alone, but in spite of the low lead concentration the reaction is slower than with sodium alone; Table 14.3 shows that the rate constant is about half the value for sodium alone. This is probably due to the accumulation of lead at the metal surface, giving a lead concentration there which is higher than that in the bulk of the liquid. The composition of the gas phase after reaction with ethyl chloride at 250°C is similar to that for sodium alone at this temperature (Table 14.4), so that apart from slowing down the reaction, the addition of 3.2 atom % of lead to sodium makes little difference to the chemistry involved. At temperatures above 250°C, the rate of the subsequent sodium–ethylene reaction increases, and at 350°C ethylene could no longer be detected in the gas phase.

In the reactions of ethyl chloride with the alloy NaPb there are two distinct temperature ranges, i.e. above and below 100°C, and the mechanism of the reaction is different within these two ranges because 100°C marks the temperature above which tetraethyl lead, and the intermediates involved in its formation, are unstable. A considerable number of investigations of the reaction have been carried out, designed largely to improve the yield of tetraethyl lead.[324] This work is not particularly relevant in the present context, since it is necessary to use solid rods of the compound NaPb; but it is of interest to compare the proposed mechanism with those discussed above for the liquid Na–EtCl reaction.

In the range 20–100°C, the reaction is dominated by the formation of tetraethyl lead. The initial reaction

$$NaPb + C_2H_5Cl \longrightarrow C_2H\dot{5} + NaCl + Pb$$

yields ethyl radicals, which can take part in the propagating step

$$C_2H_5^{\cdot} + NaPb \longrightarrow PbC_2H_5^{\cdot} + Na$$

The formation of $PbC_2H_5^{\cdot}$ allows the reaction

$$PbC_2H_5^{\cdot} + C_2H_5Cl + NaPb \longrightarrow 2C_2H_5^{\cdot} + NaCl + 2Pb$$

to occur, whereby two $C_2H_5^{\cdot}$ radicals are produced for every $PbC_2H_5^{\cdot}$ radical reacting. This reaction, together with

$$Na + C_2H_5Cl \longrightarrow NaCl + C_2H_5^{\cdot}$$

provides $C_2H_5^{\cdot}$ radicals in abundance, and the stepwise addition of these radicals to $PbC_2H_5^{\cdot}$ eventually yields the volatile product $Pb(C_2H_5)_4$.

At temperatures above 100°C, it is believed that the radical $PbC_2H_5^{\cdot}$ is unstable. The initial reaction is the same as for lower temperatures, but the

$C_2H_5^{\bullet}$ radical now reacts according to

$$C_2H_5^{\bullet} + C_2H_5Cl \longrightarrow C_2H_6 + C_2H_4Cl^{\bullet}$$

The $C_2H_4Cl^{\bullet}$ radical can take part in both the reactions

$$2C_2H_4Cl^{\bullet} \longrightarrow CH_2{=}CHCl + C_2H_5Cl$$

and

$$C_2H_4Cl^{\bullet} + NaPb \longrightarrow C_2H_4 + NaCl + Pb$$

so that the main volatile products are ethane, vinyl chloride and ethylene. As with the reaction of sodium alone, the formation of butane by combination of C_2H_5 radicals occurs to a relatively minor extent.

14.5 Boron and boron trichloride

The reaction of the liquid alkali metals with boron trichloride can be expressed by the simple equation

$$3M_{(l)} + BCl_{3(g)} \longrightarrow 3MCl_{(s)} + B$$

One of the problems encountered in attempts to develop this process commercially has been its low efficiency, owing to the difficulty in maintaining the reactants in constant contact; this difficulty arises because of the rapidity of the reaction and the highly cohesive nature of the solid film produced at the metal surface. The production of elemental boron by reduction of boron halides using liquid alkali metals was established many years ago.[326,327] In 1944, Spevack and Kurtz[328] made a serious attempt to use the process commercially by a study of the reaction of liquid sodium with boron trifluoride, but efforts to achieve a dynamic process generally failed because of stoppage of the metal flow by the solid reaction product. However, Spevack was able subsequently to patent a selection of improved designs which he claimed would allow some degree of continuous production.[329] Pichat and Forest[330] studied the liquid sodium–potassium alloy reaction with boron trichloride vapour at room temperature; they were able to isolate finely divided boron by removing excess metal with ethanol, and by vacuum distillation. They also noted that the reaction was slower at lower temperatures, and at lower boron trichloride partial pressures. This can be compared with experiments in the USSR in which a sample of boron trifluoride was simply added to a pressure vessel containing liquid sodium.[331] The reaction was conducted at temperatures of 600–850°C, when considerable sodium vapour would also be present. Experimental hazards were encountered because of the ignition of the finely divided sodium, which occurred when the vessel was opened, but the boron produced was finely divided and reasonably pure. The stated aim of the work was to devise a preparative route to isotopically enriched boron, for subsequent nuclear reactor use, and no attempt was made to extend the scope of the reaction for continuous boron production.

It is clear from the above that there is still much to be learnt about this apparently simple reaction. Adams[332] has attempted to find answers to two aspects raised by the following questions:

(a) In the liquid alkali metal–boron trichloride reaction, is the boron produced as insoluble, pure boron powder, or are there conditions under which the metal boride might be produced? As a corollary, is boron indeed quite insoluble in the liquid alkali metals, or could some solubility become detectable if boron is added in very finely divided form, e.g. as 'nascent' boron produced by the reaction, or as plasma boron?

(b) Can the complications inherent in the formation of films of reaction product be avoided by the use of liquid metal jets?

In the course of these studies some interesting and relevant observations have been made which are worth recording.

Alkali metal–boron interactions

Binary borides of some of the alkali metals are known compounds. The existence of borides of sodium and potassium has been conclusively demonstrated, and their stoichiometry established.[333,334] Both borides were first prepared by the direct synthetic method, in which the alkali metal and finely divided boron of high purity were placed in a molybdenum crucible inside a sealed iron vessel. Vigorous conditions were necessary to effect reaction, involving high temperatures and pressures of the alkali metal vapour:

$$\text{Na} + \text{B} \quad (\alpha \text{ or } \beta \text{ rhombohedral}) \quad \xrightarrow[\text{10 atm}]{1000°\text{C}} \quad \text{NaB}_6 \quad \xrightarrow{1100°\text{C}} \quad \text{NaB}_{15}$$

$$\text{K} + \text{B} \quad (\alpha \text{ form}) \quad \xrightarrow[\text{18 atm}]{1200°\text{C}} \quad \text{KB}_6 \quad \xrightarrow[\text{temps}]{\text{higher}} \quad \text{thermal decomposition}$$

Potassium requires more forcing conditions, and will only react with the less thermodynamically stable α-rhombohedral boron. On continued heating at higher temperatures under an atmosphere of argon, KB_6 reverts to boron and potassium vapour.[335] It is isostructural with the alkaline earth hexaborides such as CaB_6, and it is interesting to note therefore that a two-electron transfer from metal to the boron octahedron is not an absolute prerequisite for the formation of metal hexaborides. The structure of NaB_6 is complex, and does not resemble the KB_6 structure. However, the structure of NaB_{15} obtained by heating NaB_6 in argon is now confirmed. The boron atoms are arranged in B_{12} icosahedra, which are linked both directly and through B_3 chains.[336]

Attempts to isolate a lithium boride have produced a much wider range of possible compositions. The first identifiable binary phase was reported in 1962 by Secrist and Childs,[337] who obtained a product, tentatively assigned the formula LiB_{12}, from direct reaction of the two elements in a sealed iron chamber at $1000°C$ for 2 days, or $600°C$ for one week. A lithium boride is also obtained using boron carbide instead of boron:

$$B_4C + Li \text{ (or LiH)} \longrightarrow LiB_x + Li_2C_2 \text{ (or C)}$$

A French patent describes this as a successful method for the preparation of the tetraboride LiB_4,[338] whereas Secrist[339] also isolated LiB_{12} from the product of this reaction.

From the above brief survey it is evident that the alkali metals can form binary borides, but the published literature gives few clues as to the range of experimental conditions under which such borides can be produced. This information is relevant to all alkali metal–boron halide reactions, and a series of experiments has therefore been conducted to determine specifically whether borides are formed when boron is added to liquid lithium, sodium, potassium and some of their alloys. The boron used was either in crystalline form, as a fine powder (prepared using a Grindex agate ball-mill, Research and Industrial Instrument Co.), as plasma boron (prepared by the hydrogen reduction of boron trichloride in a radio-frequency induced argon plasma, and supplied by the Borax Research Centre, Chessington) or as nascent boron (by addition of boron trichloride to the liquid metal). The boron was added to an excess of the liquid metal in steel beakers encased in glass or steel outer containers, depending on the temperature. The excess metal was removed by distillation and the residue was examined by X-ray powder diffraction, differential thermal analysis, and chemical analysis.[332]

The results are collected in Table 14.5. Lithium reacts at $650°C$ (but not at $300°C$) to give a boride of empirical composition about $LiB_{10.5}$; X-ray powder diffraction shows this to consist of a mixture of two cubic phases, probably LiB_6 and LiB_{12}, and differential thermal analysis indicates that the lithium–boron reaction occurs in the temperature range $510–550°C$. The products of reaction with sodium–lithium alloy were the same as with lithium alone. Potassium gives a boride KB_6, but only at temperatures which are approaching its boiling point. Potassium also reacts when in solution in sodium, but the product is an indefinite one. It is of interest that there was no evidence of reaction between liquid sodium and boron, even when using plasma boron at $650°C$. The possibility existed that reaction might be stimulated by addition of barium or strontium to the sodium (as with nitrogen), but this appeared not to be the case.

Since solubility of a non-metal in the liquid metal does not normally arise unless the metal and the non-metal combine with one another, these experiments give a pointer to liquids in which some solubility of boron might occur. The use of resistivity measurements in the study of solubility has been described earlier (Chapter 8). With the results in Table 14.5 in mind, liquid lithium (at $550°C$), sodium–lithium alloy (4.5 atom % Li at $400°C$) and sodium–barium alloy (4.7 atom % Ba at $400°C$) were circulated through the capillary tube of the resistivity apparatus described in Chapter 3, for

Table 14.5 Reactivity of boron with the liquid alkali metals*

Metal	Crystalline boron	Powdered boron	Plasma boron	Nascent boron
Li	$LiB_{10.5}$[†]	—	$LiB_{10.5}$[†]	None[‡]
Na	None[§]	None[§]	None[†]	None[†]
Na with 15 and 27 atom % Li	None[§]	None[§]	$LiB_{10.5}$[†]	—
K	—	—	KB_6[†]	—
Na–K alloys	—	—	KB_{20}[†]	—
Na with Ba 4.5 atom %	None[†]	None[§]	—	—
Na with Sr 4.5 atom %	None[§]	None[§]	—	—

*'None' indicates that only boron remained after distillation of the metal.
[†]650°C for 24 h.
[‡]300°C for 24 h.
[§]300°C for 3 h.

prolonged periods. No significant change in resistivity occurred, which confirms the generally held opinion that boron is quite insoluble in the liquid alkali metals. We may conclude, therefore, that provided the temperature does not exceed 500°C the solid product resulting from the reaction of boron trichloride with any of the alkali metals will consist only of the metal chloride and elemental boron.

Boron trichloride

Earlier published work (discussed above) emphasized the problems in setting up this reaction as a continuous process. Since this is attributable to the formation of films between the reactants, Adams[332] studied the reaction under conditions in which the influence of surface films might be expected to be minimal, i.e. at the surface of a jet of the liquid metal. The apparatus described in Chapter 2 was used, in which metal circulation was maintained by an electromagnetic pump and the rate of reaction was followed by pressure changes. Using liquid sodium, with boron trichloride at an initial pressure around 200 torr and a temperature range of 150–250°C, no conditions could be found in which the reaction was reproducible. The surface film of sodium chloride and boron which is responsible for this forms instantaneously, and is remarkably cohesive. Under static conditions the film can prevent any reaction taking place for several days. When boron trichloride was added to the reaction vessel, containing the sodium jet, in a series of aliquots, complete reaction might take several hours or several seconds, without any detectable change in conditions. The reaction is obviously not a suitable one for kinetic studies, and the difficulties encountered by earlier workers in developing a continuous process are understood and appreciated.

When sodium–potassium alloy was substituted for sodium, the reaction times were more reproducible, and the pressure fell to zero in 15, 10 and 6 minutes at 70, 90 and 145°C respectively. Nevertheless, these are long reaction times for a reaction which is virtually instantaneous at a fresh metal surface, so that the reaction is still dominated by a film at the NaK alloy surface. The film is still present on sodium or NaK alloy even when the boron trichloride is diluted with argon (up to 80 per cent Ar). In contrast, the reaction with liquid lithium (at 200–250°C) was consistent and reproducible. The pressure of boron trichloride decreased smoothly to zero, without any breaks or irregularities, and the reaction rate was first order with respect to boron trichloride pressure. Reaction times of up to 1 h were necessary, so that the reaction is still hindered by the presence of a film, but the film formed on lithium is now readily penetrated by boron trichloride. The solid products of the alkali metal–boron trichloride reaction are clearly not simple mixtures of the metal halide and elemental boron, and a study of the structure of these solids would appear to be worthwhile.

When the boron trichloride is diluted with hydrogen, its rate of reaction with sodium at 150°C is slower by a factor of 10 than when diluted with argon. Analysis of the gas phase during the reaction by mass spectrometry shows that the hydrogen pressure remains constant until almost all of the trichloride has reacted. The hydrogen pressure then falls, and diborane appears in the gas phase. Diborane is probably produced by reaction between adsorbed hydrogen atoms and adsorbed trichloride molecules; the BCl_3 molecules are no doubt strongly adsorbed at the metal surface, and it is only at low boron trichloride pressures that hydrogen can compete adequately for occupancy of the surface.

Chapter 15

Hydrogen, oxygen and carbon meters

15.1 Introduction

It will have become apparent, from earlier chapters, that the main impurities in liquid sodium are hydrogen, oxygen and carbon. Because of the extensive use of liquid sodium as a coolant, methods for the measurement and control of these impurities have been developed in recent years to a high degree of precision. Methods for impurity monitoring in the other alkali metals have not been developed to the same extent. Nitrogen impurity also becomes important in liquid lithium; the degree to which methods found suitable for sodium might be applicable to impurity monitoring in liquid lithium has been considered, and this will be referred to at the end of this chapter.

When considering liquid sodium circuits, it is better to think in terms of control, rather than complete removal, of dissolved impurities. For example, the optimum oxygen and hydrogen concentrations have to be determined by consideration of such factors as corrosion and tribology. The corrosion rate of steels increases with oxygen concentration, but reducing the oxygen levels to minimize corrosion could prevent the formation of surface oxide films on the steels which has its effect on the chemical and frictional properties of the steel surface. In practice, hydrogen and oxygen levels are commonly controlled by the use of 'cold traps'. Part of the following sodium is diverted through a bypass which contains a cooled vessel, the temperature of which can be controlled. Sodium hydride and sodium oxide in solution which are in excess of their solubilities at the lower temperature of the cold trap will precipitate there, and eventually the hydride and oxide concentration in solution in the whole circuit becomes adjusted to the values corresponding to the saturation solubilities at the cold trap temperature.

The same principle has been applied to give a measure of impurity levels in flowing liquid sodium. These devices, termed 'plugging temperature indicators' or more usually 'plugging meters', have been in use for many years for impurity measurement,[343-345] and also to follow changes in impurity concentration caused by cold trap treatment. In these devices, part of the sodium flow is diverted through a bypass which carries an orifice, usually in the form of a needle-valve, and the temperature of the orifice can be controlled. As the temperature decreases, the saturation temperature of the impurity is eventually reached; precipitation occurs on the internal surface of the orifice, and this restricts the flow of liquid sodium. The change in flow rate is readily detected, and occurs at the 'plugging temperature'. By suitable calibration, and with a knowledge of the solubility of the impurity, the plugging temperature can then be related directly to the impurity concentration. For instance, if the impurity was oxide and the plugging temperature 200°C, the impurity level would be about 10 ppm by weight of oxygen.

However, it is very seldom the case that the impurity consists of one species only; oxygen is usually accompanied by hydrogen because of accidental contact with water vapour, and some carbon is also likely to be taken up from container vessels. The plugging temperature can therefore give only an overall measure of impurity levels, and cannot be specific for any one impurity. Nevertheless, the instrument is still regarded as the most reliable and rugged indicator of total impurity content in liquid sodium circuits. Attempts are being made to interpret the multiple breaks which occur in the sodium flow-rate/temperature plots in terms of separate impurities, and to study the kinetics of precipitation of the different impurity species.[344,348] If this work can be developed beyond the uncertainty which exists at present, it will add greatly to the value and the range of applicability of plugging meters.

For the satisfactory monitoring of impurities in liquid sodium systems, sampling and chemical analysis by traditional methods are no longer adequate. All such methods involve interference with the system in order to withdraw samples, which in itself can introduce impurity. Time is required for analysis, and in any case the very small impurity levels (generally parts per million) puts the solutions outside the scope of many analytical techniques. Among the many requirements for an ideal monitoring device, the following are important:

(1) It should be capable of being fitted permanently into the liquid metal circuit and not require withdrawal between measurements.

(2) It should give a continuous record of impurity levels, with minimum response time to concentration changes.

(3) It should be specific for a given impurity species, and have high precision.

There are various devices (e.g. the plugging meter) which go some way towards satisfying these ambitious demands, and they will be mentioned briefly where appropriate. Methods based on electrochemical cells come close to satisfying these requirements, and merit more detailed consideration. M. R. Hobdell, C. A. Smith and their colleagues have made valuable contributions in this field, and the discussion in this chapter on electrochemical techniques for monitoring dissolved hydrogen, oxygen and carbon in liquid sodium draws heavily on their publications on this subject.[346,347] These devices not only provide the basis for reactor instrumentation, but are finding widespread use in the research laboratory.

15.2 Hydrogen meters

The diffusion flux device

This method is based on the diffusion of hydrogen through a nickel membrane, and has the advantage that hydrogen is thereby separated selectively from all other components of the system. The nickel membrane can be positioned either in the cover gas, or (more usually) immersed in the liquid metal, and the hydrogen gas pressure within the membrane is related to the hydrogen concentration in the liquid sodium by the Sievert's law constant $K = [NaH]/P_{H_2}^{\frac{1}{2}}$ for the reaction

$$\tfrac{1}{2}H_{2(g)} + Na_{(l)} \longrightarrow NaH_{soln}$$

which has already been discussed in Chapter 4. Meters of this type have been developed in the UK, the USA, France, the Netherlands and the USSR. They all employ the same principle, and differ essentially in the technique used to measure hydrogen pressure developed within the membrane, and in the shape of the nickel membrane employed.

In a meter developed at the Argonne National Laboratories[350] (the 'ANL meter') the membrane consisted of a 12 mm diameter bellows fabricated from 0.1 mm thick nickel, giving a surface area of about 40 cm². In the equilibrium mode of operation the nickel vessel is first evacuated; hydrogen then diffuses through the nickel membrane until its pressure on the vacuum side of the membrane is in equilibrium with the partial pressure of hydrogen in the sodium. Using Sievert's law, this pressure gives a direct measure of the hydrogen content of the sodium.

This mode of operation suffers from the possible disadvantage that several minutes can be required before equilibrium is established, particularly with dilute hydrogen solutions. In the dynamic mode of operation, a steady state flux of hydrogen through the membrane is monitored. The diffusion of hydrogen from the sodium on the outside of the membrane is maintained by continuous removal of hydrogen from inside the membrane either by evacuation, or by continuous flow of an inert gas. In the ANL meter, a vacuum of 10^{-6}–10^{-8} torr is imposed on the membrane by an ion pump; the current to the ion pump depends on the quantity of hydrogen being withdrawn, and is approximately proportional to the hydrogen concentration in the sodium. Using this technique, a continuous readout of hydrogen concentration can be obtained. An online liquid sodium–hydrogen meter designed by Westinghouse[353] also employs a nickel membrane and an ion pump. It operates successfully at hydrogen concentrations in the range 0.05–5 ppm, and has an almost instantaneous (<1 s) response time.

A device designed at Dounreay[349] used the nickel membrane in the form of a coiled tube which could be positioned either in the liquid sodium or in the cover gas. The coils were about 25 mm in diameter, and 160 mm long, and were fabricated from nickel tubing of 2.5 mm outside diameter and 0.15 mm wall thickness. Hydrogen diffusing through the nickel membrane was swept out by argon, at such a rate that the hydrogen concentration in the outlet gas was near to the equilibrium partial pressure. The sweep gas was monitored continuously using a katharometer; in this instrument, the temperature of the filaments, and hence their electrical resistivity, depends on the rate at which heat is conducted away by the sweep gas. This is sensitive to the hydrogen content of the sweep gas, so that this modification also provides a continuous record of the hydrogen content of the liquid sodium.

The electrochemical hydrogen meter

Judged by the increasing extent to which these meters are being used, it would appear that electrochemical cells may become the most acceptable devices for impurity monitoring. They are specific for a given impurity, capable of high precision and rapid response time, and form the basis for meters which are equally suitable for research and plant use. There are some general aspects which apply to all electrochemical cells, and which can conveniently be discussed at this point.

The meters are, in effect, electrode concentration cells. The e.m.f. is generated between two electrodes which are separated by an electrolyte and arises from a difference in chemical activity of some species which is common to both electrodes. The electrode material is an electronic conductor which also provides chemically inert support for the species of interest, and the electrolyte contains ions which can undergo a reversible electrochemical reaction with this species at the electrode–electrolyte interface. For such a cell on open circuit, the cell e.m.f. E is related to the activities of the species concerned by the equation

$$E = (-RT/nF)\ln(a/a_{ref}) \tag{15.1}$$

where R is the universal gas constant, F the Faraday constant, T the absolute temperature, and n is the number of electrons participating in the electrode reaction per atom of the species concerned; a_{ref} is the activity at the reference electrode, and a the activity at the other electrode. If a_{ref} is known, values for a may then be obtained from measurements of E. It will be noted that these cells provide values of activity rather than concentration. The relationship between these quantities (discussed in Chapter 4) can be complex if the species concerned exists, in sodium, in more than one ionic form; this is the case with dissolved carbon which will be considered in due course. For dissolved hydrogen and oxygen, which give monatomic ions H^- and O^{2-} in solution, activity a and concentration x may be related to one another by the expression

$$a = x/S$$

where S is the saturation solubility of hydrogen or oxygen at the given temperature. This makes the reasonable assumption that Henry's law is applicable to these dilute solutions. In the simplest cases, therefore, activity represents the ratio of actual to saturation concentration, and in a saturated solution the activity is unity. In practice, solutions in sodium are usually referred to in terms of activity rather than concentration, since much of the behaviour of dissolved species (e.g. mass transfer in sodium circuits) can be related directly with their activity.

In the development of a satisfactory cell for the determination of hydrogen activity,[351,352] mixtures of molten ionic salts containing hydride ions have been considered to be most suitable as electrolytes. Eventually, taking into account melting points, chemical stability and reactivity at the electrodes, a mixture of calcium chloride with 15 mol % of calcium hydride was found to provide a suitable electrolyte. A metal membrane is necessary to separate the electrolyte from the liquid sodium, and a 0.36 mm thick iron membrane permits the necessary rapid diffusion of hydrogen through it. The reference electrode is provided by a lithium metal–lithium hydride mixture; the hydrogen-dissociation pressure of this mixture is dependent only on temperature, and this provides a fixed hydrogen activity at a fixed temperature. The dissociation pressure also lies in a convenient pressure range, since the pressure within the temperature range 380–550°C is equivalent to a hydrogen pressure over sodium containing 0.08–3 ppm of

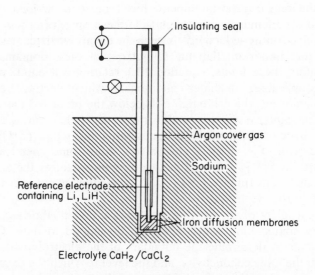

Figure 15.1 Hydrogen meter: basic features

hydrogen. The cell can therefore be represented as

$$Na(H) \,|\, Fe \,|\, CaH_2 + CaCl_2 \,|\, Fe \,|\, Li + LiH$$

and the design of the cell is shown schematically in Figure 15.1. The response of this cell to a change in hydrogen concentration is rapid, even at concentrations below 1 ppm; the determining step is the rate of diffusion of hydrogen through the iron membrane. Its shape is so designed that it can be readily inserted into a flowing sodium loop as an online instrument giving a continuous record of hydrogen activity. The rate of meter response is also rapid enough to render it suitable for use in laboratory facilities, and it has been employed in the kinetic study of the rates of precipitation and dissolution of sodium hydride in sodium, and the kinetics and mechanism of the sodium–sodium hydroxide reaction.

15.3 The electrochemical oxygen meter

This meter differs from the hydrogen meter in that a solid electrolyte is available which is compatible with liquid sodium, so that no sheath is required to separate it from the sodium. The electrolyte in common use is a solid solution of yttrium oxide (Y_2O_3) in thorium dioxide (ThO_2). Substitution of trivalent yttrium for quadrivalent thorium produces a lattice containing vacant oxygen sites, so that electrical conduction can take place by migration of oxygen ions, via the vacant sites. The electrolyte is manufactured by forming the mixed oxide powders to the required shape under very high pressures, and then firing at $2200\,^\circ C$ to produce a ceramic which is impervious to liquid sodium, and compatible with it chemically. Within certain ranges of temperature and oxygen pressure, the electrical conductivity is essentially ionic and the ionic transport number (which represents the ionic contribution to the total conductivity) is unity. Outside these ranges (termed electrolytic domains), electronic conductivity through the electrolyte becomes significant. It is therefore important that the boundaries of these domains should be known; within these limits, equation 15.1 relating cell e.m.f. with oxygen activity is applicable, but under conditions in which electronic conductivity becomes significant, the cell e.m.f. falls below the predicted thermodynamic value and the meter no longer exhibits theoretical behaviour. Measurements have been made on cells with electrolytes based on zirconia (ZrO_2) as well as thoria. However, the zirconia electrolytes show some chemical attack at temperatures above $350\,^\circ C$, and at low oxygen pressures the zirconia cells show impaired performance owing to a greater tendency towards electronic conduction in the electrolyte.

The solid electrolyte consisting of yttria-doped thoria is the most satisfactory composition which has been developed to date. Conductivity measurements to determine the boundary of the electrolytic domains have shown that the ionic conductivity, and the level to which the oxygen pressure can be increased while maintaining essentially ionic conductivity, increase with the yttria content of the electrolyte up to a maximum of about 7.5 wt % yttria, and this is the usual composition now employed. It is termed YDT by the practitioners, and its electrolytic domains are known.[354] When used for the determination of oxygen in sodium, the cell can be written

$$Na(O) \,|\, ThO_2 - 7.5 \text{ wt \% } Y_2O_3 \,|\, \text{reference electrode}$$

The choice of a suitable reference electrode system has to take into account these properties of the electrolyte. A platinum–air reference electrode has been incorporated in cells developed by Westinghouse,[355] the Central Institute for Nuclear Research, GDR,[356,357] Interatom[358,359] and Harwell.[360] Their advantages are largely of a physical nature, i.e. stability and resistance to vibration. They allow the cell to be fitted in any orientation, whereas cells with metal–metal oxide reference electrodes have to be fitted in a near-vertical position. A major disadvantage is that cells with platinum–air reference electrodes give e.m.f. values which are less than the thermodynamic equilibrium values (equation 15.1), so that recorded e.m.f. values cannot be used (without calibration) to yield activity values for dissolved oxygen. This effect is attributed to electrode polarization. Furthermore, it is difficult to reproduce with sufficient precision the surface conditions at the platinum–air–electrolyte interfaces, so that there are meter-to-meter variations in recorded e.m.f. values. These problems can be overcome by calibration of individual meters, in which case the platinum–air electrodes can probably be applied to the measurement of oxygen concentrations higher than the range to which the metal–metal oxide reference electrodes are applicable. The platinum–air electrodes are probably used to best advantage under circumstances in which detection of a change in oxygen concentration (as would occur following a leak of water into sodium) is more important than the oxygen content itself.

Reference electrodes consisting of liquid metal–metal oxide couples have been successfully adopted in meters developed by (for example) General Electric[361–363] and Harwell;[360] they include the Sn–SnO_2, Ga–Ga_2O_3 and In–In_2O_3 couples. These couples give oxygen pressures which lie within the electrolytic domain of the YDT electrolyte, so that the conductivity through the electrolyte is essentially ionic, and recorded e.m.f. is a measure of oxygen activity in the sodium. Good reproducibility has been reported for cells using the In–In_2O_3 reference electrodes[362] and these cells are being used extensively in laboratory research.[364] The full cell may be written as

$$Na(O) \,|\, ThO_2, 7.5 \text{ wt \% } Y_2O_3 \,|\, In, In_2O_3$$

The cell reaction is

$$O_2 + 4e \rightleftharpoons 2O^{2-}$$

and n, the number of electrons in the cell reaction (equation 15.1), is 4.

One design of the oxygen meter is shown schematically in Figure 15.2. The electrolyte is tubular and sealed at one end, and the perforated guard tube is

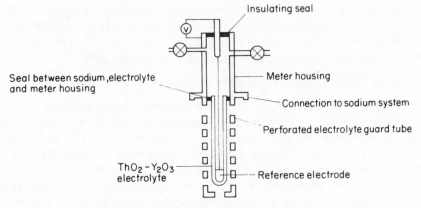

Figure 15.2 Oxygen meter: basic features

present to prevent mechanical damage. In this form it can readily be inserted into a sodium loop. Other designs employ the electrolyte in the form of a crucible, or disc. The $ThO_2-Y_2O_3$ ceramic has one grave disadvantage; it is very susceptible to thermal shock, and should not be subjected to rapid temperature changes, however small. A second purpose for the guard tube is to minimize the temperature gradient along the cell, and to this end it is fitted closely around the electrolyte tube.

15.4 Carbon meters

The two main techniques by which the carbon content of liquid sodium may be monitored are by use of electrochemical or diffusion meters. They have been developed, at an increasing pace, over the past 20 years, and have now reached a stage where each method is capable of responding to changes in the carbon content of the metal. This, in itself, is useful since rapid change in carbon content may indicate failure of some component in the system. However, the use of these meters to determine the actual carbon concentration or activity remains a major problem. Unlike hydrogen and oxygen, carbon in solution in sodium is present both as dimeric (C_2) and monomeric (C) units (see Chapter 4); and without a knowledge of the precise ratio of these two species at any given temperature, the e.m.f. for an electrochemical meter cannot be converted directly, by equation 15.1, into carbon activity. The alternative approach, by calibration of the meters against a set of carbon solutions of known activity, should eventually prove successful. At present, the preparation of such a set of standards is rendered difficult by the extremely small solubility of carbon in sodium.

The electrochemical carbon meter

No solid electrolyte possessing the necessary electrical and chemical properties has been found, and attention has centred around liquid electrolytes. It was natural that the first electrolytes to be examined should contain the acetylide (C_2^{2-}) ion, and in 1968 a cell was described which used a solution of calcium acetylide in molten calcium chloride[365] with electrodes of graphite and iron–carbon alloys. In 1971[366] this cell was shown to be applicable to the measurement of carbon in sodium, but the high melting point of calcium chloride (772°C) was a serious limitation. Other cells have been constructed which employed solutions of calcium acetylide in lithium chloride (m.p. 614°C),[367] and in other molten chlorides[358, 368] and solutions of Li_2C_2 in LiCl–KCl mixtures.[369] These electrolytes appear to be unsuitable either because of their chemical instability or because they do not exhibit reversible behaviour at the electrodes.

The meter developed at the Berkeley Nuclear Laboratories (the 'BNL meter')[346] uses a molten mixture of anhydrous sodium carbonate and lithium carbonate as the electrolyte. This equimolar Na_2CO_3–Li_2CO_3 mixture has a melting point around 500°C, and has a high chemical stability at this and higher temperatures. To most chemists, it comes as something of a surprise to see that an alkali metal carbonate can function as a source of carbon under these conditions, but intensive study of this cell appears to leave no doubt that this is indeed the case. The BNL meter may be represented schematically as follows:

$$Na(C) \mid \alpha\text{-Fe} \mid Na_2CO_3\text{–}Li_2CO_3 \mid \alpha\text{-Fe} \mid Fe_3C \quad \text{(cementite)}$$

The molten electrolyte is held between electrodes of thin carbon–permeable α-iron. The carbon content of the reference electrode is buffered to a constant carbon activity by cementite, and carbon from the sodium can diffuse through the test electrode. The e.m.f. of this cell is determined by the difference in carbon activity at the two electrodes. Tests show that as carbon is added to the sodium, in the form of carburized iron or graphite, or removed by gettering with stainless steel, the e.m.f. of the cell changes accordingly. If sodium is replaced by argon gas, the cell responds to the addition of carbon-containing gases to the argon, such as carbon monoxide or methane. This is explained on the basis of the reactions

$$2CO \longrightarrow C + CO_2$$

and

$$CH_4 \longrightarrow C + 2H_2$$

which occur at the gas–electrode interface, the carbon so produced diffusing through the test electrode to the iron–electrolyte interface.

The way in which a molten carbonate can function as an electrolyte in a carbon meter is most easily understood in terms of the electrode reactions. It is well established that, in a carbonate melt, the self-dissociation reaction is

$$CO_3^{2-} \longrightarrow O^{2-} + CO_2$$

The concentration of the oxide ion in the melt is extremely small (as is the concentration of H^+ in pure water) but it plays an important role in the

238

chemistry of the cell. As carbon diffuses through the test electrode, it is able to react with oxide ions, giving the electrode reaction

$$C + 3O^{2-} \longrightarrow CO_3^{2-} + 4e$$

which is formally equivalent to

$$C \longrightarrow C^{4+} + 4e \qquad (15.2)$$

and the operation of the cell indicates that this reaction is reversible. In this context, therefore, the CO_3^{2-} ion is regarded as a C^{4+} ion complexed by three O^{2-} ligands.

The salient features of the meter design are shown in Figure 15.3. In this form it can be inserted into flowing sodium circuits to detect changes in carbon activity. There is little meter-to-meter variation in this cell, but attempts to relate the measured e.m.f. with carbon activity (equation 15.1), using the value $n = 4$ from equation 15.2, have met with mixed success. In this connection it must be remembered that carbon is present partly as the dimer C_2 in solution in sodium, but as monomeric C units in an iron environment. It follows that a reaction $C_2 \rightarrow 2C$ must take place at the sodium-test electrode interface before the carbon can diffuse through the electrode, and this is not allowed for in the calculations. The meter is less satisfactory at very low carbon activities in sodium, and tends to record activities greater than those believed to be present. On the other hand, if the sodium is replaced by mixtures of carbon monoxide and dioxide in helium, where the carbon activity is known, the agreement with theoretical values is good.[370,371] Present difficulties therefore arise from the sodium rather than the cell, and application of the meter will develop further when the atomicity of carbon in sodium is known more precisely, and when standard solutions of carbon in sodium are more readily prepared.

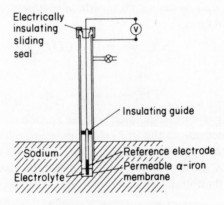

Figure 15.3 Carbon meter: basic features

The carbon diffusion meter

This instrument makes use of an α-iron membrane, usually in the form of a helically wound tube, the inner surface of which has been oxidized to form a film of iron oxide. When immersed in liquid sodium containing dissolved carbon (at an operating temperature of 490–750°C), the outer surface of the membrane achieves the same carbon activity as the liquid sodium (a_c) and carbon diffuses through the α-iron membrane. On reaching the inner surface, the carbon reacts with the iron oxide film to form oxides of carbon (CO and CO_2) which are swept away by a flow of argon carrier gas, so that at equilibrium there is a gradient of carbon activity from a_c at the outer wall of the membrane to zero at the inner wall. The rate at which carbon diffuses through the membrane (and therefore the rate at which oxides of carbon are evolved into the carrier gas) is theoretically proportional to the activity of carbon in the sodium. The rate at which oxides of carbon are produced is monitored continuously by passing the carrier gas through a gas analysis unit.

The meter was developed at UKAEA Harwell[372] and is commonly referred to as the Harwell Carbon Meter (HCM). It has been tested in various establishments, ranging from the Prototype Fast Reactor (Dounreay) to the University of Nottingham laboratories. The dimensions of the tubing forming the membrane varies widely, depending on the particular application; wall thickness is usually about 0.25 mm, but diameters range between 3 mm and 25 mm, and the length of tube in the helix can be varied to give surface areas of between 10 and 1000 cm². The carbon oxides in the sweep gas (flow rate about 25 ml per minute) are conveniently determined using a flame ionization detector; for this purpose, the carbon oxides are reduced to methane catalytically, and burnt in a hydrogen–air flame. The oxide film (FeO at temperatures above 570°C) can be regenerated by passing air through the membrane tube at its operating temperature. If the sodium contains hydrogen, this will also diffuse through the membrane and reduce the iron oxide. In this case, regeneration of the iron oxide must be carried out more frequently, but otherwise the presence of hydrogen is not a restricting factor. The speed of response of the meter to changes in carbon activity in sodium is determined essentially by the time required for the diffusion rate through the membrane to adjust to the new activity value. Theoretically, this is between about 4 minutes at 700°C and 100 minutes at 500°C, and experiments show that response times observed in practice are of this order.[373]

As with all measurements of carbon activity in liquid sodium, calibration of the diffusion meter presents a major problem. Asher *et al.*[373] have approached this problem by injecting carbon monoxide at a known rate into the system through a calibrated leak. If it is assumed that the rate-controlling process is the diffusion of carbon through the membrane, then the activity of carbon in sodium is given by

$$a_c = k \, \frac{L}{DS} \frac{F(M)}{F(L)}$$

The constant k incorporates physical features such as thickness and area of membrane. L is the calibrated leak rate, and $F(M)$ and $F(L)$ are the ionization detector readings respectively from the meter and from the leak. D is the diffusion coefficient of carbon in iron, and S the solubility of carbon in iron. Unfortunately the published values for D and S cover a wide range, so that these calculated values of a_c need independent confirmation. This has been attempted using nickel tabs.[373] These tabs (50 to 125 μm thick) were exposed to sodium in the neighbourhood of the meter, and held there until carbon distribution between sodium and nickel reached equilibrium, when the activity of carbon in the nickel should equal that in sodium. The nickel tabs were then withdrawn, washed, and analysed for carbon content. Knowing the solubility of carbon in nickel, this analysis gives the activity of carbon in the immersed nickel, and hence in the sodium. This value can then be compared with that calculated from the above diffusion equation. A correction factor x, where

$$x = a_c(\text{from Ni})/a_c(\text{calculated})$$

has been determined for a range of carbon activities. Ideally x should equal unity, and there is evidence that at high a_c values x may be close to unity. However, over most of the a_c range x has high values. For example, over the activity range $1-10^{-3}$ at 600°C, x rises to about 5, and becomes as high as 15–25 at a_c values of 2.5×10^{-4}. It is not appropriate here to examine the many experimental and theoretical reasons for this deviation,[373] but one reason may be that the solubility of carbon in nickel is itself extremely small, and difficult to determine.

At present, this carbon meter therefore provides a satisfactory means of notifying changes in carbon content of the liquid metal, but further study of calibration methods was called for to enable the meter readings to be related directly to carbon activity. Barker and colleagues at the University of Nottingham[374] have adopted a different approach, using thermodynamic standards for carbon activity. Unlike the electrochemical meter, the diffusion meter involves the continual removal of carbon from the liquid metal, so that there is a need to establish a series of solutions having both known and constant carbon activities, using a source of carbon which will continually replenish that which is removed during measurement. For this purpose 'carbide couples' have been added to the sodium, consisting of mixtures of a metal and its carbide or two carbides of the same metal. These carbide couples are added individually, in powder form; this gives a high surface/volume ratio, and rapid equilibration with the sodium. The sodium already contains some carbon, but the addition of a carbide couple establishes a 'buffered' carbon activity in solution. For example, the reaction

$$Cr + Na(_1C) \longrightarrow Cr_{23}C_6 + Na(_2C)$$

operates when the activity of the carbon initially present in sodium is higher than that which would be in equilibrium with the carbide $Cr_{23}C_6$, whereas the reaction

$$Cr_{23}C_6 + Na(_3C) \longrightarrow Cr + Na(_2C)$$

applies if the original carbon activity is lower than the value in equilibrium with $Cr_{23}C_6$. By this means, the carbon activity is automatically adjusted to the value which is in equilibrium with $Cr_{23}C_6$.

For any given carbide, the corresponding carbon activity a_c can be calculated from the relation

$$\Delta G_{carbide} - \Delta G_{Na_2C_2} = RT \ln a_c$$

and these are the 'theoretical' a_c values which are given, for the various carbon couples, in the first column of Table 15.1. In a typical experiment, the meter is placed in sodium which has been extensively purified by filtration and yttrium gettering. The carbide couple giving the lowest a_c value is added to the sodium, and the reading given by the meter is recorded until a constant reading has been obtained over several hours. The carbide giving the next lowest a_c value is then added, and the process repeated. The final carbide to be added is Fe_3C, which should give unit activity in the sodium. Alternatively, graphite may be added. In this way, a relationship is obtained between meter readings and known carbon activities. There is not a direct proportionality between meter reading and theoretical a_c values over the whole activity range; the meter readings are considered to be low for the higher a_c range ($>10^{-2}$). Reasons for this are not difficult to find. For example, at high activities the carbon flux through the membrane may not truly reflect the activity in sodium solution. Again, the proportion of carbon in sodium which is present as dimer C_2 increases with increasing carbon activity, and it could be that the carbon flux in the membrane is determined by the monomer only. For the lower a_c range (10^{-3}–10^{-4}) the procedure has been to select a couple in the middle of this range (Mo–Mo_2C) as reference, to assume the a_c value to be accurate, and to express meter readings for other a_c values in terms of this reference.

Table 15.1 Calibration of the Harwell carbon meter

Carbide couple	Carbon activity a_c (theoretical	Carbon activity a_c (experimental)	
		Expt. 1	Expt. 2
Cr–$Cr_{23}C_6$	2.6×10^{-4}	2.0×10^{-4}	1.9×10^{-4}
$Cr_{23}C_6$–Cr_7C_3	3.0×10^{-4}	2.7×10^{-4}	5.0×10^{-4}
*Mo–Mo_2C	6.9×10^{-4}	6.9×10^{-4}	6.9×10^{-4}
Cr_7C_3–Cr_3C_2	2.1×10^{-3}	1.3×10^{-3}	1.9×10^{-3}
W–WC	1.2×10^{-2}	3.1×10^{-3}	3.5×10^{-3}
316 steel	3.0×10^{-2}	5.4×10^{-3}	—
Fe_3C	1.0	0.5	—

*Reference couple.

Experimental a_c values determined on this basis from two separate experiments are given in Table 15.1; under the circumstances, reproducibility is satisfactory, and there is good agreement between experiment and theory.

The fact that meters, electrochemical or diffusion, have been developed which can respond to changes in the carbon content of sodium containing only a few parts per million of carbon is itself quite remarkable, and it is understandable that calibration should not be an easy task.

15.5 Impurity meters in liquid lithium

The application of meters for the measurement of non-metal impurities has not been developed for liquid lithium to the same extent as for sodium, but the potential requirements of nuclear fusion reactors will certainly continue to stimulate such development. The major differences arise from the greater solubility of the non-metals in lithium than in sodium (see Chapter 5), the greater stability of the binary lithium salts compared with the corresponding sodium salts, and the greater corrosive properties of lithium than sodium. These differences are best illustrated by reference to particular investigations.[375-378]

Of the non-specific methods, the use of electrical conductivity measurements has been discussed in Chapter 8. This was not included in the above treatment of impurity monitoring in sodium because of the recognized success of other methods; but it does show some promise for lithium, if only because of the higher impurity concentrations which are available. Provided that the impurities do not interact in solution, their influence in increasing the resistivity of the metal is additive, and a non-specific resistivity probe can be envisaged. However, two of the most important impurities, carbon and nitrogen, do interact, and this interaction brings about a decrease in resistivity (see Chapter 12). Interactions of this sort raise problems which may be sufficiently major to prevent the development of a resistivity monitor.

The use of plugging meters for sodium has been discussed in section 15.1; their success is a direct consequence of the very low non-metal solubilities at temperatures just above the melting point of sodium. In liquid lithium, however, the larger non-metal solubilities result in recognizable eutectics on the metal-rich side of the metal–metal salt binary phase diagram. These occur, for example, at 1372 wt ppm N for lithium nitride solutions, and at 23 wt ppm H for lithium hydride solutions, and it is not until impurity concentrations greater than these eutectic values are reached that they will be detected by a plugging meter. The interpretation of plugging temperatures in liquid lithium will therefore be much more difficult than in a sodium system.

A specific method for the determination of hydrogen in sodium involving direct pressure measurement has been discussed in section 15.2. Two main differences are encountered in applying this method to liquid lithium. Firstly, the partial pressure of hydrogen over lithium, and consequently Sievert's constant, is considerably lower for dilute solutions of lithium hydride in lithium (compare a partial pressure range of 10^{-7}–10^{-3} torr for hydrogen in lithium with 10^{-3}–10^{-1} torr for hydrogen in sodium at 500°C). This makes a much greater demand on the precision of the instruments used for pressure measurement, either in the equilibrium or the dynamic mode of operation. The second major difference concerns the membrane material. Nickel is suitable for sodium solutions, but its permeability for hydrogen at the lower partial pressures associated with lithium solutions is insufficient to produce measurable readings in the sensors. Nickel also has a relatively larger solubility in lithium, and is therefore subject to severe corrosion. These problems have been overcome by employing a niobium membrane; niobium is not only compatible with liquid lithium, but it also has an extremely high permeability for hydrogen. Oxide films on the membrane can seriously reduce this permeability; but niobium oxides are not thermodynamically stable in liquid lithium, so that the niobium surface remains free from oxide contamination.

Another device involving diffusion through a metal membrane which has proved successful in sodium, i.e. the Harwell carbon meter, has also been given some preliminary tests in liquid lithium.[378] With the iron membrane immersed in liquid lithium, carbon was added to the liquid metal in the form of Fe_3C at 490°C, and the liquid then gettered with yttrium; the meter responded to all carbon additions and carbon getters. In an experiment at 530°C, carbon was added as Li_2C_2 and the liquid gettered, and again the meter responded accordingly. However, as with the hydrogen diffusion meter, a different material will be required for the membrane on changing from liquid sodium to liquid lithium, because of the greater corrosive power of the latter. At 600°C, membrane iron was transferred from below the surface of the lithium (causing severe grain boundary grooving) to the region of the liquid metal surface, where it was precipitated as small crystals of α-iron. The transfer was so rapid that a hole in the membrane was formed within 48 h at 600°C.

So far as electrochemical cells are concerned, doubts have been expressed[376] as to the applicability of a hydrogen cell in liquid lithium, on the grounds that the partial pressure of hydrogen to be measured could be below the dissociation pressure of the conducting electrolyte. An electrochemical carbon meter might be possible, but at the time of writing no attempts at this development have been published. In each case, corrosion by lithium will call for extensive modification of the cells used in liquid sodium. Initial tests[376] have demonstrated that an oxygen meter of the type described above for sodium can also be used in liquid lithium, and that this meter shows appropriate responses to changing oxygen content of the liquid metal. For long-term use the major problem is the stability of the solid thoria electrolyte to liquid lithium. The reduction of thoria by pure liquid lithium is not favoured thermodynamically, but if dissolved nitrogen is also present, the reaction

$$Li(N) + ThO_2 \longrightarrow ThN + Li_2O$$

can occur.[379] Impurities other than oxygen can therefore influence the corrosion of the thoria electrolyte, and in this example it is the free energy of formation of ThN which is the significant factor.

15.6 Applications

The main application of these meters is in the routine monitoring of alkali metal coolant cycles, and this is the purpose for which they were originally designed. They are able to supply a constant record of metal purity, and to check on the efficiency of cold traps. An additional role played by the hydrogen and oxygen meters is to provide a rapid response to any entry of water into the liquid metal which might result from tube failure. Apart from routine monitoring, the carbon meter finds specific use in detecting leakage of oil from mechanical pumps into the liquid metal. Virtually all the carbonaceous products of the sodium–oil reaction are retained in the sodium, and tests conducted at 500–650°C in experimental loops have shown that the electrochemical meter responds in the predicted fashion on addition of oil to sodium by increasing the carbon activity.[384] The meter has also been applied with success in a number of metallurgical problems. In the absence of suitable analytical techniques for the very small amounts of carbon in sodium, it was difficult to distinguish sodium conditions which caused extensive carburization of steels from those which gave insignificant changes in the carbon content of steel. With the aid of the carbon meter it has been possible to define this relationship; for instance, a carbon activity of unity leads to extensive precipitation of carbides in austenitic 316 steels, and can increase the carbon content near to the surface of the steel to as much as 4 wt %. The carbon meter can therefore help to define carbon activities above which carburization is unacceptably high.[347,380]

The application of these meters in laboratory research has considerable potential, though the number of examples to date is very limited. Their particular value will probably be in the study of interactions between non-metals in solution, since they are able to operate at much lower concentrations than those to which electrical resistivity measurements are sensitive. The progress in this field which has been possible using resistivity measurements has been illustrated in several chapters in this book, but the next generation of equipment is likely to involve reaction vessels fitted with two or more of the meters described above.

The oxygen meter has been used to measure the oxygen concentration in equilibrium with a variety of metal oxides in liquid sodium. This is useful for defining the conditions required for the formation of ternary oxides. In research concerned with the identification of products obtained from reactions of transition metal oxides with an excess of liquid sodium,[381] it is realized that the product is in equilibrium with a solution of oxygen in liquid sodium, and it is sometimes observed that the product depends on the oxygen level in the sodium. For example, the reaction of the oxide VO_2 with sodium containing low levels of oxygen gave the ternary oxide $NaVO_2$, whereas at high oxygen levels Na_4VO_4 is obtained.[382] The oxygen level in equilibrium with the ternary oxide has now been measured for a number of oxides using the oxygen electrochemical meter.[361,383] The metal oxide is added to the sodium at the 'sample' electrode, and the mixture is contained in a nickel crucible. At equilibrium, the e.m.f. of the cell is related to the oxygen potentials of the sample and reference electrode by the expression

$$E \text{ (volts)} = \frac{\Delta \bar{G}_{O_2}(\text{sample}) - \Delta \bar{G}_{O_2}(\text{ref})}{4F}$$

so that the oxygen potential of the sample is directly available from the e.m.f. of the cell. The results may also be extended to determine the free energy of formation of the ternary oxides involved, and these measurements have been made for the ternary oxides Na_3UO_4, $NaCrO_2$,[361,383] $NaVO_2$ and $NaMnO_2$.[383] Taking the uranium compounds as an example, the product Na_3UO_4 was prepared by reaction of U_3O_8 with liquid sodium:

$$U_3O_8 + 3Na \longrightarrow Na_3UO_4 + 2UO_2$$

and since this reaction occurs at the melting point of sodium, both UO_2 and Na_3UO_4 are present at the sample electrode of the cell. The cell reaction may therefore be written

$$3Na + UO_2 + O_2 \rightleftharpoons Na_3UO_4$$

and the equilibrium oxygen potential is given by

$$\Delta \bar{G}_{O_2} = RT \ln P_{O_2} = \Delta G_f(Na_3UO_4) - \Delta G_f(UO_2)$$

where P_{O_2} is the oxygen activity, and $\Delta G_f(Na_3UO_4)$ and $\Delta G_f(UO_2)$ represent the free energies of formation of these two compounds. Knowing ΔG_f for UO_2, the value for the ternary oxide is then available. The practical value of these results is evident from the fact that ternary oxides play a significant role in corrosion of transition metals by liquid sodium; they can indicate the threshold concentration of oxygen at which ternary oxides will be formed, and thus provide a means of relating the oxygen level of sodium with the mechanism of corrosion.

A schematic diagram of an apparatus incorporating both hydrogen and oxygen meters which was used successfully in the study of the rate of decomposition of sodium hydroxide in liquid sodium:[195]

$$NaOH + 2Na \longrightarrow NaH + Na_2O$$

is shown in Figure 15.4 and is a good example of the type of equipment which is necessary. Hydrogen and oxygen in the sodium were gettered before the experiment to about 0.1 ppm using foils of yttrium and uranium respectively. The oxygen meter used an air–platinum reference electrode which allowed the meter to be fitted horizontally, and since the oxygen levels used were high

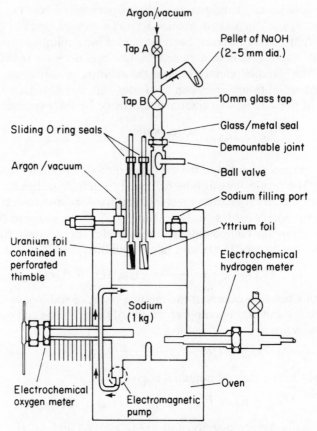

Figure 15.4   Apparatus for the simultaneous monitoring
of hydrogen and oxygen in sodium

during the reaction, it could be calibrated by addition of weighed quantities of sodium monoxide. Using this apparatus, it was possible

(a) to determine the rate law and activation energy for the dissolution of Na_2O in sodium;

(b) to follow the rate of decomposition of sodium hydroxide from both the hydride and the oxide produced in solution, and hence to determine the equilibrium constant for the association reaction $(O) + (H) \rightleftharpoons (OH)$;

(c) to measure the rates of gettering of hydrogen and oxygen by yttrium and uranium, and hence to study the mechanism of gettering.

It is obvious that apparatus of this type, using more than one meter simultaneously, could be of considerable value in many other areas of alkali metal chemistry.

The carbon meter has been used to give a direct indication of the proportion of carbon, in solution in sodium, which reacts with dissolved lithium.[129] The carbon activity of sodium at the test electrode was controlled by judicious use of carburized iron rods, and lithium (up to 2 atom %) was added in stages at 600°C. The carbon activity in the sodium was recorded continuously; it was found to decrease on addition of lithium, but not to the extent expected if carbon was being completely removed from solution as solid Li_2C_2. The results have been explained on the basis of an equilibrium

$$Li_2C_2 \text{ (solid)} \rightleftharpoons 2Li \text{ (in solution)} + 2C \text{ (in solution)}$$

Knowing the amount of carbon remaining in solution in sodium, it is then possible to evaluate the equilibrium constant for the reaction, and the free energy of formation of Li_2C_2. The carbon meter has also been used to gain additional insight into the nature of the species formed when carbon is dissolved in sodium.[346, 384] Evidence for the presence of the dimeric species in solution was discussed in Chapter 4, and further information based on the relation between carbon concentration and carbon activity, as determined by the carbon meter,[384] supports the presence of the C_2 species.

Chapter 16

Surface chemistry and wetting

16.1 Introduction

The surface tensions of the liquid alkali metals, usually determined against an inert gas, have been given in Chapter 1. The values are low compared with those for most other metals, which is consistent with bulk properties such as melting and boiling points, and reflects the relatively low attractive forces between the alkali metal atoms in the liquid. In contrast to molecular liquids and molten salts where studies of the surface have led to considerable progress, our understanding of the surfaces of liquid metals remains rather poor. The surface tension values themselves are well characterized, but as Evans[391] among many others has pointed out, many of the theories which have had some success with other liquids are not applicable because of the free-electron nature of the liquid metal. The surface tensions of liquid mixtures of two alkali metals have been discussed in Chapter 6. One important conclusion arising from those results is that, to varying extents depending on the binding energies of the atoms towards themselves and towards one another, the composition at the surface will differ from that in the bulk liquid; this has been recognized in several earlier chapters where reactions at liquid metal surfaces are discussed.

Whether or not materials exist which can act as surface active agents in liquid metals (corresponding to soaps in water), or indeed whether there can be such a phenomenon as surface activity in this medium, is at present unknown. Solution of non-metals (e.g. oxygen in sodium[392]) causes small changes only, and the deviations do not represent surface activity as it is understood in molecular liquids. Basically, molecules of substances which can bring about the profound decrease in the surface tension of water are characterized by an ability to take up a position across the surface; a hydrophilic end of the molecule, which often carries a charge, is capable of solution in the liquid, while the hydrophobic end (usually a long carbon chain) remains outside the surface. Attempts to devise surface active agents for the alkali metals, based on analogy with aqueous solutions, have been quite unsuccessful, since it is difficult to envisage molecules having both metallophilic and metallphobic properties. For the same reason, no additives

248

have been found which will give stable foams on the alkali metals. Their surface tensions are sufficiently high that, even after vigorous shaking, no foam is formed on any of the liquid alkali metals. Mixtures of sodium and potassium will foam on shaking, but the foaming is feeble and breaks almost immediately.

The surface property which has the most extensive technical significance is that of wetting. The terminology used in this area of study is not always consistent, so that a brief survey of the quantities measured, and their practical significance, is relevant here. Suppose that a liquid L rests on a solid surface S with an inert gas G as the third phase (Figure 16.1(a)). At equilibrium, the liquid surface will meet the solid surface at an angle θ, such that

$$\gamma_{GS} = \gamma_{LS} + \gamma_{GL} \cos \theta$$

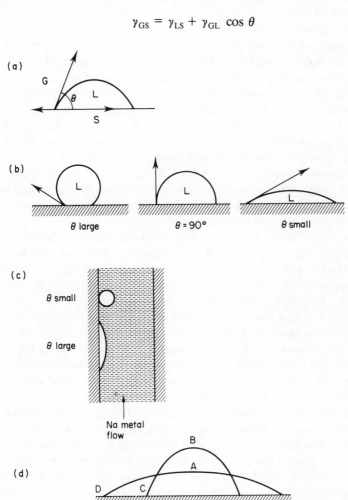

Figure 16.1   Contact angles

where γ_{GS}, γ_{LS} and γ_{GL} represent the surface tensions, or the surface energies, at the three interfaces. The angle θ is termed the contact angle, and is always measured in the liquid. Thus when the liquid sits as a near-spherical droplet on the surface, θ is large, and in the limit reaches $180°$ (Figure 16.1(b)). If the liquid tends to spread out on the solid surface, θ is small and in the limit reaches $0°$. These two limits represent conditions termed complete non-wetting and complete wetting respectively. Good examples of these extremes are provided by water on glass, where the water is said to wet the glass, and mercury on glass, where no wetting occurs. Confusion can arise unless it is remembered that the contact angle is always measured in the liquid, whatever the physical environment.

Imagine, for example, a vertical steel tube up which liquid sodium is flowing, and a bubble of entrained gas meets the tube wall (Figure 16.1(c)). If the liquid sodium continues to wet the steel, the gas bubble will retain a spherical shape, with small contact angle; whereas if the gas bubble spreads, pushing back liquid sodium from the steel wall, the contact angle is then large. If we continue the same definition of contact angle as above then at first sight the physical picture appears to be exactly the opposite of that illustrated in Figure 16.1(b). The technical significance of the contact angle now becomes fairly obvious. When liquid sodium is used as a coolant, high thermal conductivity, and hence heat transfer, across the solid metal–liquid metal interface is desirable; this implies intimate contact between solid and liquid metal, and a small contact angle. When the liquid metal is used as the mobile phase in a magnetohydrodynamic device, maximum electrical conductivity (i.e. minimum 'contact resistance') across the interface is desirable, and complete wetting is again necessary.

When contact angles are measured, values intermediate between $0°$ and $180°$ are very often obtained. It is a common convention to say that wetting occurs when θ is less than $90°$, and that there is no wetting if θ exceeds $90°$. Some authors prefer to express these intermediate contact angles below and above $90°$ as representing partial wetting, and partial non-wetting respectively, and this does give a better picture of the real situation. Under ideal conditions of perfect purity, these intermediate angles would truly represent the balance of the three surface energies involved (Figure 16.1(a)). In practice such conditions are never achieved; the solid metal may carry an oxide film, the liquid may react with this film, the gas–solid interface may carry an adsorbed film of the liquid, and so on.

Another complication results from the two different directions from which equilibrium can be reached. Using the diagram shown in Figure 16.1(d), suppose that a drop reaches an equilibrium state B from an earlier position A. At the contact point C, any chemical reactions which may occur between the liquid metal and the solid surface (such as a reaction between liquid sodium and an oxide film) have had an opportunity to take place as the contact point moves from D to its final position at C. Under these conditions the equilibrium angle at C is termed the receding contact angle. If, on the other hand, equilibrium is achieved by a readjustment of the drop shape from B to A, the gas–solid interface at the equilibrium point D has had no opportunity to react with the liquid phase. A contact angle derived in this way is termed an advancing angle.

Since we are therefore dealing with two potentially different gas–solid interfaces, it is not surprising to find that there is often a wide difference between the advancing and receding contact angles for the same system. This alone makes the study of contact angles a very imprecise science. Consequently, there have been very few attempts to relate contact angles with any physical properties of the liquid and solid metals. An early attempt was made by Hildebrand and his coworkers[394] to relate wettability with a fundamental property. Jordan and Lane[393] attempted a correlation with the atomic radius ratio of the metals; they met with only limited success, although the ideas put forward may well have greater relevance in experimentally ideal conditions. Angles near 0° and 180° have a very clear meaning, but it is difficult to attach much in the way of fundamental significance to equilibrium angles in the intermediate range. Nevertheless, some interesting correlations have been found, particularly between contact angles and the reactivity of liquid alkali metals with oxide films on the surface of solid metals, and these are discussed in the following sections.

Other techniques have been used to study the wetting process of one metal by another. Winkler and Vandenburg,[395] for example, employed molybdenum, nickel and steel specimens in the form of ribbons of known electrical resistance. A wetting film of sodium acts as a low-resistance conductor in parallel, and its presence is revealed by change in electrical resistance. However, measurements of contact angles remains the most generally applicable technique for the study of wetting.

16.2 Measurement of contact angles

Two main methods are in common use. The 'vertical plate' method, first suggested by Wilhelmy as early as 1863,[396] involves the suspension of a plate of the specimen so that it is partly immersed in the liquid metal. The various forces involved can be measured, and the contact angle then calculated. In the second method (the 'sessile drop' technique) a drop of the liquid metal rests on the surface of the solid metal, as shown in Figure 16.1(a), and the contact angle determined by direct observation.

The vertical plate method

The application of this method to the study of wetting by liquid sodium at the University of Nottingham,[397–399] at the UKAEA (Risley) Engineering and Materials Laboratory,[400,401] and at AERE Harwell,[402] have followed the same general pattern, with varying degrees of sophistication.

The apparatus used at Nottingham is illustrated in Figure 16.2. Molten

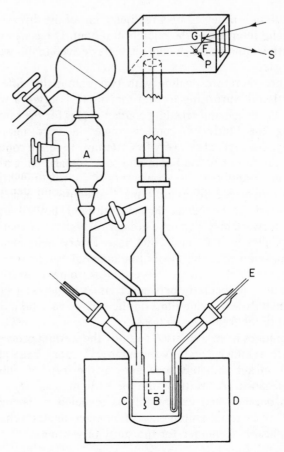

Figure 16.2   Apparatus for measurement of contact
angles

sodium was forced, under argon pressure, through the sintered glass filter A
(maintained at 110°C to reduce dissolved impurities to a minimum) and into
the stainless steel beaker B. The beaker was itself contained in the Pyrex glass
vessel C, which was surrounded by a furnace D. The temperature of the
sodium was measured by the thermocouple E, which also served to control
the furnace temperature. The metal plate was suspended by a Pyrex glass
thread from an arm attached to a torsion wire F. The vertical position of the
plate could be varied by rotating a pointer P attached to the torsion wire. The
force on the torsion wire was amplified and converted to a reading on a scale
S by means of the mirror G also mounted on the torsion wire. Calibration was
achieved by adding a series of known weights to the plate, in the absence of
sodium. When a metal plate is suspended and partly immersed in liquid
sodium, the surface tension γ and contact angle θ which the sodium subtends

at the plate at equilibrium is related to other quantities by the equation

$$\gamma \cos \theta = g \ \frac{tx \ \mathrm{d}l + (F - W)}{2(t + x)}$$

where t, x and W are the thickness, width and weight of the plate, l is its depth of immersion, d is the density of sodium and F the force exerted by the suspension wire. All the terms on the right-hand side of this equation represent measurable quantities, and since the surface tension γ is known, θ can be calculated.

At a chosen temperature, the plate was about half immersed in sodium, then drawn upwards through the liquid metal surface in steps of 1 mm at a time. At each step, the receding contact angle was recorded. When the plate had been raised about 5 mm through the surface, the contact angle became independent of further movement, and this represented the true receding angle. By repeating the measurements at different time intervals and different temperatures, it is possible to follow the variation in contact angle as the reaction between the liquid metal and (say) the oxide film on the plate surface proceeds. It is also possible to obtain advancing contact angles by pushing the plate downwards stepwise into the liquid metal. Buoyancy problems sometimes arise because of the upthrust on the plate, but this can be overcome by attaching small calibrated weights to the bottom of the plate. However, very few such measurements have been made, because of their more limited theoretical and technical value.

A number of modified versions of this apparatus have been described. Kerridge and Ford[403] substituted a bending beam for the torsion balance, and measured the deflections of the beam by a proximity meter, which registered small changes in capacity as two plates moved relative to one another. With this apparatus it was necessary to achieve movement of the plate through the liquid surface by raising or lowering the crucible containing the liquid. Longson and Thorley[401] replaced the torsion wire by a sensitive magnetic balance; they also incorporated a simple sodium loop, so that contact angles could be measured during and after purification of the sodium. The use of a steel torsion wire has one minor disadvantage, in that it may show some fatigue after long periods under high tension, so that recalibration is necessary. Matthews[402] overcame this difficulty by suspending the metal plate on a coiled spring made from beryllium–copper alloy.

The sessile drop method

This method has both advantages and disadvantages when compared with the vertical plate method. It involves direct observation of the contact angle, and is therefore free of any complications arising from calibration and calculation. It is understandable that it should have been one of the first techniques to be employed to study the wetting of transition metals and steels by liquid sodium.[404,405] More recently, in the hands of Hodkin *et al.*[406] the method has been raised to a high degree of sophistication. The specimens were polished with diamond dust, ultrasonically washed in methyl alcohol, and dried in a blast of warm air. The sodium drops (about 0.03 g) were extruded on to the specimen plates in a glove box filled with argon which contained less than 1 ppm of oxygen and 5 ppm of water, and was then further purified by passing over zirconium turnings at 800°C. The drop profiles were photographed using back-illumination, and an accuracy of ±2° in the contact angle is claimed.

Although the angle can be measured accurately, its significance is less certain. The method is supposed to yield advancing contact angles, which involves movement from shape B (Figure 16.1(d)) which the drop took up when first placed on the surface, to an equilibrium shape A. However, the force which is responsible from this movement is gravity. Since this is small, there is a danger that the equilibrium shape A is never reached, and very small drops have a tendency to retain the shape which they take up on first application to the surface. For this reason, approach to equilibrium is inevitably slow, and it is not advisable to use the method for rate measurements.

Other more obvious disadvantages include evaporation of the liquid metal at higher temperatures, which reduces the volume of the drop, and lowers the contact angle. If the solid substrate is porous, as can be the case with ceramics, loss of liquid from the drop again lowers the contact angle. These disadvantages are not encountered in the vertical plate method, since movement of the liquid with respect to the solid surface is achieved by application of an external force to the plate. The sessile drop method is perhaps most useful when evaluating and comparing the wetting behaviour of a range of solid substrates, such as ferrous materials[406] or ceramics.[407]

16.3   Factors responsible for the wetting of transition metals

Surface films on the solid metals, and the nature of their reactions with liquid sodium, are by far the most important factors controlling wetting. The guiding principles have now been worked out for liquid sodium; experimental work on wetting by the other alkali metals is limited, but from a knowledge of the composition of the films and the thermodynamic quantities relevant to their interactions with the alkali metals it is possible to make a reasonable prediction of their wetting properties. We are excluding from the present discussion metal pairs in which the solid metal has an appreciable solubility in the liquid metal, since wetting is only a temporary state on the way to complete solution. For example, cadmium dissolves in liquid sodium to the extent of 4.9 atom % at 200°C. A thin plate of cadmium placed in liquid sodium (at 100°C and above) was wetted immediately. Within a few minutes the plate thickened because of the formation of the intermetallic compounds Cd_6Na and Cd_2Na, some of which also dissolved in the sodium.

One of the first indications that surface films on the solid metal could influence wetting arose from a study of the liquid sodium–zinc system.[408] Abraded and degreased zinc plates were completely wetted within a few minutes at temperatures just above the melting point of sodium. No oxide of zinc is stable in contact with liquid sodium, so that the wetting can be attributed to the reduction, by sodium, of the surface layer of zinc oxide on the surface of zinc plates abraded in air. However, when zinc is electropolished in an alcohol–phosphoric acid solution, the metal surface becomes endowed with an invisible film of zinc phosphate about 50 Å thick. The wetting of plates polished in this way is very slow below 160°C, but is rapid above 160°C. This temperature corresponds to that at which liquid sodium reacts with zinc phosphate.

During the period 1955–65 an extensive study was made, at the University of Nottingham, of the influence of temperature, time and experimental conditions on the wetting of many of the transition metals by liquid sodium, using the vertical plate technique. Although this was early work, the general conclusions are still valid, and can be summarized as follows:

(a) The observed changes of θ with time and temperature are direct reflections of the rate and nature of the chemical reactions taking place between sodium and the oxide film on the metal surface.

(b) In the absence of any film, pure sodium (and presumably also the other alkali metals) will always spread to give a zero contact angle θ on any pure solid metal.

(c) Transition metals, when abraded in air to a high gloss, still carry an invisible surface film of oxide.

(d) There are several types of reaction which can occur between the liquid metal and the solid film, and each type give rise to a different wetting pattern. Where the free energy of formation of the metal oxide is much less than that of sodium monoxide (e.g. Fe, Co, Ni) complete reduction to metal usually occurs, and it is the clean metal surface which is wetted by sodium. Where the transition metal oxide and sodium oxide have similar $-\Delta G$ values (e.g. V, Nb, Ta) the bridge between the liquid and solid metals can be built up by the formation of ternary oxides; and where the solid metal is able to withdraw dissolved oxygen from the sodium (e.g. Zr) it is the wetting of the oxide surface which is significant.

(e) The wetting behaviour can be influenced considerably by the concentration of oxygen in the liquid sodium.

Table 16.1 shows how the sodium–metal oxide reaction products vary with the position of the metal in the periodic table; this correlates with the general trend towards lower free energies of formation of the metal oxides on moving from left to right. These results were obtained by examining the bulk products of reaction of transition metals with sodium containing dissolved oxygen, but would appear to provide a useful guide to what products are formed when transition metal oxides and sodium react on the solid metal surface. This is a

Table 16.1   Reactions of transition metals (M) in liquid sodium containing oxygen

Na + M oxide	Ternary oxides (Na–M–O)				M + Na [O]				
Sc	Ti	V	Cr	Mn	Fe	Co	Ni	Cu	Zn
Y	Zr	Nb	Mo						
Lanthanides	Hf	Ta	W						
	Ce								

general classification and cannot be taken too literally, but it is an attempt to show which metals are able to form ternary oxides which are stable in 'pure' sodium containing only a few parts per million of oxygen.[409] The position is blurred somewhat by the fact that some ternary oxides can only exist in sodium in the presence of high concentrations of oxygen. Thus, no ternary oxides are stable in sodium for the elements from cobalt to zinc; the known ternary oxides $NaNiO_2$ and Na_2NiO_2, and the ternary oxides of cobalt $NaCoO_2$, $Na_{10}Co_4O_9$, Na_4CoO_3 and Na_4CoO_4, are all reduced to metal by liquid sodium. The ternary oxide Na_4FeO_3 has been observed in liquid sodium, but only at saturation levels of oxygen; the compound $NaFeO_2$ is reduced to iron metal at 260°C, whereas the corresponding chromium compound $NaCrO_2$ is stable at all oxygen levels. In the vanadium system at 750°C Hooper and Trevillion[410] demonstrated the formation of the oxide VO on vanadium wires exposed to sodium at an oxygen activity of 10 ppm, which changes to the compound $NaVO_2$ at 23 ppm. At the lower left of Table 16.1, there is a group of elements (outlined by a dotted line) whose oxides are stable to liquid sodium, but which can form ternary oxides at saturation levels of oxygen in the sodium.

Many of the metals which are of value for constructional purposes are also the metals which can form ternary oxides with sodium. The binary and ternary oxides of these elements, which are stable to liquid sodium, are collected in Table 16.2, it being stipulated that in some cases a high oxygen level in the sodium is necessary for their stability.[411,412]

So far as evidence is available, the position with respect to liquid potassium is not greatly different from that for sodium, and this is illustrated by some examples given in Table 16.3. However, the ΔG_f, value for lithium oxide is so much more negative than that for the sodium and potassium monoxides (-441 kJ mol^{-1} at 725°C compared with -277 and -211 for Na_2O and K_2O respectively) that the oxides of many of the metals which yield ternary oxides in liquid sodium or potassium are reduced to the metal in liquid lithium. This is apparent in Table 16.3, and only in the case of those metals whose oxides have the highest ΔG_f° values (i.e. the lanthanides) can ternary oxides with lithium be prepared.[414,415]

For many of the systems mentioned in Tables 16.1, 16.2 and 16.3,

Table 16.2   Binary and ternary oxides stable in liquid sodium*

Ti	V	Cr	Mn	Fe
TiO	VO	$NaCrO_2$	MnO	Na_4FeO_3
$NaTiO_2$	$NaVO_2$		$NaMnO_2$	
Na_4TiO_4	Na_4VO_4		$Na_4Mn_2O_5$	
Zr	Nb	Mo		
ZrO_2	Na_3NbO_4	Na_2MoO_4		
Na_2ZrO_3				
	Ta	W		
	Na_3TaO_4	Na_2WO_4		
		Na_3WO_4		

*Many of these compounds have been identified as corrosion products, and are discussed in more detail in Chapter 17.

information is also available on the temperatures at which reactions occur. However, it has to be remembered that the reactivity of a metal oxide in the form of a thin film in intimate contact with the metal substrate may well differ from that of the bulk oxide. It is reasonable to look for correlation between wetting and type of reaction at the surface; there are also examples (as in the zinc phosphate reaction quoted above) where wetting and reaction

Table 16.3   Products of reaction of alkali metals with some transition metal oxides[409,413]*

Oxide	Reaction product in:		
	Liquid lithium	Liquid sodium	Liquid potassium
Fe_2O_3	$Fe + Li_2O$	$Fe + Na_2O$	$Fe + K_2O$
NiO	$Ni + Li_2O$	$Ni + Na_2O$	$Ni + K_2NiO_2$
VO_2	$V + Li_2O$	$NaVO_2$	KVO_2
V_2O_3	$V + Li_2O$	$NaVO_2 + VO$	$KVO_2 + VO$
Nb_2O_5	$Nb + Li_2O$	$Nb + Na_3NbO_4$	$Nb + K_3NbO_4$
NbO	$Nb + Li_2O$	$Nb + Na_3NbO_4$	—
Ta_2O_5	$Ta + Li_2O$	$Ta + Na_3TaO_4$	$Ta + K_3TaO_4$
CrO_2	$Cr + Li_2O$	$NaCrO_2$	—
Cr_2O_3	$Cr + Li_2O$	$NaCrO_2 + Cr$	—
TiO_2	$Ti + Li_2O$	Na_4TiO_4	—
ZrO_2	$Zr + Li_2O$	Na_2ZrO_3	—
Ce_2O_3	$LiCeO_2$	$NaCeO_2$	—
L_2O_3†	$LiLO_2$	$NaLO_2$	—

*Many of these reactions are involved in corrosion mechanisms, and are discussed in more detail in Chapter 17.
†L indicates one of the lanthanide elements.

temperatures can also be correlated, but a widespread agreement of this nature is not to be expected.

16.4 Wetting rates and wetting temperatures

Iron, cobalt, nickel and steel

A plate of cobalt was partially immersed in liquid sodium and the receding contact angle measured by the vertical plate method at various immersion times and temperatures; the results obtained are shown in Figure 16.3[399] and a similar set of curves was obtained with iron and nickel plates. The rate at which θ falls to zero increases with increasing temperature. There is a characteristic temperature (190°C in the case of cobalt) above which θ always falls to zero, and below which complete wetting does not occur. The critical nature of this change in behaviour is shown in Figure 16.4.[398] With nickel, for example, the equilibrium value of θ is zero at all temperatures above 195°C; the behaviour changes sharply within a few degrees, and at lower temperatures θ exceeds 90°. Iron and cobalt show the same phenomenon, but at critical temperatures of 140°C and 190°C respectively. This 'critical wetting temperature' is associated with that group of metals (Table 16.1) where the oxides are reduced fully to the metal by liquid sodium, and represents the temperature at which reaction occurs at the surface. Above this temperature, the clean transition metal surface is completely wetted by sodium. Below the critical wetting temperature no reaction occurs, and the

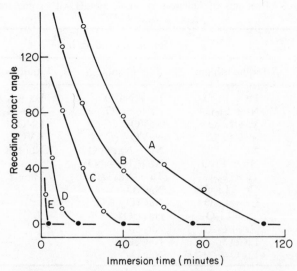

Figure 16.3 Receding contact angles for molten sodium at a cobalt surface. A, 190°C; B, 200°C; C, 230°C; D, 300°C; E, 400°C

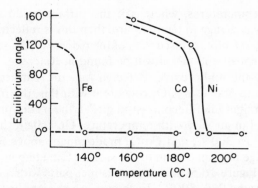

Figure 16.4   Critical wetting temperatures

high contact angles observed are typical of a metal oxide in contact with liquid sodium of low oxygen content.[416]

These metals provide the simplest and most direct correlation between wetting and surface reaction, and there are other observations which support this simple interpretation. In Figure 16.5, the times t_w for complete wetting are taken to be a measure of the rate at which the surface reaction occurs, and $\log t_w$ is plotted against the inverse of the absolute temperature T. The surface reaction is not a reversible one, and the lines in Figure 16.5 are not intended as Arrhenius plots; however, this method of correlating the results is of considerable value. A single line is obtained where the same surface oxide is

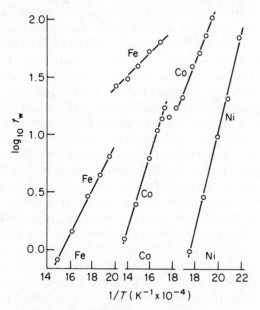

Figure 16.5 Influence of temperature on times for complete wetting by liquid sodium

present at all temperatures, whereas if the surface oxide is converted to another oxide by a change in temperature, then this is reflected by a break in the $\log t_w$ against $1/T$ plot. A discussion of the oxides which are formed on the surface of iron, cobalt and nickel will be found in Reference 398.

In summary, the only oxide detected on the surface of nickel at temperatures up to 500°C is NiO; consistent with this, the results for nickel lie on a single straight line. In contrast to nickel, the oxides present on cobalt and iron abraded in air vary with temperature. On cobalt, the oxide CoO is predominant below 300°C, and Co_3O_4 predominant above 300°C. There is therefore a change in composition of the surface film at about 300°C and the plot for cobalt (Figure 16.5) shows two distinct parts with a discontinuity at temperatures around 260–300°C. Each part is a straight line, but the slopes differ and this can be attributed to different surface reactions in which the oxides CoO and Co_3O_4 are reduced by sodium at different rates. The separation of the iron results into two parts is even more obvious, the break in the line occurring at around 220–240°C. Below 225°C the film on iron consists of α-Fe_2O_3, γ-Fe_2O_3 and Fe_3O_4. At about 225°C there are transitions from γ- to α-Fe_2O_3, and from α-Fe_2O_3 to Fe_3O_4, and this agrees well with the change in rates of wetting. Thus, the wetting of iron below 220°C is considered to involve reduction of all three oxides to metal, and above 240°C the wetting rates depend on the reduction of Fe_3O_4 only.

Another important piece of evidence supporting the wetting-surface reaction correlation involves solutions of barium and calcium in liquid sodium.[398] It was to be expected that, if the wetting process reflects oxide film reduction, the presence of metals dissolved in the sodium which have a greater oxygen affinity than sodium would accelerate wetting. Values of $-\Delta G_{298}$ for CaO, BaO and Na_2O are 600, 530 and 376 kJ mol^{-1} respectively, and the addition of small quantities of calcium and barium to the sodium has been found to influence wetting rates profoundly. The rates of wetting of nickel and iron plates by calcium solutions are compared in Table 16.4 with the rates for sodium alone, and the behaviour of barium solutions was very similar. There is a threshold concentration of calcium of about 0.07 wt % below which calcium does not appear to influence wetting by sodium, but above this concentration its effect is very marked; iron plates which required almost an hour to be wetted by sodium alone at 150°C were wetted immediately by sodium containing 0.082 per cent of calcium. The solubility of calcium in sodium falls below the threshold value at the critical wetting temperature for iron, so that it is not possible to study the effect of calcium in solution on this critical temperature. However, at the critical temperature for nickel (195°C) the calcium solubility is above the threshold value. A solution containing 0.095 per cent of calcium reduces the critical wetting temperature for nickel from 195°C to below 160°C, which supports the postulate that the critical wetting temperature is that below which reduction of the oxide film cannot occur.

The wetting of iron and nickel has been studied also by Longson and

Table 16.4   Wetting of nickel and iron by solutions of calcium in sodium

Solid metal plate	Concentration of Ca in Na (wt %)	Temperature (°C)	Equilibrium θ in Ca soln	Equilibrium θ in pure Na	Immersion time for complete wetting (minutes) in Ca soln	Immersion time for complete wetting (minutes) in pure Na
Nickel	0.062	160	>90°	>90°	—	—
	0.062	200	0°	0°	29	31
	0.095	160	0°	>90°	Immediate	—
	0.095	200	0°	0°	Immediate	31
	0.095	250	0°	0°	Immediate	4.5
	0.252	230	0°	0°	Immediate	10
	0.252	250	0°	0°	Immediate	4.5
Iron	0.055	140	0°	0°	65	63
	0.055	220	0°	0°	28	26
	0.082	150	0°	0°	Immediate	55
	0.082	250	0°	0°	Immediate	6
	0.200	200	0°	0°	Immediate	33
	0.200	280	0°	0°	Immediate	4

Prescott at the UKAEA Risley Engineering and Materials Laboratory;[417] this follows on earlier work by Longson and Thorley, also at UKAEA, Risley.[401] They were able to use a wider range of experimental conditions, particularly with respect to surface treatment of the solid metal plates. In all cases, iron and nickel plates were readily wetted at 250°C and above, and it will be noted that this is above the critical wetting temperatures already established. A remarkable feature of their results is the influence of oxygen content of the sodium on the rates of wetting. This was studied by using sodium which had been purified by cold-trapping at two different temperatures; cold-trapping at 250°C gives sodium containing about 400 ppm of oxygen, whereas the oxygen content of sodium cold-trapped at 120°C is only of the order of 1 ppm. The unexpected feature was that the rate of wetting was much greater at the higher oxygen content. This was shown by individual experiments using sodium containing two different oxygen levels, but also by oxygen addition. For instance, during the course of an experiment in which contact angle was being plotted against time using low-oxygen sodium, oxygen was added to the system; the rate of wetting increased immediately. No explanation of this phenomenon was offered, but it may be that ternary oxides of these metals, although unstable in sodium, have a part to play in the reduction mechanism. If we continue this speculation, it may be that ternary oxides such as Na_4FeO_3 and Na_2NiO_2 are more readily reduced by sodium than are the simple oxides, and it has been mentioned already that these ternary oxides can only exist in sodium at high oxygen levels.

The behaviour of type M.316 stainless steel was quite different from that of pure iron or nickel.[417] No matter how vigorous the wetting treatment, the contact angle did not fall below 30°. The rate of change of θ was not increased by increase in the oxygen content of the sodium, and in some experiments oxygen addition had the opposite effect. The only property of type 316 steel (17% Cr, 12% Ni, 2.5% Mo) likely to contribute to this behaviour is its chromium content. The high equilibrium contact angle suggests that it is an oxide film, rather than clean metal, which is being partially wetted by sodium, and it will be recalled that chromium forms a ternary oxide $NaCrO_2$ which is not reduced by sodium.

The sessile drop technique has also been applied to the study of the wetting of a range of ferrous materials by liquid sodium.[406] Correlations were found between the chromium content of the solid metal and wettability, which was also influenced by the chromium–oxygen ratio at the surface. Among pure metals, chromium was found to be more readily wetted than was iron or nickel, which at first sight is not in accord with the vertical plate results. However, it has to be remembered that this technique gives advancing angles; in the temperature range 100–300°C where the vertical plate method is recording angles at, or approaching, zero, most sessile drop angles were in excess of 90°, and only at the highest temperatures (500–600°C) did θ fall towards 20°. Because of this hysteresis, it is unwise to attempt to coordinate results by the two methods in any detail.

Chromium, molybdenum and tungsten

These metals are of special interest, since they are included in that group of metals which give ternary oxides stable to sodium (Table 16.1), yet they have wetting characteristics which are typical of those metals (e.g. Fe, Co, Ni) whose oxides are reduced to metal by sodium. Taking molybdenum as an example,[418] the variation in receding contact angle with time at various temperatures gives a family of curves similar in general pattern to those shown for cobalt in Figure 16.3. The contact angle reached zero at all temperatures above 160°C, at a rate which increased with increasing temperature, and wetting occurred immediately at temperatures above 300°C. There is a critical change in wetting behaviour at 160°C, and below this temperature the contact angle remained above 90°. A curve for molybdenum could therefore be added to those shown in Figure 16.4 for iron, cobalt and nickel, with a critical wetting temperature of 160°C. Again, when $\log t_w$ is plotted against $1/T$, as in Figure 16.5, the results for molybdenum lie on a straight line over the 160–300°C range, indicating that the same reaction is responsible for wetting throughout this range, and that the nature of the oxide film does not change with temperature.

All this evidence is consistent with reduction of the oxide film on molybdenum by liquid sodium to the metal above 160°C, but a study of the reaction of sodium with bulk molybdenum dioxide (which has been identified on the surface of the metal) suggests that this is not entirely true.[419] Differential thermal analysis showed that an exothermic reaction occurred within the range 277–304°C, to give a black powder which consisted of two distinct phases; one molybdenum metal, and the other a ternary oxide phase. The latter was originally represented as $Na_3Mo_2O_6$, but later work would favour the composition Na_2MoO_4, and the reaction is then

$$2Na + 2MoO_2 \longrightarrow Mo + Na_2MoO_4$$

If this reaction also occurs at the metal surface during wetting, then there is no inconsistency in the experimental results; the reaction does produce a ternary oxide stable to liquid sodium, but also free molybdenum metal which can account for the wetting behaviour. It is interesting that the reaction of the bulk oxide occurs at 277–305°C, since this is the temperature above which wetting occurs immediately. It is also reasonable to assume that this reaction can occur (albeit more slowly) at the metal surface at temperatures between the critical temperature of 160°C and that at which the bulk dioxide reacts. The evidence from corrosion tests on molybdenum metal (Chapter 17) may also be relevant here.

The behaviour of tungsten metal is closely similar to that of molybdenum. There is again a critical wetting temperature at 160°C, wetting is immediate above 300°C, and reactions between tungsten oxides and sodium produce a mixture of tungsten metal and a ternary oxide stable to liquid sodium. It is unfortunate that the wetting of chromium has not been examined to the same extent as molybdenum and tungsten, largely because of the difficulty of fabricating thin plates from pure chromium metal. From visual observation of plates on which chromium had been electrodeposited and then abraded, it appeared that a critical wetting temperature existed, but at a somewhat higher temperature than for molybdenum and tungsten. The oxide Cr_2O_3 has been identified on the surface of chromium;[420] this oxide reacts with sodium according to the equation[421]

$$3Na + 2Cr_2O_3 \longrightarrow 3NaCrO_2 + Cr$$

to give the very stable ternary oxide $NaCrO_2$, and chromium metal to account for the wetting pattern.

Titanium, zirconium and vanadium

These metals show a wetting behaviour quite different from that discussed above.[399] There is no sudden change in contact angle with temperature, and thus no 'critical wetting temperature'. At a sufficiently high temperature (200°C, 180°C and 170°C for Ti, Zr and V respectively) the contact angle does fall to zero, but below these upper limits the equilibrium angle has a positive value which gradually decreases with increasing temperature. This is shown in Figure 16.6 for all three metals, and this figure also illustrates the entirely different character of the two wetting patterns. Some curves showing the variation of contact angle with time for vanadium at a series of temperatures are collected in Figure 16.7, and a very similar set of curves

264

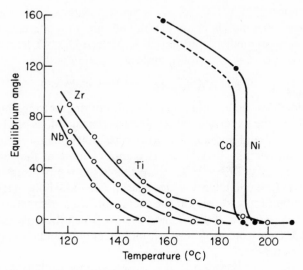

Figure 16.6   Wetting temperatures

were obtained for titanium and zirconium. At the lowest temperatures (curves A and B) the contact angle is high, and near its initial value when the plate was first immersed; the surface processes responsible for wetting occur at a negligible rate, so that little change occurs in the angle, and these curves have no real significance. Between 130°C and 250°C, however (curves C–K), the curves fall into a well organized pattern. As the temperature increases, the rate of wetting increases and the equilibrium angle decreases until, at 170°C (curve G), complete wetting is achieved. Above 170°C complete wetting is reached more rapidly, until at temperatures above 250°C wetting is almost immediate.

It is clear, particularly from Figure 16.6, that a different mechanism is involved in the wetting of these three metals, and that this should not involve the reduction of the surface oxide film to metal. Any mechanism proposed at present can only be tentative, but it is not difficult to accept the concept that any reaction between the liquid metal and the oxide film at the surface must encourage intimate contact, and stimulate wetting. In such cases the surface film would remain, even though its composition will have changed, and would form a bridge between the liquid and solid metal. From separate investigations of reactions between liquid sodium and the bulk metal oxides, it seems likely that the surface reaction responsible for enhanced wetting is the conversion of the surface oxide to a ternary oxide by liquid sodium, or the conversion of one ternary oxide into another, and that these reactions may also involve oxygen dissolved in the sodium. With titanium, for example,[422] the main oxide film on the surface of the metal abraded in air is TiO_2. In sodium, the only stable oxide is TiO, and reduction to titanium metal does not occur. Both oxides, TiO_2 and Ti_2O_3, react with sodium to give the very stable ternary oxide Na_4TiO_4, which can readily be formed at low oxygen levels in

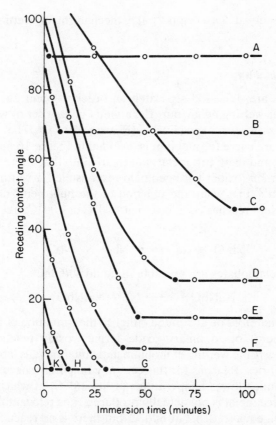

Figure 16.7   Receding contact angles for molten
sodium at a vanadium surface. A, 110°C; B,
120°C; C, 130°C; D, 140°C; E, 150°C; F,
160°C; G, 170°C; H, 200°C; K, 250°C

the sodium. Reduction to titanium metal would be possible if we considered
only the $-\Delta G$ values for the separate oxides, but it is the high free energy of
formation of Na_4TiO_4 which prevents reduction to metal. The postulate is
therefore that it is the formation of the ternary oxide from the binary oxide
which is responsible for wetting, and that the rate of this reaction is sensitive
to temperature. Similarly, with zirconium, the surface oxide ZrO_2 can be
converted to the ternary oxide Na_2ZrO_3 at sufficiently high oxygen levels, and
the ternary oxides Na_4TiO_4 and Na_2ZrO_3 have been observed on titanium and
zirconium plates wetted by liquid sodium. The oxide V_2O_3 has been identified
on the surface of abrated vanadium. Sodium is not able to reduce this to
vanadium metal; instead, provided some dissolved oxygen is present, the
binary oxide VO together with the ternary oxide $NaVO_2$ are formed, and at
sufficiently high oxygen levels these may be converted to another ternary
oxide Na_4VO_4.[382] It is unwise to draw too close an analogy between the
reactions of sodium with bulk oxides and with surface films, but the bulk

oxide reactions do at least support the mechanism of wetting postulated above.

Niobium and tantalum

These metals are discussed separately in order to point out an apparent anomaly in their wetting behaviour. Each metal gives a set of wetting curves resembling those for vanadium (Figure 16.7). The equilibrium angle/temperature curve for niobium is included in Figure 16.6; the tantalum curve is similar, and these two metals fall clearly into that group for which the formation of ternary oxides has been held responsible for wetting. However, it is now known[423] that when the reaction of the bulk pentoxides Ta_2O_5 or Nb_2O_5 with sodium reaches equilibrium at 400°C or 600°C, free metal is also formed, e.g.

$$Nb_2O_5 + Na \longrightarrow Nb + Na_3NbO_4 \qquad (16.1)$$

and that the metaniobates are unstable in liquid sodium:

$$NaNbO_3 \longrightarrow Nb + Na_3NbO_4$$

Wetting could therefore be attributed either to the formation of free metal (as with molybdenum) or to binary oxide–ternary oxide reactions (as with vanadium). Clearly the wetting of niobium and tantalum falls into the latter of these two categories. Reasons for this can only be purely speculative. It may be that at the much lower temperatures (100–200°C) at which wetting has been studied, free metal is not actually produced; it is relevant that the oxide NbO reacts as in equation 16.1 at 600°C, but there is no reaction at 400°C.

16.5 Wetting of ceramics by liquid sodium

Very little work has been carried out in this area, in spite of the fact that ceramics are considered for use as structural materials in industrial units using liquid sodium as coolant. Hodkin and Nicholas[407] made a broad survey of the wetting behaviour of thirty ceramics, using the sessile drop technique and sodium containing about 20 ppm of oxygen. Wetting was found to vary widely from one ceramic to another, but there was some evidence that oxides (MgO, Al_2O_3, Cr_2O_3, Y_2O_3, ThO_2, ZrO_2) were less readily wetted than carbides (B_4C, SiC, TiC, ZrC, HfC, VC, NbC, TaC, Cr_3C_2, Mo_2C, WC, UC); this conclusion was based on a direct comparison of the behaviour of related pairs such as Cr_3C_2/Cr_2O_3, UC/UO_2 and ZrC/ZrO_2. There were insufficient data to draw any general conclusions about the wettability of nitrides (BN, TaN, UN and Si_3N_4). None of these ceramics was wetted below 200°C (i.e. the contact angle remained about 110–150°C). In the range 200–465°C, depending on the particular ceramic, a sudden change in wetting behaviour occurred. The contact angle decreased sharply, falling to about 30–50° at 500°C, but in no case was a zero contact angle observed. There is no doubt that the conclusions which can be drawn from this work are restricted by the sluggish response to changing conditions which is characteristic of the sessile drop, and the study needs to be repeated using the vertical plate method.

Observations on uranium dioxide are of particular interest because of the earlier comments on wetting which can result from the 'bridging' effect of ternary oxides. No reaction occurs between sodium and stoichiometric $UO_{2.00}$ in bulk.[419] Consistent with this, UO_2 is not wetted by sodium having a low oxygen content; the contact angles at 200°C and 500°C were found to be 117° and 60° respectively.[407] This is also in agreement with observations made in an entirely separate investigation. Because of its use as a nuclear fuel, suspensions of UO_2 in sodium–potassium alloy were studied as part of a programme to develop a fluid fuel.[425] The stability of the slurry is dependent on the wetting of the particulate matter. In a loop containing circulating NaK alloy, pressure variations were measured against changes in flow velocity; this relationship changes if the particulate matter flocculates and clear evidence was obtained for flocculation of suspended UO_2 in the slurry. The introduction of oxygen can have a dramatic effect on the wetting of UO_2, irrespective of the way in which the oxygen is introduced. In sessile drop measurements on the non-stoichiometric oxides $UO_{2.05}$ and $UO_{2.08}$, Hodkin and Nicholas[407] found that the contact angle for sodium fell to zero at 500°C, and the sodium spread completely over the substrate surface. Using the same technique, Bradhurst and Buchanan[416] observed that, when sufficient oxygen was present, the wetting of UO_2 by liquid sodium could be brought about at temperatures above about 300°C. Matthews[402] used the vertical plate technique. Contact angles of pure sodium against plates of stoichiometric $UO_{2.00}$ did not fall below about 30–40°, but when oxygen was introduced into the argon cover gas complete wetting was achieved. This behaviour is readily explained on the basis of ternary oxide formation. Although the pure dioxide is not reactive, the higher uranium oxides (e.g. U_3O_8 and UO_3) do react with sodium, and a prominent reaction product is the ternary oxide Na_3UO_4.[426] The presence of oxygen beyond the stoichiometry $UO_{2.00}$ makes it possible for this or some other ternary oxide to be formed at the surface, with consequent wetting by sodium.

Preliminary work[407] suggests that somewhat similar compositional effects may occur with uranium carbide (UC), though no explanation has yet been offered. The wetting temperature of the carbide was increased from 145°C to 425°C when the carbon content was increased from 4.67 to 4.89 wt % carbon, and subsequent addition of oxygen caused a further rise in the wetting temperature.

16.6 Deposition from solution

There is another area of surface chemistry concerned with the liquid metal–solid metal interface which has considerable technological significance as well as academic interest, namely the deposition of metals and non-metals dissolved in sodium, in very dilute solution, on to the walls of metal containers. The technical aspect which has stimulated interest and research is the possible consequences of leakage of uranium and plutonium fission products from failed fuel pins in a sodium-colled reactor. For the fission products rubidium, caesium, strontium, barium, antimony, tellurium, bromine, iodine and some transition and lanthanide metals, measurements have been made of the extent to which they remain in sodium solution, or deposit on the walls of the steel containers. The behaviour of caesium is particularly important because of the volatility of metallic caesium, the long half-life of the ^{137}Cs isotope, and the relatively high fission yield of caesium.

Treatment in detail of this topic will not be attempted here. It is the subject of current research,[427] but there is not as yet general agreement on the extent of partition of fission product between sodium solution and steel surface. Some of the problems have arisen because the radiochemical techniques used to determine the deposition are complicated by the activity of decay products, such as the daughter nuclide ^{137m}Ba formed by decay of ^{137}Cs (Reference 428). There is a lack of consistency in the deposition measurements, but it would appear that the extent of deposition decreases with increasing temperature, though barium is exceptional in showing deposition above 500°C. The pretreatment of the steel surface, and the presence of oxide and hydride at the surface, also play important roles in the deposition process. Readers wishing to explore this topic in greater detail will find that References 427–435 provide a useful introduction to the subject.

Chapter 17

Corrosion of transition metals by the liquid alkali metals

17.1 Introduction

The general principles governing corrosion by liquid metals bear very little resemblance to those involved in corrosion of metals in aqueous media, so that the extensive research which has been carried out over many years on corrosion in moisture-containing environments can contribute little or nothing to our understanding of liquid metal corrosion. Although corrosion in a liquid metal environment is a relatively new science, there is now a voluminous literature on the subject, and this can be related to changing industrial demands. Aqueous corrosion is an expensive inconvenience, but its onset is usually all too obvious. In contrast the major applications of liquid alkali metals are in the field of nuclear reactors; corrosion is not so obvious, and can lead to major hazards.

The interactions between liquid alkali metals and constructional materials, and the consequences of corrosive action on flowing liquid metal systems, are being studied in most developed countries throughout the world. A reader new to the subject and wishing to gain a picture of the countries and the research workers involved, the range of research interests and an up-to-date overview of the field should consult the published proceedings of several recent conferences, including the first and second Topical Meetings on Fusion Reactor Materials, Miami (1979), and Seattle, Washington (1981),[385] the seminar on Materials Behaviour and Physical Chemistry in Liquid Metal Systems, Karlsruhe (1981),[386] the international conferences on Liquid Metal Technology in Energy Production held at Champion, Pennsylvania (1976) and Richland, Washington (1980),[388] and the international conference on the Liquid Alkali Metals organized by the British Nuclear Energy Society, Nottingham (1973).[389]

It is not possible, in the present context, to attempt a review which would do justice to the large number of technically difficult studies now being carried out on corrosion. Although some of the work is in the nature of direct corrosion rate measurements, many of the most important features of liquid metal corrosion are now being put on a logical and thermodynamic basis, and the aim of this chapter is to define these features. There are a few general aspects which merit mention at this stage; these aspects are sufficiently general that references to sodium can be taken to refer to the other alkali metals too.

(a) Corrosion can be defined as the wastage of metal in contact with liquid sodium, leading to reduction in thickness of metal in immersed samples or in the walls of containing vessels. This can be caused by solution of the metal in liquid sodium (or solution of one of the constituents in the case of an alloy such as steel), and this process can be stimulated by intergranular attack of the sample metal. Also, corrosion can occur as a result of chemical attack by non-metals dissolved in the liquid metal. Oxides, nitrides or carbides of the sample metal are then formed as surface layers. This layer may adhere, or may be swept away continually (if in a stream of flowing sodium), either in solution or in the form of suspended particles.

(b) A very large proportion of the published investigations has been concerned with corrosion of stainless steel and the nickel-based alloys, in sodium. This is to be expected since these alloys are favoured construction materials, and their properties can be varied quite widely by change in composition. A relatively smaller amount of work has been carried out using other transition metals, which are usually metals of the titanium, vanadium and chromium groups of the periodic table. Corrosion by liquid lithium is receiving increasing attention, but corresponding experiments using liquid potassium, rubidium or caesium are almost negligible.

(c) Corrosion rates are small, and are commonly recorded in terms of thickness of metal corroded per year. The unit 'mil' appears frequently in US publications (1 mil = 0.001 inch) but is not in common usage in the UK. More surprising, however, is the fact that the definition of a year also differs. The US definition of a year is 8766 h, representing 365 days plus 6 h per year to average out the effects of leap year; one UK definition is 8064 h, representing 48 weeks. This variation introduces an error of about 8 per cent, making the UK data slightly high compared with the remainder. However, the spread of experimental data is such that a difference of this magnitude is small enough to be disregarded.

Actual corrosion rates are dependent on a combination of many factors, and little is to be gained, in this chapter, by quoting rates without at the same time being able to define precisely all the relevant experimental conditions. Instead, we shall concentrate on the corrosion process, since it is only with this knowledge that it is possible to influence and control corrosion rates.

(d) The phenomenon known as 'mass transfer' is one of the most important, and also one of the most troublesome, effects encountered in all systems in which a liquid metal is confined in a metal or alloy container, the constituents of which have some solubility in the liquid metal. The best known examples relate to sodium loops used as coolant cycles; one part of the loop (the hot leg) is receiving heat, and at another region (the cold leg) heat is being withdrawn. The temperature differences in these two legs can amount to several hundred degrees, and if the solubility of the structural metal in sodium varies with temperature, it will dissolve in the hot leg and deposit in the cold leg; this is mass transfer. The phenomenon is therefore closely related to solubilities, but is also greatly influenced by the presence of other elements (e.g. oxygen) in the liquid metal. Mass transfer is so closely related to corrosion that in some accounts the two terms are used as though they were synonymous, which can be confusing. Mass transfer can only occur following some type of corrosion; it requires some finite solubility of the corrosion product, which may be the corroded metal itself or one of its compounds (e.g. an oxide). If a sodium circuit is constructed from two different alloys, having a given component in two different compositions, then its chemical potential in the neigbourhood of each alloy will differ. The component will tend to migrate from one alloy to the other to achieve equilibrium, and this can provide another driving force for mass transfer in addition to the temperature gradient.

(e) There are several terms which are used in the corrosion literature in such a way as to imply that they may represent individual effects which are divorced from the normal chemistry of the systems, whereas the use of separate titles is only a matter of convenience. For example, if a corrosion test specimen is inserted into a stream of flowing sodium, and a second plate is inserted in a downstream position, the corrosion at the second plate is usually less than for the first plate. This is the so-called 'downstream effect', and is consistent with the simple chemistry of the system. Thus, if corrosion is due to the presence of oxygen, this will be decreased as the sodium passes downstream and corrodes each successive specimen, with consequent reduction in corrosion rate. Again, if corrosion is a solution effect, the concentration of metal (say nickel) dissolved from the specimens will increase as the sodium flows downstream, and solution rate decreases as the concentration of solute metal increases. There is also the 'rate of flow effect', whereby corrosion of specimens in flowing sodium is generally found to increase at higher flow rates. In cases where corrosion results from the formation of a separate phase of corrosion product at the solid surface, the increasing eroding action of sodium at the higher flow rates will more effectively remove the corrosion product from the surface, exposing more solid metal surface to corrosion. On the other hand, if corrosion involves solution of one of the components of the solid metal, a higher flow rate will reduce the concentration of the solute metal in the sodium in the neighbourhood of the surface, and thus enhance further solution. Corrosion rates can therefore only be compared directly if the same sodium flow rates are used.

(f) It will be obvious that wetting is a necessary precursor to corrosion. If the solid specimen is not wetted by the alkali metal, it will not be corroded by it. The same factors which cause wetting (i.e. solution of solid metal, reduction of oxide films, formation of ternary compounds etc.) are also responsible for corrosion, so that there is much common ground between this and the previous chapter.

17.2 Thermodynamic aspects

If we can imagine a system in which a pure solid metal is completely wetted by a pure liquid metal, then the only factor responsible for corrosion would be solution of the solid metal in the liquid metal. In practice, however, it is the non-metals (mainly oxygen, hydrogen, nitrogen and carbon) dissolved in the liquid metal which cause corrosion. Ideally, corrosion should be studied by following the reactions which take place at the actual surface, separation of reaction products from the surface, and so forth. Although physical methods for studying the corroding surface (such as electron-probe microanalysis, secondary-ion mass spectrometry and X-ray diffraction) are being developed, data are not always available to provide a complete picture of processes occurring at the surface during corrosion. It is fortunate, therefore, that we can often anticipate the nature of the corrosion process from separate experiments involving reactions of the solid metal, or its binary oxides, nitrides and carbides, with liquid metal containing various dissolved non-metals, and such experiments will be discussed where appropriate.

At the solid metal–liquid metal interface, the fundamental feature is the partition of the non-metal between solid and liquid metal, which is thermodynamically controlled. This is illustrated in Figure 17.1.[409] The high value of the free energy of formation of lithium oxide ensures that solutions of oxygen in liquid lithium will not react with the surface of transition metals such as niobium. Indeed, any oxygen present on the metal surface, or in solid

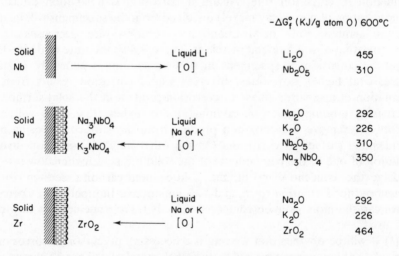

Figure 17.1 Partition of oxygen between a liquid alkali metal and a solid metal

solution, will be removed and will dissolve in the liquid lithium. This would suggest that oxygen in lithium should not act as a corrosive impurity, and this is found to be the case in practice. Resulting from the lower free energies of formation of sodium oxide and potassium oxide, solutions of oxygen in sodium or potassium are more reactive towards metal surfaces. In the example shown in Figure 17.1, oxygen is removed from solution and is transferred across the solid–liquid interface, and diffuses into the solid metal forming initially a solid solution. As more oxygen diffuses across the interface, a surface film is formed, consisting of a ternary oxide in the case of niobium or of a binary oxide in the case of zirconium. Whether or not the oxygen remains in solution in sodium or reacts with the solid metal surface then depends essentially on the free energies of formation of the binary or tertiary oxides of the solid metal. Eventually equilibrium is reached when the activities of oxygen in the alkali metal and in the surface film are balanced. This type of behaviour has been observed for many metals, and is summarized in the previous chapter (Tables 16.1, 16.2 and 16.3).

The corrosion of stainless steel in sodium provides an interesting development of this theme. The components iron, chromium and nickel react towards sodium containing oxygen in different ways. Chromium reacts with oxygen to form a ternary oxide

$$Cr + Na\,[O] \longrightarrow NaCrO_2$$

and this reaction takes place at virtually all levels of oxygen in sodium. Iron will only form a ternary oxide at very high oxygen levels (i.e. the saturation value at $600°C$), and any iron oxide film on the surface would be reduced by sodium at temperatures above the critical wetting temperature (see Chapter 16). Nickel shows no interaction with oxygen in sodium, but has a relatively high solubility in the liquid metal. The net result of all these interactions is that a surface layer of $NaCrO_2$ forms on the stainless steel on exposure to sodium containing dissolved oxygen. The formation of the outer chromite layer leaves an underlying ferritic layer, and this is enhanced by the additional loss of nickel by direct solution in the liquid metal. In time, the system reaches equilibrium and the rate of corrosion decreases.

Free energy values are also the most important property when we consider the corrosion behaviour of the various non-metals which may be dissolved simultaneously in a single liquid, i.e. liquid lithium. This is illustrated in Figure 17.2. The values of $\Delta G°$ at $625°C$ for Li_3N and Li_2C_2 are only -45 and -178 kJ mol^{-1}, compared with -455 for Li_2O;[436] therefore, while oxygen usually remains in the liquid lithium, nitrogen and carbon are more readily transferred from solution to the solid metal surface. In practice, nitrogen in lithium plays the same corrosive role as does oxygen in sodium. The value of $\Delta G°$ for the niobium nitride NbN is -161 kJ mol^{-1} at $600°C$, and when niobium metal undergoes corrosion in liquid lithium containing dissolved nitrogen, the binary nitrides Nb_2N and NbN are formed as surface products with the added possibility of ternary nitride formation at high nitrogen

274

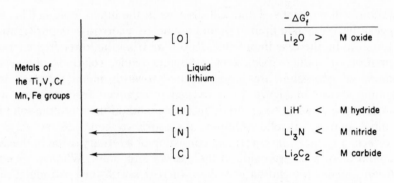

Figure 17.2 Transition metals (M) in liquid lithium containing dissolved non-metals

levels.[436] Nitrides therefore feature prominently in the corrosion chemistry of liquid lithium. Many metals are attacked by nitrogen and by carbon dissolved in liquid lithium, and Table 17.1 lists a number of known compounds which might possibly be formed during corrosion in lithium. As will be seen below, many (but not all) of these compounds have actually been identified on the metal surfaces.

Values of the free energy of formation of many of the binary compounds which will be discussed in this chapter are collected in Table 17.2. These values are being constantly refined as research continues and techniques develop, and values for all the compounds are not available at the same temperature. Nevertheless, they provide a useful basis for comparison. Note that the absolute scale of temperature is used in Table 17.2, as this is normally employed in books of thermodynamic data.

Table 17.1   Some known binary and ternary nitrides and carbides of transition metals*

Metal	Nitrides	Carbides
Fe	Li_3FeN_2	Fe_3C
Cr	Li_9CrN_5	$Cr_{23}C_6$, Cr_7C_3, Cr_3C_2
V	V_3N, VN, Li_7VN_4	V_2C
Nb	Nb_2N	Nb_2C
Ta	$TaN + Ta_2N$	Ta_2C
Ti	Ti_2N	Ti_2C
Zr	$ZrN + Li_2ZrN_2$	—
Mo	MoN	Mo_2C

*This list is restricted to the constructional metals.

275

Table 17.2 Some free energies of formation of binary compounds (kJ mol^{-1})

Compound	$-\Delta G_f^\circ$	T(K)	Compound	$-\Delta G_f^\circ$	T(K)	Compound	$-\Delta G_f^\circ$	T(K)
Li_2O	441	1000	TiO	435	873	Cr_2O_3	870	1000
	469	800	TiO_2	796	800	CrN	58	873
	497	600	ZrO_2	942	800	Cr_2N	52	1000
	523	400	ZrN	280	900	$Cr_{23}C_6$	387	1000
Na_2O	274	1000	ThO_2	1072	800	Cr_7C_3	190	1000
	302	800	Ce_2O_3	511	1000	Cr_3C_2	96	1000
	333	600	CeO_2	459	1000	MoO_3	529	900
	362	400	VO	344	1000	Mo_2C	56	900
K_2O	260	673	V_2O_5	1123	1000	WO_3	613	900
Li_3N	45	900	VN	131	1000	MnO	312	1000
Li_2C_2	178	900	Nb_2O_5	1550	900	FeO	198	1000
			NbN	161	900	Fe_2O_3	288	500
			Ta_2O_5	1650	900	NiO	149	1000
			TaN	181	900			

17.3 Competitive reactions

Liquid lithium is particularly interesting in this respect, since a wide range of non-metals show reasonable solubilities in the liquid metal, and oxygen, nitrogen and carbon are difficult to remove from the liquid. Once again, experiments show that when two or more dissolved non-metals are in competition in the corrosion processes, it is the relative values of the free energies of formation of the compounds formed with the liquid and solid metals which determine the nature and composition of the corrosion product. The competition between oxygen and nitrogen is illustrated by the reactions of vanadium. The ΔG° values of the vanadium oxides are of the order of -300 kJ/g atom O at 625°C; and when vanadium monoxide is added to pure lithium, vanadium metal is produced, the oxygen being dissolved in the liquid lithium. However, vanadium metal immersed in liquid lithium containing nitrogen is rapidly nitrided; the ΔG° values for VN and Li_3N are -130 and -45 kJ/g atom N at 625°C respectively. In fact, the extent and nature of the nitriding is dependent on the ratio of vanadium to the nitrogen present in solution. Competition between oxygen and nitrogen in solution was set up by adding vanadium monoxide to liquid lithium containing various concentrations of dissolved nitrogen. The results (Table 17.3) show that a clear distinction can be made between the behaviour of oxygen and nitrogen in liquid lithium.[437]

If the $-\Delta G^\circ$ value for the solid metal oxide is similar to, or exceeds, the value for Li_2O, then a situation arises in which both oxygen and nitrogen are capable of being partitioned across the solid–liquid interface. This is the case with thorium,[438] where ΔG° at 625°C is -525 kJ/g atom O. Thorium metal reacts with liquid lithium containing dissolved oxygen to give thorium dioxide as a surface product:

$$Th + Li\,[O] \longrightarrow ThO_2 + Li \qquad (17.1)$$

Similarly, solutions of nitrogen in lithium react with thorium metal to give a surface layer of thorium nitride:

$$Th + Li\,[N] \longrightarrow ThN + Li \qquad (17.2)$$

Since ΔG° for thorium nitride at 625°C exceeds -200 kJ/g atom N, the free

Table 17.3 Reaction of vanadium metal with nitrogen and oxygen in liquid lithium at 250–600°C

Nitrogen content of liquid lithium	Solid phase	Solution
0	V	Li_2O
0.04–0.06	$V + V_3N$	Li_2O
0.1–0.8	V_3N	Li_2O
0.9–1.7	$V_3N + VN$	Li_2O
1.8–2.0	VN	Li_2O
>2.6	Li_7VN_4	Li_2O

energy change in reaction 17.2 is much greater than in reaction 17.1; so that under competitive conditions, when thorium metal is in contact with liquid lithium containing both oxygen and nitrogen in solution, only ThN is produced so long as nitrogen is in excess:

$$Th + Li [N]_{excess} + Li [O] \longrightarrow ThN + Li [O]$$

The dioxide ThO_2 can only be formed when nitrogen in solution has been removed by the formation of ThN:

$$Th + Li [N] + Li [O]_{excess} \longrightarrow ThN + ThO_2 + Li [O]$$

This order of reactivity has also been observed in reactions of cerium dioxide with liquid lithium.[415] In the absence of nitrogen, the dioxide reacts with lithium to give a ternary oxide:

$$Li + CeO_2 \longrightarrow LiCeO_2$$

Traces of nitrogen dissolved in the lithium cause decomposition of the ternary oxide:

$$Li [N] + LiCeO_2 \longrightarrow Li_2O + CeN + LiCeO_2$$

at nitrogen levels as low as one weight ppm of nitrogen in the lithium. Given sufficient nitrogen, all the ternary oxide is decomposed:

$$Li [N] + LiCeO_2 \longrightarrow CeN + Li_2O$$

and, in an excess of nitrogen, a ternary nitride is formed:

$$Li [N]_{excess} + CeO_2 \longrightarrow Li_2O + Li_2CeN_2$$

In an almost identical series of experiments, reactions were carried out using lithium containing dissolved carbon. The ternary oxide $LiCeO_2$ was again unstable relative to carbide formation, and complete decomposition to the carbide could be achieved;

$$Li + Li_2C_2 + CeO_2 \longrightarrow Li_2O + CeC$$

Molybdenum metal provides an example of extreme selectivity.[439] No interaction was observed at 600°C over 16 days between molybdenum and oxygen dissolved in sodium or in lithium, or between molybdenum and nitrogen dissolved in lithium, but introduction of carbon into the system resulted in the ready formation of the carbide Mo_2C. Carbides formed as corrosion products in this way are also included in Table 17.1.

17.4 Ternary oxides

In order to attempt a complete correlation between reactivity in these systems and the thermodynamic quantities involved, values for the free energies of formation of the many ternary oxides formed should also be available; in fact, very few values are known at the time of writing. An estimate of their values can sometimes be made by consideration of the various reactions by which they may, or may not, be formed. Three examples are given in Table 17.4. With the vanadium compounds, and using pure reactants, reaction (4) does not occur. This sets a top limit to $- \Delta G_f^{\circ}$ for NaVO$_2$, since if it exceeded $- \Delta G_f^{\circ}$ for 2VO, we would expect the reaction to go forward. Knowing the values of $- \Delta G_f^{\circ}$ for the binary oxides, a top limit can therefore be estimated. Reactions (1)–(3) do proceed, and can provide lower limits for $- \Delta G_f^{\circ}$. Thus in reaction (3), $- \Delta G_f^{\circ}$ (NaVO$_2$) should exceed $- \Delta G_f^{\circ}$ (VO + Na$_2$O) for the reaction to be favourable. It is obviously desirable that as many reactions as are available should be used in determining this lower $- \Delta G_f^{\circ}$ limit; in this case reaction (2) provides the most negative value, and is taken as the lower limit of the range within which the true value lies. In the three examples given in Table 17.4, sufficient experimental evidence is available to make it possible to estimate free energy values within a narrow range, and experimental values (see below) fall within these ranges.

It is also possible to take advantage of reactions which do, or do not, proceed according to the temperature. The ternary oxide Na$_3$NbO$_4$ is formed by reaction (7) at 600°C, but not at 400°C.[423,424] Using $- \Delta G_f^{\circ}$ values for NbO at the appropriate temperatures, upper and lower values for $- \Delta G_f^{\circ}$ (Na$_3$NbO$_4$) are obtained, and the lower limit is supported by reactions (5) and (6). Similar arguments have recently been applied to determine the $- \Delta G_f^{\circ}$ range for the ternary oxide LiCeO$_2$.[449] This example illustrates how any reaction which involves the ternary oxide may, in principle, be used, and shows that it is desirable to use as many reactions as possible to define upper, as well as lower, limits.

Values for the free energy of formation of ternary oxides can also be determined experimentally using the oxygen electrochemical cell discussed in Chapter 15.[455] All oxides which are stable in liquid sodium exist in equilibrium with a solution of oxygen in the sodium. By measuring the activity of oxygen in solution, the free energy of formation of the coexisting oxides can then be determined. In a typical experiment, the oxide V$_2$O$_3$ was added to liquid sodium in the sample side of the cell. At 600°C, a reaction takes place according to

$$V_2O_3 + Na \longrightarrow VO + NaVO_2$$

The cell reaction may therefore be written as

$$Na_{(l)} + VO_{(s)} + \tfrac{1}{2}O_{2 \text{ dissolved}} \longrightarrow NaVO_{2(s)}$$

and at equilibrium

$$\Delta G_f^{\circ}(NaVO_2) = \Delta G_f(VO) + \tfrac{1}{2} \Delta \bar{G}_{O_2}$$

The e.m.f. of the cell leads directly to the activity of oxygen in solution, and hence to the chemical potential and free energy of formation of the solution

Table 17.4 Estimated $-\Delta G_f^\circ$ values for some ternary oxides (kJ mol^{-1})

Reaction	Temperature (K)	$-\Delta G_f^\circ$	
		for NaVO$_2$	
(1) VO$_2$ + Na → NaVO$_2$	1000	>547	Estimated range 637–682;
(2) V$_2$O$_3$ + Na → NaVO$_2$ + VO	1000	>637	experimental value 661
(3) VO + Na$_2$O → Na + NaVO$_2$	1000	>595	
(4) 2VO + Na → NaVO$_2$ + V	1000	<682	
		for Na$_3$NbO$_4$	
(5) 4Nb$_2$O$_5$ + 15Na → 3Nb + 5Na$_3$NbO$_4$	873	>1308	Estimated range 1329–1400;
(6) 2NbO$_2$ + 3Na → Nb + Na$_3$NbO$_4$	873	>633	experimental value 1346
(7) 4NbO + 3Na → 3Nb + Na$_3$NbO$_4$	873	>1329	
(8) 4NbO + 3Na → 3Nb + Na$_3$NbO$_4$	673	<1400	
		for LiCeO$_2$	
(9) Li + CeO$_2$ → LiCeO$_2$	873	>917	Estimated range 996–1022
(10) LiCeO$_2$ + Li(N) → 2Li$_2$O + CeN	873	<1137	
(11) Li + 2CeO$_2$ → Ce$_2$O$_3$ + Li$_2$O	873	>996	
(12) Li + 2Ce$_2$O$_3$ → 3LiCeO$_2$ + Ce	873	<1022	

$\Delta \bar{G}_{o_2}$. Knowing ΔG_f for VO, the value for the ternary oxide is then available. The experimental values quoted in Table 17.4 were determined in this way.

17.5 Survey of corrosion experiments

Sections 17.1–17.4 have been concerned with the general principles governing corrosion of metals by the liquid alkali metals, and a few solid metals have been discussed as examples. A brief survey of corrosion studies on some other metals will now be given. The emphasis varies considerably from metal to metal, depending on their potential value as constructional materials; but it will be convenient to group the solid metals according to their position in the transition block of the periodic table. Since there is a good correlation in almost all cases between the products obtained on corrosion of metal plates, and the separate reactions of alkali metal and solid metal oxides and nitrides, the later will continue to be referred to as appropriate.

Titanium, zirconium, hafnium and thorium

The results of corrosion studies on titanium and zirconium metals in liquid sodium are collected in Table 17.5.[422] Only one ternary oxide corrosion product, Na_4TiO_4, is formed on titanium at oxygen concentrations in the sodium varying from 100 to 12,000 ppm, indicating its high stability in a liquid sodium environment. Table 17.5 shows that titanium monoxide is only formed in experiments having the longest reaction times. This suggests that formation of Na_4TiO_4 is the initial reaction, followed by absorption of oxygen into the metal lattice until the composition TiO is reached at the surface. Since Na_4TiO_4 can be formed readily at low oxygen levels, it is likely to be present on titanium metal exposed to flowing sodium containing low concentrations of oxygen. Titanium so exposed will thus be corroded both by the formation of Na_4TiO_4 and its erosion, and embrittled by oxygen absorption up to the composition of the monoxide.

Table 17.5 Corrosion of titanium and zirconium in liquid sodium

Metal	Temperature (°C)	Time (days)	Oxygen content of sodium (ppm)	Corrosion product
Ti	600	3	3,400	Na_4TiO_4
Ti	600	8	12,000	Na_4TiO_4
Ti	600	14	100	Na_4TiO_4 + TiO
Ti	600	14	1,600	Na_4TiO_4 + TiO
Ti	700	14	7,000	Na_4TiO_4 + TiO
Zr	600	2	3,000	ZrO_2
Zr	600	14	2,300	ZrO_2 + βNa_2ZrO_3
Zr	550	24	3,000	ZrO_2 + αNa_2ZrO_3

The first stage in the corrosion mechanism for zirconium would appear to be the penetration of oxygen from liquid sodium into the solid metal, and the eventual formation of a film of zirconium dioxide. Once this layer has been formed, a much slower reaction then occurs to give the ternary oxide corrosion product Na_2ZrO_3. This compound has a phase transition temperature of 570°C, and the production of both the α and β forms has been observed. It is reasonable to postulate that the corrosion rate should depend on the rate of diffusion of oxygen through the ZrO_2 layer; in flowing liquid sodium circuits containing low concentrations of oxygen, any Na_2ZrO_3 layer would be eroded; but since its formation is a slow process the dominating reaction would still be the formation of the ZrO_2 layer and the penetration of oxygen through it. No corresponding studies on the surface of hafnium and thorium metals are known at the time of writing, but a behaviour similar to that of zirconium would be expected.

The mechanism of corrosion of these metals by liquid lithium is dominated by the readiness with which pure liquid lithium can reduce the metal oxides. Relevant free energy values are given in Table 17.6. The reactions of the metal dioxides with liquid lithium have been examined,[438] and we may take the products of the MO_2 + Li reactions as giving a good guide to the behaviour of the Li [O] + M system. Titanium dioxide reacts with liquid lithium, at all temperatures at which lithium is liquid, to give titanium metal, which is in accord with free energy data (Table 17.6). The situation for zirconium and hafnium dioxides could be more complex, since the free energy change for the reactions

$$4Li + MO_2 \longrightarrow M + 2Li_2O$$

changes from a small, negative value at low temperatures to a small, positive value above 600°C; these are the circumstances in which ternary oxides are often formed. In the event, zirconium dioxide is reduced by lithium to zirconium metal. Hafnium dioxide is also reduced to the metal, but on isolation of the reaction product (Hf + Li_2O) a small quantity of the ternary oxide $LiHfO_2$ was also identified. As mentioned earlier, thorium dioxide is not reduced by liquid lithium, so that the behaviour of the titanium group metal dioxides follows closely the predictions which could be made from the free energy data in Table 17.6. Extending this argument, it seems likely that any oxide films on the surface of titanium, zirconium or hafnium would be removed by lithium, and that corrosion of the solid metals by very pure

Table 17.6 Free energies of formation of the dioxides of the titanium group (kJ mol^{-1})

Oxide	Li_2O	TiO_2	ZrO_2	HfO_2	ThO_2
$-\Delta G_f^\circ$ at 27°C	560	888	1036	1032	1168
$-\Delta G_f^\circ$ at 527°C	468	796	942	956	1072

lithium would proceed largely by dissolution of the metal into the liquid. This, however, has little practical significance, since corrosion problems usually arise when large quantities of liquid lithium are being used; the liquid metal will contain nitrogen impurity, and attack of the solid metal surface by the nitrogen species in solution in lithium will carry the main responsibility for corrosion.

The products of reaction of the dioxides with lithium containing dissolved nitrogen can be expressed as follows:

$$Li\,[N] + TiO_2 \longrightarrow Ti \text{ metal}, Ti_2N, Li_2O$$

$$Li\,[N] + ZrO_2 \longrightarrow Zr \text{ metal}, ZrN, Li_2ZrN_2, Li_2O$$

$$Li\,[N] + HfO_2 \longrightarrow Hf \text{ metal}, HfN, Li_2HfN_2, Li_2O$$

$$Li\,[N] + ThO_2 \longrightarrow ThN, Li_2ThN_2, Li_2O$$

the relative amounts of the various products depending on the concentration of nitrogen in solution. The formation of the nitrides of titanium and zirconium corresponds to almost complete removal of nitrogen from the lithium, and free energy changes calculated for the reactions

$$Li + Li_3N + M \longrightarrow MN + 4Li$$

indicate that this is a very favourable process. It is interesting to note that the addition of thorium dioxide to liquid lithium containing nitrogen gave visible signs of reaction, in that sparks and small flames were produced, whereas its addition to pure lithium showed no such effects. Some experiments which are also closely related to corrosion have been carried out by reacting lithium nitride with hafnium and thorium metals in the solid state.[440] The reactions proceed by formation of the mononitrides, with the ternary nitrides Li_2MN_2 as final products.

Two additional points deserve passing mention. Occasionally, a phase corresponding to the composition Th_2N_2O was identified in the reaction products, and this oxide nitride could possible feature as a corrosion product. The corresponding compound of cerium, Ce_2N_2O, is also known.[440] Secondly, in liquid lithium systems containing carbon, titanium forms the carbide Ti_2C as corrosion product.[412]

Vanadium, niobium and tantalum

The general corrosion pattern for these metals is similar in principle to that described for the metals of the titanium group, i.e. reduction by liquid lithium of oxides, in the form the surface films or bulk oxides, to metal, which is then available for attack by dissolved nitrogen or carbon, and reaction of oxides in liquid sodium or potassium to give ternary oxides.

All the vanadium oxides are reduced to metal in nitrogen-free lithium. The reactions of vanadium oxides in liquid sodium are shown in Table 17.7. Only

Table 17.7 Reactions of the vanadium oxides with liquid sodium

Oxide	$-\Delta G_f^{\circ}$ at 600°C (kJ/g atom O)	Reaction temperature (°C)	Reaction products at equilibrium
Na_2O	292	—	—
V_2O_5	236	100	$NaVO_2 + Na_4VO_4$
V_2O_4	282	150	$NaVO_2$
V_2O_3	337	400	$VO + NaVO_2$
VO	341	(No reaction up to 600°C)	

three compounds are stable in liquid sodium at 600°C, namely the monoxide VO and the ternary oxides $NaVO_2$ and Na_4VO_4.[382,441] The stabilities of the compounds VO and $NaVO_2$ depend on the oxygen content of the sodium in that both these compounds are oxidized to the ternary oxide Na_4VO_4 in sodium containing high concentrations of oxygen. Although this compound has a high stability in liquid sodium, it is relatively unstable when isolated and heated under vacuum. At 290°C it loses sodium, with the formation of the orthovanadate Na_3VO_4. This conversion could take place while sodium was being distilled from the reaction product so that it is of great advantage to be able to identify the product, by X-ray techniques, while it is still immersed in sodium. In corrosion experiments on vanadium plates in static sodium at 600°C, at oxygen contents ranging from 300 to 3000 ppm and for immersion periods up to 5 days, the corrosion product was identified as Na_4VO_4.[442] The behaviour of liquid potassium, so far as reactions with the vanadium oxides is concerned, closely resembles that of liquid sodium; it seems likely that corrosion of the metal will also be similar, though no definitive corrosion experiments designed to identify corrosion products appear to have been made. There are some minor differences in the behaviour of the compounds formed. Thus, the orthovanadate K_3VO_4 is stable in liquid potassium, whereas sodium orthovanadate Na_3VO_4 is reduced to the compound Na_4VO_4 in liquid sodium. Again, vanadium monoxide, VO, although stable to both sodium and potassium, differs in its behaviour towards dissolved oxygen. In liquid potassium, the tendency is towards the formation of oxygen-rich compositions of the monoxide, i.e. in excess of $VO_{1.0}$, rather than the formation of ternary oxides, as occurs in liquid sodium.

These results are in good agreement with studies of the surface of vanadium sheets exposed to static liquid sodium under various experimental conditions. After immersion, the surface was examined by X-ray diffraction analysis, with sodium still present on the metal surface.[443] In experiments at 600°C, the X-ray diffraction pattern confirmed that no ternary oxide other than Na_4VO_4 was present. This compound is decomposed in water, so that its X-ray pattern disappeared when the sheets were washed in water; other patterns appeared representing various underlying oxide phases formed during corrosion. The most prominent of these was the sub-oxide V_9O, with VO in smaller amounts. The formation of the ternary oxide is dependent on the supply of oxygen to the surface of the metal. The amount of ternary oxide decreased in successive experiments at 600°C in which the initial oxygen concentration was reduced from 3000 to 100 ppm oxygen; the X-ray pattern for Na_4VO_4 was scarcely detectable at 100 ppm oxygen, and at 20 ppm only the oxide phases resulting from oxygen penetration into the metal lattice could be detected. The first stage in the corrosion mechanism of vanadium must be the solution of oxygen in the metal. As this continues, layers of oxide phases such as V_9O and VO are formed in the region of the surface. At sufficiently high concentrations of oxygen in the sodium, the compound Na_4VO_4 is formed; this is a competing process, and it would appear that at lower oxygen levels the dissolution process becomes thermodynamically favoured compared with the formation of the ternary oxide.

In nitrogen-free liquid lithium, the oxides of niobium and tantalum are reduced to the metals, which is consistent with the wide difference between the free energies of formation of the respective oxides. The products of reaction in liquid sodium[423] and liquid potassium[444] are given in Table 17.8. The remarkable feature here is that, in spite of the free energy values, the oxides are reduced, even by liquid potassium, to the free niobium and tantalum metals, together with a ternary oxide. An acceptable interpretation is that an atomic rearrangement of the oxide lattices takes place to give two lattices (metal and MO_4^{3-}), which is brought about by the exceptionally high stability of MO_4^{3-} compounds. It is therefore the formation of Na_3NbO_4 and Na_3TaO_4 which is responsible for the appearance of free metal in the reaction products, since the two lattice together are energetically more favourable in the liquid metal environment. It is possible that this disproportionation may involve the meta-compound as an intermediate phase, since it is known (for example) that sodium metaniobate rearranges to $Na_3NbO_4 + Nb$ in liquid sodium.

Some interesting measurements carried out at an early stage in liquid metals research[446,447] showed that the corrosion of niobium and tantalum is dependent on the temperature, oxygen content and flow rate of the liquid

Table 17.8 Reactions of niobium and tantalum oxides with liquid sodium and liquid potassium

Oxide	$-\Delta G_f^\circ$ at 600°C (kJ/g atom O)	Reaction products at equilibrium		
		in sodium (400°C)	in sodium (600°C)	in potassium (400°C)
K_2O	260 (400°C)	—	—	—
Na_2O	292	—	—	—
Nb_2O_5	310	$Nb + Na_3NbO_4$	$Nb + Na_3NbO_4$	$Nb + K_3TaO_4$
NbO_2	315	$Nb + Na_3NbO_4$	$Nb + Na_3NbO_4$	No reaction
NbO	332	No reaction	$Nb + Na_3NbO_4$	No reaction
Ta_2O_5	334	$Ta + Na_4TaO_4$	$Ta + Na_3TaO_4$	$Ta + K_3TaO_4$

Table 17.9   Weight loss for the corrosion of niobium and tantalum in liquid sodium

Metal	Temperature ($^\circ$C)	Oxygen level (ppm)	Weight loss ($mg/cm^2/month$)
Niobium*	500	5	15
	500	15	50
	550	5	25
	550	15	85
	600	5	190
	600	15	575
Tantalum†	500	10	0.7
	500	40	1.8
	550	10	0.9
	550	40	3.8
	600	10	1.4
	600	40	9.3

*Reference 446: sodium flow rate 760 cm/s.
†Reference 447: sodium flow rate 6 cm/s.

metal, and the effect of some of these variables is shown in Table 17.9. The weight loss figures do not have any individual significance, as they represent loss in weight after the samples have been washed in water, which may also remove corrosion product; but they do give a good indication of the extent of attack. The corrosion rate is clearly increased by a rise in the oxygen level of the sodium, and by increases in temperature and flow rate of the liquid sodium past the solid metal specimen. These effects, in themselves, suggest that corrosion of niobium and tantalum involves the formation of a binary or ternary oxide on the solid metal surface. The surface of plates of these two metals after immersion in sodium have now been examined by X-ray diffraction, and the corrosion products identified.[445] At temperatures below 300°C, no corrosion product was found on the surface of either metal, irrespective of the oxygen level in the sodium, and this has been confirmed by other workers. Above this temperature the corrosion rate increased significantly, and the corrosion products consisted of the compounds Na_3NbO_4 and Na_3TaO_4, in either cubic or orthorhombic forms, and no other compounds.

Perhaps the most interesting feature is that these ternary oxides are formed even at oxygen levels as low as 5 ppm, which is in marked contrast to the behaviour of vanadium. The first stage in the corrosion mechanism must again be solution of oxygen in the solid metal, but the absence of monoxide in the corrosion product indicates that the ternary oxides are more stable. Oxygen dissolution will therefore not proceed up to the composition of the binary oxide, and the formation of the ternary oxide becomes the primary corrosion process.

In Figure 17.1 we illustrated the general point that dissolved oxygen in the solid niobium–liquid lithium system will tend to concentrate in the liquid metal, whereas in liquid sodium oxygen migrates to the niobium surface. This accounts for the formation of the ternary oxide corrosion product discussed above, but further confirmation has now been obtained by examination of the metal surface underneath the layer of corrosion product;[436] this contained between 1.15 and 2.08 atom % of oxygen in solid solution in the metal, which is much greater than the amount (0.06 atom %) originally present in the niobium metal before exposure to sodium. When corresponding samples of niobium metal were exposed to liquid lithium, no oxide film is formed, but X-ray diffraction of the metal surface showed that the lattice parameter had decreased during immersion. A microhardness traverse along a cross-section of the sample showed a smooth decrease in hardness towards the edge, for a zone of thickness 200 μm. These results are consistent with the removal, by liquid lithium, of the oxygen in solid solution in the niobium to a depth of 200 μm, with the edge of the sample being almost totally depleted of oxygen.

The formation of binary and ternary nitrides of vanadium and niobium at the surface of these metals immersed in liquid lithium containing nitrogen has been referred to in sections 17.2 and 17.3, and in Table 17.1.

Chromium, molybdenum and tungsten

There has been particular interest in the corrosion of chromium, which stems from its incorporation into stainless steels, the favoured materials for containing the liquid alkali metals. When the oxide Cr_2O_3 is added to pure lithium, oxygen is transferred to the liquid lithium, leaving chromium metal, and this proves to be a useful starting point for further reactions involving nitrogen. The chromium–nitrogen reaction occurs readily, and can be followed by change in electrical conductivity of the liquid metal. In 1980, Calaway[448] showed that the resistivity of liquid lithium containing dissolved nitrogen, at 500°C, decreased whenever a sample of chromium metal was lowered into the liquid, and a dark, crystalline material rich in nitrogen was formed at the solid metal surface. Calaway suggested that this might be the ternary nitride Li_9CrN_5, but gave no confirmation of this. In 1983, results were published from the Nottingham laboratories[278,129] which showed conclusively that the corrosion product was indeed Li_9CrN_5, and that this is the most important compound in the Li–Cr–N system. Small successive quantities of nitrogen were added to a suspension of chromium metal in liquid lithium, and the resistivity changes in the liquid were monitored. At 475°C, nitrogen was absorbed rapidly, firstly into solution in the lithium, then on to the chromium surface as corrosion product. The resistivity was unchanged until a point was reached (which depended on the amount of chromium present) at which it suddenly began to increase rapidly at a rate characteristic of the known resistivity coefficient for nitrogen in lithium (see Chapter 8). Both the position at which this break occurred in the resistivity/nitrogen addition curve, and subsequent isolation and examination of the corrosion product, showed it to be the ternary nitride Li_9CrN_5, so that the complete reaction can be represented by the equation

$$5Li_3N + Cr \longrightarrow Li_9CrN_5 + 6Li$$

At 475°C the formation of Li_9CrN_5 decreases the nitrogen content of the lithium to below 140 ppm which is equivalent to a nitrogen activity of 2.5×10^{-3} in the liquid lithium. This is much lower than the concentrations necessary for the formation of the binary nitrides Cr_2N and CrN, so that the ternary nitride is a more thermodynamically stable compound than either of the binary nitrides. Some relevant activity values are given in Table 17.10.[449,278]

The presence of carbon in the liquid lithium also influences the processes occurring at the metal surface. The binary carbides $Cr_{23}C_6$, Cr_7C_3 and Cr_3C_2 have been obtained from the reaction of the oxide Cr_2O_3 with liquid lithium containing added carbon. These carbides are all more stable than Li_2C_2, and hence will be expected to form at very low carbon levels in the liquid lithium (Table 17.10).

So far as the corrosion of chromium in liquid sodium is concerned, nitrogen and carbon are less important because they have very low solubilities in liquid sodium; oxygen now becomes the significant impurity, and the chemistry of the Na–Cr–O system is dominated by the ternary oxide $NaCrO_2$ (sodium chromite). Another compound Na_4CrO_4 is formed in the presence of large quantities of oxygen (as in the solid-state reaction of chromium metal and sodium oxide), but this compound dissociates to give $NaCrO_2$ in an excess of liquid sodium.[450] All the chromium oxides yield chromite when added to liquid sodium:[421]

$$2Cr_2O_3 + 3Na \longrightarrow 3NaCrO_2 + Cr$$

$$CrO_2 + Na \longrightarrow NaCrO_2$$

$$CrO_3 + 3Na \longrightarrow NaCrO_2 + Na_2O$$

Most investigations involving corrosion of chromium have dealt with the metal as a constituent of an alloy, and this is referred to again below. In static sodium, the rate of corrosion of chromium is very small. In experiments in

Table 17.10 Activities of carbon or nitrogen with which carbides or nitrides of chromium are in equilibrium at 627°C with solutions of the non-metal in liquid lithium

Compound	Activity
Cr_2N	0.11
CrN	0.20
$Cr_{23}C_6$	1.3×10^{-3}
Cr_7C_3	2.9×10^{-3}
Cr_3C_2	4.4×10^{-2}

which chromium was exposed to sodium containing 110 ppm of oxygen for 1000 h, the metal showed either slight or no weight gain.[451] Discs of chromium metal were immersed at 600°C in sodium containing 6200 ppm oxygen for 2 days, or at an oxygen level of 20 ppm for 7 days, and the surface of the chromium discs examined by X-ray diffraction while still covered by a layer of sodium.[452] In each case the only corrosion product present was $NaCrO_2$. This chromite was much more adherent than corresponding compounds on other metal surfaces; it was difficult to remove by water washing, though rapidly flowing liquid sodium can cause its removal by erosion. There was no significant change in the lattice constant of the chromium metal underlying the chromite layer, which is consistent with the very low solid solubility of oxygen in chromium. No binary oxides were observed on the metal surface during the corrosion experiment, and it would seem that the direct formation of sodium chromite:

$$2Na_2O + Cr \longrightarrow NaCrO_2 + 3Na$$

is the only corrosion mechanism associated with this metal in liquid sodium. Thermodynamic calculations[453] show that $NaCrO_2$ can form in sodium containing as little as 5 ppm of oxygen and at temperatures up to 850°C, and there is experimental evidence that the product is also stable at all oxygen concentrations up to saturation.

It will be convenient to discuss tungsten next, since the behaviour of molybdenum is unusual. In liquid lithium, the oxide WO_3 is reduced to tungsten metal. Research on the reactivity of the metal with nitrogen and carbon dissolved in the liquid lithium is very limited; but it would appear that, in contrast to chromium, tungsten is reluctant to form a binary or ternary nitride under these conditions. Carbide formation, however, does occur.[454] In liquid sodium, the reactions of tungsten oxides again give a good guide to the nature of the corrosion products formed at the metal surface. The reaction scheme can be represented as follows:

$$WO_3 + Na_{(l)} \xrightarrow{\;<400\,°C\;} W + Na_2WO_4$$

$$700°C \text{ (vacuum)} \qquad \text{in Na } (600°C)$$

$$WO_3 + Na_{(l)} \xrightarrow{\;600°C\;} W + Na_3WO_4$$

Free energy values would not support the reduction of the trioxide by sodium to free tungsten metal. As discussed above for niobium and tantalum, this is probably brought about as a result of a disproportionation for which the high thermodynamic stability of the WO_4^{n-} group is responsible. Two ternary oxides can be formed, and are interrelated as shown. Additionally, the solid-state reaction between sodium monoxide and tungsten powder at 360°C and above yields another ternary oxide Na_6WO_6. Corrosion experiments have been carried out in which the corrosion product on the surface was identified by its X-ray diffraction pattern recorded through a matrix of sodium; the corrosion is strongly influenced by the concentration of oxygen in the liquid metal. After immersion of a plate of tungsten in a flowing sodium loop at 600°C for 5 days, at an initial oxygen level of 50 ppm, the surface film consisted of the cubic ternary oxide Na_3WO_4. In static sodium at very high (>3000 ppm) oxygen levels, again at 600°C for 5 days, the surface film consisted of orthorhombic Na_6WO_6.[439] At the tungsten surface, it seems likely that the reaction of dissolved oxygen to give a ternary oxide, which acts as a protective layer, is a mechanism which is preferred to that which involves initial penetration of oxygen into the tungsten lattice to give a solid solution. The lattice parameter of the metal immediately below the corrosion product layer was little changed after the corrosion experiments, which also agrees with the non-formation of sub-oxides and oxygen solid solutions.

The behaviour of molybdenum is exceptional in that there is not the close correlation between corrosion and the known chemistry of oxygen and nitrogen impurities which has been such an important feature of the metals discussed above. The partition of oxygen between molybdenum oxides and liquid lithium lies entirely on the lithium side, so that on immersion in pure lithium the oxides are reduced to metal. In lithium containing nitrogen no nitride is formed even at nitrogen concentrations up to saturation, which is in marked contrast to the behaviour of chromium. In lithium containing carbon, however, the carbide Mo_2C forms readily, even in the presence of traces of carbon only, and the corrosion chemistry of molybdenum in both lithium and sodium can be said to be dominated by the compound Mo_2C. The existence of ternary oxides in the Na–Mo–O system is well established, and can be represented as follows:

$$MoO_2 + Na \xrightarrow{400°C} NaMoO_2 \xrightarrow{600°C} Mo + Na_4MoO_5$$

$$MoO_3 + Na \xrightarrow{400°C} Mo + Na_2MoO_4 \xrightarrow{600°C} Mo + Na_4MoO_5$$

In spite of this, there is no evidence for the corrosion of molybdenum by the formation of any oxide films, in sodium containing oxygen levels from 5 ppm up to saturation. In corrosion tests[430] in both static and flowing liquid sodium, corrosion of molybdenum was independent of oxygen concentration. Weight loss measurements over a 2-week period at 600°C indicated no measurable weight change, and the molybdenum metal lattice parameter was virtually unchanged. However, when the liquid sodium was doped with carbon, the carbide Mo_2C was identified on the solid metal surface. It seems possible that this carbide, which forms so readily, creates a barrier at the molybdenum surface, and prevents the reaction of dissolved oxygen which would otherwise take place.

Manganese

This is an instance in which the reactions of the bulk oxides with liquid sodium might be expected to give a fairly clear indication of the nature of the corrosion products, although no specific corrosion experiments appear to have been carried out. Table 17.11 shows the reaction products (which have been identified by X-ray powder diffractometry) obtained when the various oxides of manganese react with sodium under different conditions.[411] We note that at low oxygen levels, the same products are formed at 400°C and 600°C, and that the ternary oxide $NaMnO_2$ is prominent. In sodium saturated with oxygen, the oxide MnO is converted to $NaMnO_2$, and at 600°C a further ternary oxide $Na_4Mn_2O_5$ appears in the product. Given the appropriate oxygen level, the compounds MnO, $NaMnO_2$ and $Na_4Mn_2O_5$ are each stable at 600°C in sodium. It seems reasonable to expect that when the surface of manganese metal is corroded by sodium, the corrosion film will consist of $NaMnO_2$, whatever oxide is originally present at the manganese surface, and that $Na_4Mn_2O_5$ will only be formed at the highest temperatures and oxygen concentrations. Using the approach outlined in section 17.4 and Table 17.4, the estimated range for $-\Delta G_f^\circ$ of $NaMnO_2$ is 652–667 kJ mol^{-1} at 627°C.

Iron

All the oxides of iron are reduced to the metal by liquid lithium. Any oxide film on the surface of an iron specimen will therefore be removed on immersion in lithium, and corrosion of iron in liquid lithium will be due to solution of iron in the liquid metal. The presence of dissolved nitrogen in the liquid can result in nitriding of the iron surface; there is a known ternary nitride Li_3FeN_2 (Table 17.1) which might appear as a corrosion product, but experiments to date[411] have not identified this product, and the iron surface is not severely affected even at high nitrogen levels. There is also a negligible uptake of carbon from solution in liquid lithium on to the pure iron surface. This is consistent with the very low stability of the carbide Fe_3C, which is often used as an alternative to graphite as a carbon source in the preparation of solutions of carbon in liquid lithium.

When considering the corrosion of pure iron by liquid sodium, we have available a full range of experimental evidence, i.e. laboratory experiments on oxide reactions, laboratory tests on corrosion products and large scale experiments using flowing sodium in mild steel loops. The interest in this system arises mainly because the corrosion of stainless steel can be better understood if the behaviour of pure iron is already known. All the oxides of iron are reduced by liquid sodium to the metal and sodium oxide, and these are the only products formed at temperatures below 550°C. Above 550°C, a further reaction takes place in sodium having a high oxygen content with the formation of a ternary oxide of composition Na_4FeO_3. This closely resembles the compound $FeO(Na_2O)_2$ which Horsley[456] first identified in 1956 as a

Table 17.11 Products of reaction of manganese oxides with liquid sodium

Temperature (°C)	Oxygen content of sodium	Reaction product with			
		MnO	Mn_3O_4	Mn_2O_3	MnO_2
400 and 600	5–10 ppm	MnO	$NaMnO_2 + 2MnO$	$NaMnO_2 + MnO$	$NaMnO_2$
400	Saturated	$NaMnO_2$	$NaMnO_2$	$NaMnO_2$	$NaMnO_2$
600	Saturated	$NaMnO_2$	$NaMnO_2 + Na_4Mn_2O_5$	$NaMnO_2 + Na_4Mn_2O_5$	$NaMnO_2 + Na_4Mn_2O_5$

brown scale on iron which had been immersed at 800°C for 7 days in sodium having a high oxygen content. Horsley's compound has been at the centre of discussions on mass transfer of iron in flowing sodium circuits even since that date; its role in corrosion is acknowledged as important, but is still not understood. It would appear that the formation of this compound in sodium depends essentially on oxygen concentration. Calculations based on enthalpy and free energy values indicated that an oxygen concentration of 1000 ppm was needed for the formation of Na_4FeO_3 as a separate phase at 600°C,[457] and this agrees with experimental evidence. Below 550°C the solubility of oxygen in sodium is too small, and the most stable phases thermodynamically are sodium oxide and iron metal. Above 550°C the oxygen level at saturation rises above 1000 ppm, and Na_4FeO_3 is formed. No other ternary oxides of iron have been observed in this system so long as excess sodium is present, so that the only oxidation state of iron which is stable in the sodium environment is Fe^{2+}. There is another ternary oxide $NaFeO_2$ (sodium ferrite) which can be prepared by the solid-state reaction of sodium oxide with iron metal, so that under these conditions oxidation to the Fe^{3+} state can occur, but $NaFeO_2$ has no significance in liquid sodium.

Results of laboratory corrosion experiments[452] were in full agreement with the above reactions. Samples of pure iron were exposed to static sodium at 600°C for 14 days, and the surfaces then examined by X-ray diffraction. At oxygen concentrations in the range 0.4–3.3 wt % the compound Na_4FeO_3 was identified as a dark-brown layer on the iron plates; at 50 ppm oxygen in the sodium, no corrosion product formed on the plate surfaces. Oxygen has a very small solid solubility in iron, and this was reflected in the lattice parameter measurements, which remained virtually unchanged during the corrosion.

It is clear, then, that at high oxygen levels in sodium, a product is formed in which there is strong chemical bonding between iron and oxygen. However, effects have also been observed which suggest that some form of iron–oxygen bonding may also be important at low oxygen concentrations, and this has been highlighted by some recent definitive experiments carried out on a technical scale by Thorley at UKAEA, Risley.[458] Pure iron specimens were inserted into the well of a loop constructed from mild steel (i.e. iron containing some carbon but no alloying elements), through which sodium flowed at a rate not less than 15 feet per second, and at an operating temperature between 624°C and 650°C. By this means the corrosion in both flowing and static sodium could be observed. After 7 days, in sodium containing less than 100 ppm of oxygen, the surfaces of the specimens and interior walls of the loop on the 'hot' side was clean, and typical of surfaces corroded by dissolution of iron into the sodium. The small amount of debris on the 'cool' side contained single crystals of iron, but no binary or ternary oxides; this is therefore an example of mass transfer by solution of iron at the hot surface, and subsequent deposition at the cooler regions.

However, it is the role of oxygen which cannot be interpreted in purely physical terms, and any explanation must take into account the following observations. Firstly, the rate of corrosion is strongly influenced by oxygen concentration; the rates of loss of iron were found to vary from 1×10^{-3} inches/year at 10 ppm of oxygen to 100×10^{-3} inches/year at 100 ppm oxygen. Secondly, the solubility of iron in sodium (measured simultaneously) showed no increase with increasing oxygen content. Thirdly, there was no decrease in the oxygen concentration in sodium during the experiments, so that none was being taken up by the iron surfaces.

These observations, which are apparently contradictory, can be reconciled if we assume that in solution the iron atoms are solvated by oxygen ions, to give a weak solvation complex, so that dissolved oxygen acts as a carrier for iron. It has also been pointed out[458] that, according to the law of mass action, increasing the chemical activity of oxygen in the sodium to form the complex would lower the activity of iron in the sodium, and thus increase the driving force for transport of iron from the metal surface.

17.6 Stainless steel

This subject is of major importance technologically, since steels of various compositions constitute the main materials used for containment of the liquid alkali metals. There has therefore been extensive research covering such aspects as variation in corrosion rates with proportions of major constituents, the pronounced influence of minor metal additives, mass transport, the composition of corrosion films, tribology, liquid metal flow rate, temperature and so on. Many of these aspects are mechanical in nature and outside the scope of this volume, and in the present context we will only point to some of the chemical principles which have been found to influence the corrosion of steel. The proceedings of the international conferences mentioned at the beginning of this chapter provide an up-to-date overview of the present state of the art, and ample references. Most of the steels used in these investigations are represented by types 316 and 321, which contain (in wt %) 70–75 Fe, 15–16 Cr, and 8–12 Ni, together with 0.5% Ti (type 321) or 2.0% Mo (type 316) and small quantities of silicon, manganese and carbon.

In liquid lithium

Because of its potential use as a blanket and heat transfer material in the fusion reactor, the compatibility of liquid lithium with steel has been examined under a variety of experimental conditions. In comparing the many published accounts of corrosion by lithium, it is important to remember that observations made on steel surfaces after exposure to flowing lithium are likely to be different, at first sight, from those obtained using static liquid lithium. In the latter case, corrosion products can collect on the surface and be identified (and are therefore more useful in studying mechanisms of corrosion), whereas in flowing lithium these products are swept away by erosion, but the corrosion rates are more important from a technological point of view.

In a general review, DeVan[459] has commented on some of the features observed in flowing lithium which has been gettered to reduce nitrogen and carbon levels. Corrosion is due essentially to dissolution of the major constituents into lithium, and can reach a steady state of about 12 μm per year at 600°C. Chromium and nickel are depleted from the hotter steel surfaces, with consequent enrichment of iron, and a steady-state surface composition is eventually set up when the iron, chromium and nickel solution rates are proportional to their concentrations in the bulk steel. At the same time, there is a transformation from austenite to ferrite (α-iron) in the depleted surface layer, which can reach a steady-state thickness of 20–60 μm within 3000 h at 600°C. Chromium has been found to deposit in the colder limb of the loop in the form of needle-like crystals, which is consistent with the fact that its solubility in lithium is much smaller than that of nickel. The presence of nitrogen in the liquid lithium increases both the rate of weight loss of steel and the intergranular attack by lithium. The addition of small amounts of calcium effectively neutralizes this nitrogen-induced corrosion, so that the solvation of the N^{3-} ion in solution is apparently able to reduce its chemical reactivity. Aluminium also shows promise as an inhibitor; as well as acting as a nitrogen scavenger, aluminium reacts with nickel to form a protective intermetallic coating on the steel surface.

Even after the most stringent purification procedures, lithium still contains significant levels of nitrogen and carbon impurities, and during the operation of a technical scale plant these impurities could become appreciable. The effects of carbon and nitrogen on the corrosion resistance of type 316 steel to liquid lithium have therefore been examined with the object of elucidating the corrosion mechanism.[279] Steel plates were immersed in liquid lithium contained in crucibles of the same steel, which were then sealed in a steel vessel and heated to 600°C. The nitrogen level in the liquid lithium was determined by addition of lithium nitride. In the case of carbon, unit activity (i.e. saturation) was achieved by addition of either graphite or the carbide Fe_3C; carbon activities less than unity were established by the use of a metal–metal carbide couple (Mo/Mo_2C, W/W_2C or $Cr/Cr_{23}C_6$) which set up an equilibrium level of carbon in the lithium.

Specimens exposed to nitrogen-contaminated liquid lithium became coated with a fine brown powder which gave the distinctive X-ray powder diffraction pattern for the ternary nitride Li_9CrN_5. This compound was the only nitride present at the surface; it was formed at very low nitrogen levels, and was formed in preference to the other possible ternary nitride Li_3FeN_2. On analysis of this surface film, the Cr:Fe ratio was found to be 400:1, so that the film is formed as a layer quite separate from the underlying metal. Metallurgical examination of steel specimens also confirmed the enhanced grain boundary penetration which was observed with flowing lithium. Steel surfaces exposed to lithium containing carbon at an activity of 0.02 showed the formation of a chromium-rich carbide $M_{23}C_6$ and a ferritic layer coexisting with the austenite. The degree of carbide formation increased with both time and carbon activity, and after exposure to unit carbon activity for 672 h the surface became a mixture of ferrite and carbide only.

The two main products which are formed in the corrosion of stainless steel type 316 by liquid lithium containing nitrogen or carbon are the ternary nitride Li_9CrN_5 and the chromium-rich carbide $M_{23}C_6$. Both carbon and nitrogen diffuse into the steel, but they differ in that the carbides can coexist with the austenite and ferrite at the surface, and precipitate in the grain boundaries behind the surface, while Li_9CrN_5 appears to form only as a surface layer. As a result, chromium is depleted from the steel surface in nitrogen-containing lithium, but chromium is enhanced near the surface of the steel in carbon-containing lithium owing to the precipitation of chromium-rich carbides.

In liquid sodium

This is the area in which there has been the greatest amount of research effort, in view of its obvious and immediate application in fast nuclear reactors. Much of the work preceded that on corrosion by liquid lithium, but further research is necessary, and important contributions were still being published in 1982.[460] In contrast to lithium, nitrogen now plays a negligible role in the corrosion of steel in view of its insolubility in liquid sodium, and oxygen is the important impurity. Carbon is also important, in spite of its very low solubility in liquid sodium. It is again possible to compare the results of laboratory experiments using static sodium, and technical scale experiments with flowing sodium.

It is largely true that the corrosion of steel by liquid sodium is the sum of the separate processes already discussed for the single elements iron, chromium and nickel; what is important, especially in the reactor environment, is the relative speeds at which these processes take place. In 1974, Barker and Wood[452] published the results of experiments in which steel specimens were immersed in static liquid sodium having a high (0.7–4.3 wt %) content of oxygen. After 14 days at 700°C, the surfaces were examined by X-ray diffraction, when the ternary oxides $NaCrO_2$, Na_4FeO_3 and α-iron were identified at the surface. The relative positions of the layers of these materials were such as to indicate that sodium chromite was formed first, and this was followed by an underlying layer of sodium ferrate which was produced by reaction of oxygen with steel which had been denuded of chromium. Below this, there was evidence of a phase transformation from γ- to α-iron. At low oxygen levels (10–15 ppm) Na_4FeO_3 is not stable in sodium, so that only $NaCrO_2$ is formed at the surface. We can imagine, then, that in flowing sodium and at low oxygen levels (which are the conditions normally attained in sodium coolant assemblies) the sodium chromite layer would be wholly or partly removed by erosion, leaving an iron–nickel alloy at the surface. Nickel is then removed differentially by dissolution in the sodium, leaving an outer ferritic layer which would corrode in a manner similar to that described above for pure iron.

Different steels vary in their carbon content, and mass transfer of carbon from one to another is a major technological problem. Carbon will transfer from an alloy having high carbon potential to one of lower carbon potential, using liquid sodium as the transfer medium, and this alters the physical properties of the steels as they become carburized or decarburized. Type 316 stainless steel shows similar behaviour towards carbon in both sodium and lithium, in that the carbide $M_{23}C_6$ is formed in both liquids. Carbon transfer is controlled kinetically as well as thermodynamically. Thus, when addition of carbon to a steel is occurring, an equilibrium carbide is formed at the surface, but the diffusion of carbon into the steel matrix will be affected by the microstructure of the steel. Similarly, the removal of carbon from high-carbon steels is time-dependent in that, after the surface has been depleted of carbon, the process is diffusion controlled. It has been possible to maintain a constant carbon activity in liquid sodium by the addition of metal–metal carbide couples such a W/W_2C, and hence to determine diffusion coefficients for the diffusion of carbon into steel.[436]

The several processes discussed above which contribute to the overall corrosion will not take place in an orderly or consecutive fashion, and will certainly interfere with one another, so that the nature of the corroded surface will depend on the physical and chemical environment of the surface. It is difficult to test this on a laboratory scale, but in a loop of flowing sodium the steel surface can be submitted to a wide range of conditions. In a paper which makes a major contribution to our understanding of the behaviour of corrosion products in a sodium loop, Thorley et al.[460] describe experiments in which 321 and 316 type steel samples were inserted into a sodium loop at various points, and their surfaces examined after 30 and 100 weeks. The liquid metal was circulated at a velocity of 3 m/s at temperature ranges of between 675°C and 410°C, and the oxygen level maintained at either 10 or 25 ppm. They found that the loop may be divided roughly into zones where corrosion, oxidation and deposition occurs. Corrosion (meaning the removal of corrosion products) occurs in the high-temperature zone, and ferrite layer formation is a general feature. The depths to which these layers form depends on temperature, oxygen level, sodium velocity and material composition. Away from this high-temperature corrosion zone, surface oxides of the sodium chromite type (together with carbide) are observed; both surface and sub-surface oxidation can occur and this causes some break-up of the surface, when mechanically unstable corrosion products are transported to the cold zone. At 10 ppm oxygen level, and in the lower velocity sections of the loop, the oxidized surface layer was still intact after two years' operation. When the oxidized layer has been removed, the transformation of the exposed austenite to ferrite depends on the rate of release of nickel at the high-temperature zone upstream. There is therefore a pleasing consistency between results obtained in laboratory and technical scale investigations.

Corrosion studies have been extended to alloys containing higher proportions of nickel (the 'nickel-based' alloys). Inconel 600, for example, contains the major constituents in the proportions 77 Ni, 16 Cr and 6 Fe (wt %). Corrosion of these alloys is controlled by the same chemistry as for the iron-based alloys, but the different composition is reflected in corrosion rates. Whereas iron transfer is influenced by oxygen content of the sodium, this is not the case with nickel; on the other hand the higher dissolution rate of nickel becomes important in the nickel-based alloys. From a survey of such alloys containing 10–77 wt % Ni, it has been established that the corrosion rate increases with nickel content. However, when the nickel content of the alloy is reduced below about 45 wt % the nickel content plays a less significant role in determining the corrosion rate. The corrosion rate then becomes increasingly sensitive to the oxygen level in the sodium as the iron content is increased, and a stage is reached when the effects of changing oxygen level in the sodium outweigh the effects of further reduction in nickel concentration.[461]

Chapter 18

Modern applications of the liquid alkali metals

In 1956 Marshall Sittig[480] gave an account of some actual and potential applications of metallic sodium, based on the unique physical properties of the metal. Since that time there has been considerable development in the uses and applications of the alkali metals; this chapter describes some modern applications, selecting in particular those which involve the use of large quantities of liquid metal.

18.1 Liquid sodium in the fast nuclear reactor

There are two aspects to the development of fast nuclear reactors which are of interest here. Firstly, the use of liquid sodium on a very large scale has illustrated how liquid metals can be applied successfully to purposes for which no other fluid could be used. Secondly, this development has provided a welcome stimulus for the study of the properties of this and other liquid alkali metals, which in turn will provide the scientific background for other major applications in the future.

In the first generation of nuclear reactors, neutrons emitted from the fuel are slowed down to thermal energy levels by the use of moderators such as graphite, and heat is carried away from the reactor core by a stream of carbon dioxide gas. As the name implies, the neutrons in a fast nuclear reactor are not slowed down by moderators, and when the possibility that such a reactor might be technically feasible was first considered in Britain in the early 1950s, it quickly became apparent that higher reactor temperatures would be involved, that the output of heat per unit volume of core would be vastly greater than in thermal reactors, and that gas cooling could not cope with the higher rate of heat removal. The basic requirements for a fast reactor coolant are therefore that it should be capable of removing a large quantity of heat from a small volume and be reasonably cheap, that it should not act as a moderator or absorb neutrons to a significant extent, that it should not react chemically with other reactor components, and that it should not involve the need to introduce a pressurized system. Water is unacceptable on many of these counts, and only a liquid metal can satisfy all requirements. Table 1.2 gave some relevant physical properties of water and mercury for comparison, and it is clear that mercury also fails to qualify, if only because of its low boiling point and high density.

Sodium, however, does appear to possess all the necessary properties. It has a wide liquid range (98–883°C) and a vapour pressure of only 24 torr at 600°C, which was the expected reactor operating temperature. It has also the typical metallic properties of low specific heat, and high thermal and electrical conductivity. The effects of radiation are greatly simplified by the fact that sodium has only one naturally occurring isotope, ^{23}Na. Under neutron bombardment this is converted to ^{24}Na, which decays by β emission to the stable ^{24}Mg with a half-life of only 15 h; so that although sodium would become radioactive in a reactor, this activity would die away quickly if and when sodium was removed from the reactor. Liquid sodium would have to be circulated rapidly over the reactor core, and its low density and low viscosity were further advantages. The only obvious difficulty in the use of liquid sodium as coolant arises from its high chemical reactivity, particularly towards oxygen, hydrogen and water, and there was no doubt that much fundamental research and technical development would be necessary before non-metallic impurities, which are largely responsible for corrosion, could be controlled reliably at a low level.

To test this and other aspects of fast reactor operation, an experimental fast reactor, known as the Dounreay Fast Reactor (DFR), was built for the UK Atomic Energy Authority in the north of Scotland during the 1950s, and first went critical in 1959. As this was the first time that a liquid alkali metal had been handled on such a large scale, it was envisaged that the use of sodium alone, which solidifies at 98°C, might cause difficulties. However, mixtures of sodium and potassium have lower melting points, and at the eutectic composition the alloy is liquid down to −12°C. This alloy was therefore chosen for the experimental reactor and DFR contained 120 t of liquid sodium–potassium alloy.[462] By 1963 this reactor (which was fuelled with uranium metal highly enriched in ^{235}U) had been operated at all available power levels; experience confirmed that the sodium–potassium alloy (termed NaK) performed satisfactorily as a reactor coolant and that its effect on reactor components could be controlled. The initial fears that a liquid alkali metal at high temperature and in large quantity might be altogether too chemically reactive to give a stable system proved to be unfounded. NaK is more reactive than is sodium alone, and can introduce the risk of explosions. After several years' experience in its manipulation, the higher melting point of sodium was no longer a problem, and the decision was taken that future reactors would employ pure sodium as coolant. As Rudolf Dehn, special assistant to the head of the reactor division at Dounreay, commented at the time, 'we haven't merely learned to live with sodium, we've really grown to like it. In its liquid state it's like whisky—lighter than water and altogether better.'

The next stage was the construction of a prototype for the commercial fast reactor.[463] This prototype (termed PFR) was also built at Dounreay. It uses 1150 t of liquid sodium as coolant, and was completed in 1974. Because of the higher temperatures involved (up to 650°C) metallic uranium was no longer suitable as fuel, and ceramic uranium dioxide enriched with plutonium dioxide is used instead; a combination of uranium and plutonium carbides is also feasible. The mixed oxides have low thermal conductivities, and to effect heat transfer to the sodium the fuel is charged, in tablet form, into thin steel tubes, or 'pins'. The PFR core contains 50 sub-assemblies, each consisting of 325 such pins, and the sodium flows through the annular spaces in the pin assembly.

Figure 18.1 shows a diagram of the reactor.[464] There are two separate sodium circuits. In the primary circuit the reactor core, control devices and sodium pumps are suspended in a large pool of liquid sodium; the sodium is circulated through the core and the intermediate heat exchangers (marked A in Figure 18.1). In the secondary circuit (of which PFR has three) sodium is heated in the heat exchangers within the pool, circulated through steam generators, and the cool sodium is then returned to the heat exchangers. By keeping the sodium circuits separate, the radioactive sodium within the pool is isolated from the steam generator circuit. The sodium temperature in the primary pool ranges from about 370°C to 540°C in the region of the core. In the secondary circuits, sodium leaves the heat exchanger at about 510°C, and returns at 340°C. Because of the low density and low viscosity of liquid sodium, very high circulation velocities are possible. The sodium in the primary pool is circulated at about 3000 litres/second, and travels through the

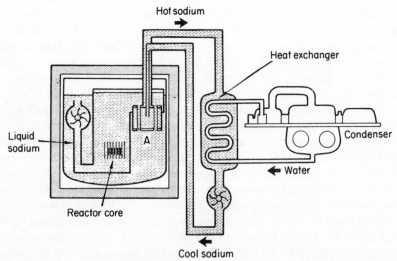

Figure 18.1 Sodium cooled reactor

core channels and heat exchangers at velocities up to 6 m/s. In the secondary circuits, sodium velocities can be as much as 8 m/s. The main constructional material is stainless steel, and argon is used as the cover gas.

Many of the problems encountered in reactor operation stem from the peculiar solvent properties of sodium. With a few exceptions, liquid sodium dissolves most elements to a very small extent (see Chapter 5), but these extremely dilute solutions are of considerable chemical interest. Some aspects have been discussed in earlier chapters, and more detailed accounts will be found elsewhere.[464] Thus, flowing sodium has the ability to transfer alloying elements in steels (notably nickel, carbon, nitrogen and boron) from one, often high-temperature, region to a region where the chemical potential of the alloying element is lower. Oxygen is also an important element because of the role it plays in corrosion (see Chapter 17), and one of the most important lessons learned from the operation of the experimental reactor was that the oxygen level in sodium must be maintained at or below a few parts per million.

One of the universally known properties of sodium is its vigorous reaction with water, and this inevitably draws attention to the steam generators (Figure 18.1) consisting essentially of tubes which have hot liquid sodium on one side, and water or steam on the other. High integrity in the construction of the generator components is obviously called for, but in fact the dangers are not so great as might be imagined, for reasons which have been discussed in Chapter 10. The existence of a leak can be recognized from the presence of hydrogen, which can now be detected both in solution and in the argon gas by quick response meters (see Chapter 15).

Successful operation of the PFR over a period of several years has encouraged the belief that much larger, commercial scale, reactors are possible which can retain most of the design features of PFR. The next stage is therefore the construction of a commercial scale reactor, and planning and design of the first Commercial Demonstration Fast Reactor (CDFR) is well advanced. This reactor will incorporate up to 7000 t of liquid sodium.

18.2 Liquid lithium in the fusion reactor

An increasing effort is being devoted throughout the world to means of tapping the essentially limitless supply of energy which is potentially available from the thermonuclear fusion of light nuclei. This effort is stimulated by the ever-increasing cost, and diminishing supply, of fossil fuels, and by the hope that the eventual introduction of fusion reactors will remove the environmental objections which arise from disposal of nuclear waste from fission reactors. Increasingly sophisticated fusion experiments are being carried out, notably in the USA, USSR, Japan and Europe, and in each case lithium plays an important role. To date, the emphasis has been heavily on the plasma physics aspects of fusion, but in the operation of an actual fusion reactor many of the problems are chemical in nature.

In a plasma at a sufficiently high temperature, a number of fusion reactions are possible involving deuterium, tritium or helium-3 atoms. The reaction which requires the lowest plasma temperature $(40–50 \times 10^6$ K), and therefore the reaction most likely to be exploited commercially, is the deuterium–tritium fusion:

$$^2_1H + {}^3_1H \longrightarrow {}^4_2He \ (+3.5 \ \text{MeV}) + {}^1_0n \ (+14.1 \ \text{MeV})$$

Deuterium, comprising 0.015 per cent of hydrogen in water, is in ample supply, but sufficient tritium for this purpose is not available in nature. However, it can be produced when the ^6Li isotope of lithium is irradiated with thermal neutrons:

$$^6_3Li + {}^1_0n \longrightarrow {}^3_1H + {}^4_2He$$

This isotope is present to the extent of 7.4 per cent in natural lithium metal, and the reaction provides a convenient method of producing the initial tritium content for a fusion reaction.

The key to successful energy production lies in the ability to contain the high-temperature plasma, and to retain within it a high density of fusing particles for a sufficiently long time for the energy of fusion to exceed the energy input required to produce and maintain the plasma. Much of the effort to date has been directed towards magnetic containment of the plasma, i.e. the use of large superconducting magnets for the heating and confinement of the plasmas. The leading magnetic confinement experiment is the Tokamac, which employs a toroidal plasma. More recently, interest has been developing in an alternative approach, the inertial confinement of the plasma in which pellets of fuel are converted to plasma by pulses of laser radiation.

In this rapidly developing field on which the future of the world's energy supply may well eventually depend, it would be pointless to include here a description of any particular model reactor; the interested reader may refer to the many available texts on technical aspects.[465,466,467,481,482] However, some basic features are common to all toroidal reactors, and Figure 18.2 represents part of a cross-section of the reactor.[467] The plasma, at $10^7–10^8$ K, is held within a vacuum, which in turn is contained within the 'first wall', or vacuum wall. This wall consists of carbon (together with other structural material) to moderate neutrons emitted from the fusion reaction down to the thermal energy level. This core lies within the blanket of liquid lithium coolant; outside the blanket is a neutron and thermal shield, which also protects the cryogenic superconducting magnet if magnetic containment of the plasma is involved.

It is with the blanket material that we are concerned here. Materials other than liquid lithium, such as the alloy Pb_4Li and the molten salt Li_2BeF_4, are still under consideration, but liquid lithium is the standard against which other candidates are assessed and continues to be a favoured blanket fluid. Some difficulty arises from the fact that circulation of liquid metal through the strong magnetic fields of the plasma confinement system is impeded by

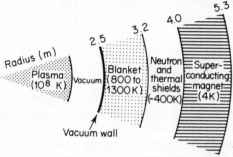

Figure 18.2 Sector of section through breeder blanket (not to scale). (Reproduced by permission of British Nuclear Energy Society)

electromagnetic interactions. Large magnetic fields are not present, however, in the inertial-confinement type of fusion reactor, so that liquid lithium is more attractive as a coolant in this type of reactor than in the magnetic containment type. Because of its wide liquid temperature range, liquid lithium can be used at a peak temperature of about 600°C if linked directly to a steam generator (as is liquid sodium in the fast nuclear fission reactor), or at 1000°C and above if an intermediate potassium topping cycle is employed. A fusion reactor might contain around 1300 t of liquid lithium as blanket and the energy used in pumping this quantity is minimized as a result of its low density and viscosity and its high thermal conductivity (Table 1.2). The very low vapour pressure, even at high temperatures, is also an advantage.

Most of the energy from the plasma is transmitted to the lithium blanket in the form of kinetic energy of the emitted neutrons, so that heat is generated homogeneously by collision with nuclei in the blanket. In addition, however, the blanket serves as a tritium breeder, to provide a continuous supply of tritium for the plasma. The tritium breeding reaction with the $^6_3$Li isotope has already been mentioned, and other reactions involving the more predominant $^7_3$Li isotope can also contribute to the formation of tritium. The unique property of lithium is therefore that it can serve both as a tritium breeder and a heat transfer medium.

Much of the technology developed in the past for the large-scale handling of liquid sodium is also applicable to liquid lithium; but there are some important differences which result from the smaller atomic size of lithium, and the greater stability of many of its compounds with non-metals. In turn, this influences the solubility of non-metals in liquid lithium (see Chapter 5), the methods to be used in purification of the metal (Chapter 3), and the compatibility of liquid lithium with structural materials (Chapter 17). Efforts to produce a viable fusion reactor has stimulated research in many of these areas, and most of these aspects have been referred to in earlier chapters. The equilibrium between lithium and hydrogen, deuterium and tritium is of special importance, since the operation of the reaction requires that the tritium produced by breeding within the liquid metal should be continuously and efficiently removed.

18.3 The potassium topping cycle

In most power stations in which heat energy (e.g. from the burning of fossil fuels, or from nuclear reactors) is converted into electrical energy, the heat source is connected, directly or indirectly, to a steam generator and thence to a turbine. The only practicable materials from which to construct the water–steam cycles are the Fe–Cr–Ni steels, which set an upper temperature limit of about 600°C. Above this temperature the strength of many steels falls away; more serious, however, is the fact that water begins to decompose into its components:

$$H_2O \longrightarrow H_2 + \tfrac{1}{2}O_2$$

above 600°C, particularly at the metal surface. Hydrogen diffuses into the steel, causing embrittlement, and the oxygen produced sets up corrosion. Considerations such as these, as much as the energy balance within the reactor, are responsible for the upper limit of 600°C at which sodium in the secondary circuits leaves the fast reactor (Figure 18.1). This limitation is a serious one, since of course the thermal efficiency of a heat cycle depends *inter alia* on the peak temperature achieved in the cycle. The problem can be overcome by the use of an intermediate cycle, the 'topping' cycle, which is stable at a higher peak temperature. Ideally, the fluid circulating in the topping cycle should have a boiling point in the 600–800°C range, be non-corrosive with respect to iron alloys, and be chemically stable. Potassium, with a boiling point 754°C under atmospheric pressure, is highly suited to this purpose, and a potassium topping cycle might well find use in the fusion reactor. The physical properties of potassium vapour at 850°C are roughly similar to those of steam at 120°C.[468]

If caesium became available in sufficient quantity it could replace potassium in topping cycles with some advantage. It has a lower boiling point (671°C), and thus a higher vapour pressure at any given temperature, than has potassium. The condensing temperature of the metal vapour can be reduced, which again improves the efficiency of the cycle. ('Bottoming' cycles are designed to reduce the condensing temperature.) The higher vapour pressure, combined with the greater atomic weight, of caesium can also lead to simplification of turbine design.

18.4 Sodium and lithium in rechargeable batteries

As a result of the increasingly high cost of fossil fuels, particularly oil, there is little doubt that electrochemical reactions in cells and batteries will play an increasingly important role in the energy economy of the advanced nations. If the fraction of nuclear-based electric power increases, it may also become attractive to store large quantities of electricity in electrochemical batteries. One of the most striking applications of this, which seems likely to become a practical proposition in the not-too-distant future, is the development of motor vehicles and trains powered by electric batteries which would be charged overnight.[483] Such vehicles are silent and free from pollution, and use electric power efficiently.[469] Since 1967 the research division of British Rail have been developing a new generation of batteries specifically for rail traction.[470] So far as road vehicles are concerned, there are already in the UK, for example, about 100,000 electrically propelled industrial trucks and about half this number of commercial road vehicles. These vehicles are slow and designed for short journeys only, and use lead–sulphuric acid batteries. The use of lead presents several problems when considered in terms of the electric car; its electrochemical equivalent is high, and sufficient of the metal may not be available to meet a worldwide demand. The weight, bulk and cost of the lead–acid battery are likely to prove prohibitive, and it has been calculated that for the average family car performance, a lead–acid battery weighing about one ton would be required.

Among the various alternatives, electrochemical cells involving the reaction between a highly electropositive element (such as lithium or sodium) and a strongly electronegative element (such as sulphur or chlorine) show considerable promise, and high energy densities may be achieved. Most attention to date has been paid to the sodium–sulphur and the lithium–chlorine batteries, and these will be discussed here.

The liquid sodium–sulphur battery was first described by Kummer and Weber in 1967,[471] and since then has been under development by Ford and GEC in the USA, in Japan, and at the British Rail laboratories and the Electricity Council Research Centre in the UK. One version of the cell is shown schematically in Figure 18.3. In this 'central sodium' cell, an inner tube contains liquid sodium, and the annular space between this and the outer steel tube contains liquid sulphur. Since the latter is a non-conductor the sulphur phase is made up of sulphur-impregnated carbon felt, which acts as current carrier. The two phases are isolated from one another using an appropriate insulating gasket. The inner tube forms the diaphragm or electrolyte separating the two liquid phases and is permeable to Na^+ ions. Early cells used polycrystalline β-alumina for the diaphragm; this has the formula $Na_2O.11Al_2O_3$, and has a layer structure. Blocks of Al^{3+} and O^{2-} ions in spinel formation alternate with layers of sodium ions, and it is the 2-dimensional layer of Na^+ ions which provides the sodium ion conduction through the diaphragm. Later cells use a second form of β-alumina, of composition $Na_2O.5Al_2O_3$, which has a similar structure but is less stable. However, if doped with Li_2O or M_gO, the structure can be stabilized and a high-density polycrystalline ceramic can be prepared. This has a lower resistivity, and is the preferred diaphragm for sodium–sulphur cells. The stability of this diaphragm under repeated charge and discharge cycles is one

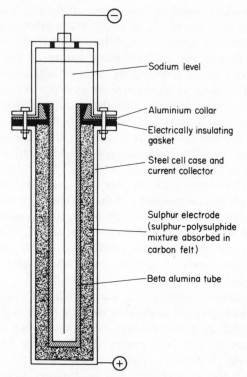

Figure 18.3   Tubular sodium–sulphur cell

of the major factors governing the lifetime of these cells, and is one of the main areas of research and development. Where short-circuits occur, they have been found to be due to penetration of the ceramic by sodium metal along flows, and microcracks or grain boundaries in the diaphragm.

In the 'central sodium' cell, a problem arises from the corrosion of the outer steel tube by sodium polysulphides, with which the steel is in contact for long periods at high temperatures. This can be avoided if the sulphur phase is located inside the diaphragm tube (i.e. the 'central sulphur' cell). This need not be described in detail here; both types of cells, and also a flat-plate model, have been discussed by Sudworth[472] and their relative merits compared. The cells are about 25 cm in length, and the cell e.m.f. lies between 2.08 V and 1.78 V. For rail traction, the required energy content can be obtained by connecting many cells in parallel, and the operating voltage obtained by connecting appropriate groups of cells in series.

The cell is operated at 300–350°C, at which temperature both sodium and sulphur are liquid. The cell reaction is written as

$$2Na + 3S = Na_2S_3$$

On discharge, sodium is oxidized to Na^+ ions, which migrate through the β-alumina diaphragm and combine with sulphide ions produced by reduction of sulphur at the β-alumina–sulphur interface. The first product is the pentasulphide Na_2S_5, but as the sodium concentration increases, the whole of the sulphur phase can be converted to the trisulphide Na_2S_3 while maintaining this phase in the liquid state. This is the discharge limit, as any discharge beyond this stage produces the disulphide Na_2S_2 which is insoluble at operating temperatures and adversely affects the performance of the electrode.

The specific energy (or energy density) of sodium–sulphur cells is high. The theoretical value is 750 W h kg^{-1}, and cells of specific energy in the range 100–200 W h kg^{-1} have been constructed; this compares with values of around 30 W h kg^{-1} for the lead–acid battery. Against this, there is the problem of maintaining an operating temperature of 350°C, and the danger that the actual specific energy might be reduced by the weight of temperature control equipment. Clearly, the battery should be kept at the high operating temperature for long periods of time, and this is why rail traction, rather than intermittent automobile use, is the more attractive application in the first instance.

Corresponding cells containing liquid lithium and sulphur have not been developed to the same extent, in spite of the higher specific energy which is theoretically possible. The cell reaction

$$2Li + S \longrightarrow Li_2S$$

for which ΔG°_{298} is -434 kJ mol^{-1}, gives a theoretical specific energy of 2624 W h kg^{-1}, but there are many more problems in handling liquid lithium than with sodium, owing to its higher reactivity. This applies particularly to the diaphragm, and in a typical liquid lithium–sulphur cell the lithium is absorbed in spongy iron, the sulphur is absorbed in carbon felt, and these two phases are separated by a molten salt eutectic mixture of LiCl/KCl. The latter is immobilized in an inert ceramic powder of magnesium oxide; the cell is operated at 420–460°C, at which temperature the vapour pressure of sulphur is uncomfortably high (0.243 atm). Somewhat lower temperatures are possible using ternary electrolytes such as LiCl/KCl/KI, or LiCl/LiBr/LiI. The severity of the practical problems can be reduced by reducing the reactivity of the lithium and the sulphur. To this end, cells are being developed in which the lithium is alloyed with aluminium or with silicon, and the sulphur is combined with a metal to form sulphides such as FeS or FeS_2.[473] With such cells the cell e.m.f. is not changed appreciably, though the specific energy may be considerably reduced.

Lithium–selenium batteries, operating at 375°C, and lithium–tellurium batteries, have been under investigation by the Argonne National Laboratory.[474] Though rather more speculative at present, they show promise for medical as well as vehicle propulsion uses. One big advantage is that they can be recharged in minutes rather than hours.

The lithium–chlorine battery is light in weight, and is unsurpassed in the high level of energy and power density that can be achieved.[475] Available power densities of 1000 W kg^{-1}, i.e. 20 times that of the lead–acid battery, are regarded as possible. The cell can be written

$$Li_{(l)} \,|\, LiCl \text{ at } 650°C \,|\, \text{porous graphite} \,|\, Cl_{2(g)}$$

and a diagram of the cell, which is purely schematic, is shown in Figure 18.4. In earlier models, liquid lithium floated on the surface of molten lithium chloride. In later models the liquid lithium is supported by a stainless steel or nickel wire mesh, both of which are wetted by the liquid metal. The chlorine gas electrode is made from porous graphite. When lithium chloride alone is used as electrolyte, the operating temperature must be above its melting point (614°C). The fact that seals and insulators are being attacked by elements on each end of the periodic table simultaneously and at this high temperature raises many problems and is delaying the adaption of this cell as a practical proposition.

One problem of immediate chemical interest arises from the solubility of lithium metal in the lithium chloride melt. This dissolved metal can reach the graphite electrode, where it forms the acetylide Li_2C_2. This induces wetting of the graphite by the normally non-wetting melt; the melt then penetrates the graphite electrode, and restricts the diffusion of chlorine. In a more recent development,[476] a mixture of alkali halides (LiF/LiCl/KCl) is used as electrolyte. This allows the cell to be used at 450°C, and reduces the solubility

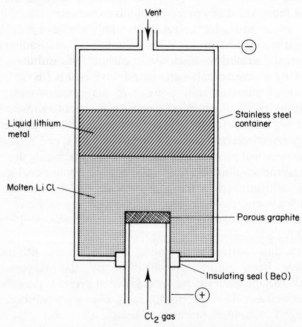

Figure 18.4   Lithium–chlorine cell

of lithium metal in the electrolyte. Such cells have been operated successfully for over 600 h. In a further modification, the liquid lithium has been replaced by a lithium–aluminium alloy.[477] Owing to the low activity of lithium in this alloy, the available concentration of lithium metal in the electrolyte is reduced by a factor of 10^2 over that in liquid lithium. Using LiCl/KCl eutectic mixture as the electrolyte, the operating temperature can be reduced to the electrolyte melting point (352°C), with corresponding reduction in corrosion.

The concept of a lithium–fluorine cell is a simple and attractive one, but its realization as a practical proposition is even further in the future. Nevertheless, fluorine handling is now highly developed and methods of handling lithium are improving. Lithium fluoride would be satisfactory as an electrolyte because of its high electrical conductivity, and the use of eutectic fluoride mixtures could bring the operating temperature below 500°C. Optimism is justified also by the high open-circuit voltage of the lithium–fluorine cell (5.26 V compared with 3.5 V for the lithium–chlorine cell), and the Li–F cell might well prove useful for power storage in industrial situations.

18.5 Sodium as a heavy-duty conductor

Sodium has many properties which make it attractive as a heavy-duty conductor of electricity. Nyholm[478] has compared some of these properties with corresponding properties of the more traditional conductors, copper and aluminium (Table 18.1). On a weight basis, sodium is more than three times better as a conductor, and now that the technical problems of handling active metals such as sodium are being overcome, it provides an attractive alternative conductor. As early as 1966, Union Carbon and Carbide Corporation experimented with the use of sodium, in a suitable sheath, as a heavy-duty current carrier.[479] An obvious advantage is cheapness and availability, since sodium is a by-product of the alkali industry for which outlets are always being sought. In 1967 it was calculated that the cost of sodium for the same current carrying capacity was only one seventh of that of copper, and with the passage of time this ratio is improving still further in favour of sodium.

Pliable polyethylene or silicone tubing is suitable for containing the sodium. The tubing should be first warmed under vacuum to remove any reactive

Table 18.1 Comparison of sodium, copper and aluminium as conductors

	Na	Cu	Al
Specific conductivity at $(\Omega|^{-1}m^{-1} \times 10^6)$ at 0°C	23	64.5	40
Relative conductance per unit weight	3.31	1.00	2.05
Density at 0°C	0.97	8.96	2.70
Melting point (°C)	98	1083	232

gases, and the liquid sodium then poured into the tube under an inert atmosphere. Tubes with about 1 cm bore have been found convenient, and connections can be made by pressing a short rod of the connector metal into the relatively soft sodium. Any break in the sodium can be rejoined merely by heating the tube. Conductors several feet in length are easily prepared; in this form the solid sodium has very little rigidity, so that the plastic tubes charged with sodium can still be bent, twisted or tied in knots if necessary.

18.6 Other applications

There are many other applications of the alkali metals, usually involving smaller quantities than in the examples given above. These include the use of lithium in organic reduction processes, its incorporation into aircraft alloys, and its aerospace applications; the use of lithium and sodium in polymerization reactions; tetraethyl lead production, which uses about 80 per cent of the world output of sodium; and the use of the alkali metals as reducing agents for the production of other metals, as in the reduction of titanium tetrachloride by sodium.

In many of these the liquid phase is involved, and the chemistry of (and in) the liquid metal, its solvent properties and surface chemistry, are important. For example, the efficiency of the sodium discharge lamps used for street lighting is influenced by whether the sodium forms droplets on the inner glass walls of the lamps, or whether it spreads as a film, and this is related to the calcium content of the sodium. Neither rubidium nor caesium have found any large scale use, but may well do so when their properties are better understood. They have been studied for use in vacuum tubes and photoelectric cells, ion propulsion engines, and power generation by means of magnetohydrodynamics and thermionic conversion. These applications take advantage of the very low ionization potentials of the rubidium and caesium atoms.

References and notes

1 See, for example:
   (a) S. Z. Beer (Ed.), *Liquid Metals Chemistry and Physics*, Dekker, New York, 1972.
   (b) P. D. Adams, H. A. Davies and S. G. Epstein (Eds), *The Properties of Liquid Metals*, Proc. 1st Int. Conf. Brookhaven Nat. Lab., New York, 1966. Taylor and Francis, London, 1967.
   (c) S. Takeuchi (Ed.), *The Properties of Liquid Metals*, Proc. 2nd Int. Conf., Tokyo, 1972. Taylor and Francis, London, 1973.
   (d) R. Evans and D. A. Greenwood (Eds), *Liquid Metals*, Proc. 3rd Int. Conf., Bristol, 1976. Institute of Physics Conference Series 30, London, 1976.
   (e) Mitsuo Shimoji, *Liquid Metals*, Academic Press, London, 1977.
2 (a) *Liquid Metals Handbook*, NAVEXOS P–733 (Rev.), 2nd Edn, 1952. *Sodium–Na/K Supplement* TID 5277, Washington, – US Atomic Energy Commission, 1955. *Sodium–NaK Engineering Handbook* (Ed. O. J. Foust), Gordon and Breach, New York, 1972.
   (b) *Physical and Thermodynamic Properties of Sodium* (2nd Edn), Ethyl Corporation, Michigan, 1955.
3 M. C. Ball and A. H. Norbury, *Physical Data for Inorganic Chemists*, Longman, 1974, p. 48.
4 R. J. Zollweg, *J. Chem. Phys.*, 1969, **50**, 4251.
5 *Handbook of Chemistry and Physics*, Chemical Rubber Co., Cleveland, Ohio, 53rd Edn, 1972.
6 C. S. G. Phillips and R. J. P. Williams, *Inorganic Chemistry*, vol. 2, p. 31.
7 R. W. Wood, *Physical Review* (B), 1933, **44**, 353; J. Monin and G. A. Boutry, *Physical Review* (B), 1974, **9**, 1309.
8 BOC Ltd., 'Special Gases' Divison.
9 Air Products (UK) Ltd.
10 C. C. Addison, E. Iberson and J. B. Raynor, *Chemistry and Industry*, 1958, p. 96.
11 *Glass*, GEC Wembley, London, 1959.
12 T. I. Barry G. S. Schajer and F. M. Stackpool, Joint Conference of German Welding Society and German Ceramic Society, Baden-Baden, Dec. 1980.
13 M. G. Barker, private communication.
14 L. R. Kelman, W. D. Wilkinson and F. L. Yaggee, *Resistance of Materials to Attack by Liquid Metals*, Argonne National Laboratory Report ANL 4417, July 1950.
15 W. Hume-Rothery, *The Structures of Alloys of Iron*, Pergamon Press, London, 1966, p. 257.

312

16 H. U. Borgstedt, *Materials Chem.*, 1980, **5**, 95.
17 D. L. Smith and K. Natesan, *Nuclear Tech.*, 1974, **22**, 392; also I. O. Cowles and A. D. Pasternak, *Lithium Properties Related to Use as Nuclear Reactor Coolant*, UCRL Report 50647, 1969.
18 M. R. Hobdell and F. J. Salzano, *Nucl. Appl. Tech.*, 1970, **8**, 95.
19 D. J. Hayes, M. R. Baum and M. R. Hobdell, *J. Brit. Nucl. En. Soc.*, 1971, 93.
20 R. J. Pulham, *Pure Appl. Chem.*, 1977, **49**, 83.
21 C. C. Addison, G. K. Creffield, P. Hubberstey and R. J. Pulham, *J. Chem. Soc.*, 1971, 1393.
22 R. J. Pulham, P. Hubberstey, M. G. Down and A. E. Thunder, *J. Nucl. Materials*, 1979, **85**, 299.
23 R. J. Pulham, *J. Chem. Soc.*, 1971, 1389.
24 C. C. Addison, P. Hubberstey, J. Oliver, R. J. Pulham and P. Simm, *J. Less-Common Metals*, 1978, **61**, 123.
25 C. C. Addison, *Proc. Roy. Inst.*, London, 1971, **44**, 317.
26 C. C. Addison, *Proc. Brit. Assoc.*, Section B, 1971, 23.
27 J. Laithwaite *et al.*, *Proc. IAEA Symp. on Sodium-Cooled Fast Reactor Engineering*, Monaco, 1970, p. 319.
28 J. W. Mausteller, F. Tepper and S. J. Rodgers, *Alkali Metal Handling and Systems Operating Techniques*, AEC Monograph, Gordon and Breach, New York, 1967.
29 R. M. Lintonbon, Ph.D. thesis Nottingham University, 1968.
30 A. Hatterer, *The Alkali Metals*, Chemical Society Special Publication 22, London, 1967, p. 317.
31 A. W. Thorley and A. C. Raine, *ibid.*, p. 374.
32 Reviewed in Reference 28, pp. 17–22.
33 R. J. Pulham and M. G. Down, Ph.D. thesis by M. G. Down, University of Nottingham, 1975.
34 C. C. Addison, M. G. Barker and D. J. Wood, *J. Chem. Soc.* (Dalton), 1972, 13.
35 M. G. Barker, Conference on Fusion Reactor Materials, Seattle, 1981.
36 P. Hubberstey, P. F. Adams and R. J. Pulham, *Proc. Int. Conf. on Radiation Effects and Tritium Technology for Fusion Reactors*, Gatlinburg, 1975, **3**, 270.
37 M. R. Hobdell, P. Simm and C. A. Smith, *CEGB Research*, 1979, no. 10, p. 15
38 R. J. Pulham and S. Hill, Ph.D. thesis by S. Hill, University of Nottingham, 1979.
39 Reference 28, p. 58.
40 US Bureau of Mines, Bulletin 603, 1963.
41 E. E. Hoffman, Oak Ridge National Laboratory, Report ORNL-2894, March 1960.
42 N. E. Cusack, 'The electronic properties of liquid metals', *Rep. Prog. Phys.*, 1963, **26**, 370.
43 J. R. Wilson, 'The structure of liquid metals and alloys', *Metallurgical Rev.* Inst. of Metals, 1965, **10**, 486.
44 H. M. Feder, *Chemistry in Liquid Metal Solvents*, 39th Annual Priestley Lectures, Penn. State University, 1965, p. 111.
45 P. W. Kendall, *J. Nucl. Materials*, 1970, **35**, 41.
46 C. C. Addison, 'The chemistry of liquid metals', *Chemistry in Britain*, 1974, **10** (9), 331.
47 E. Veleckis, S. K. Dhar, E. A. Cafasso and H. M. Feder, *J. Phys. Chem.*, 1971, **75**, 2832.
48 K. Thormeier, *Nucl. Eng. Des.*, 1970, **14**, 69; *Atomkernenergie*, 1969, **14**, 449.
49 H. Shimojima and N. Miyaji *J. Nucl. Sci. Eng.*, 1975, **12**, 658.

313

50 R. A. Blomquist, F. A. Cafasso and H. M. Feder, *J. Nucl. Materials*, 1976, **59**, 199.

51 E. Veleckis, F. A. Cafasso and H. M. Feder, *J. Chem. Eng. Data*, 1976, **21**, 75.

52 H. C. Schnyders and H. M. Feder, USAEC Reports ANL 7750 and 7823, 1971.

53 H. Slotnick, S. M. Kapelner and R. E. Cleary, Pratt and Whitney Aircraft Report PWAC-380 to USAEC, 1965, p. 25.

54 A. C. Whittingham, *J. Nucl. Materials*, 1976, **60**(2), 119.

55 M. N. Arnol'dov, M. N. Ivanovskii, V. A. Morozov and S. S. Pletenets, *Tep. Vys. Temp.*, 1970, **8**(1), 88.

56 S. Hill, Ph.D. thesis, University of Nottingham, 1979.

57 G. Parry and R. J. Pulham, *J. Chem. Soc.* (Dalton), 1975, 446.

58 G. Parry and R. J. Pulham, *J. Chem. Soc.* (Dalton), 1975, 1915.

59 M. R. Hobdell and A. C. Whittingham, 'Liquid alkali metals', *Proc. Int. Conf. BNES*, Nottingham, 1973, p. 5.

60 P. Hubberstey, R. J. Pulham and A. E. Thunder, *J. Chem. Soc.* (Faraday I) 1976, **72**, 431.

61 P. Hubberstey, P. F. Adams, R. J. Pulham, M. G. Down and A. E. Thunder, *J. Less-Common Metals*, 1976, **49**, 253.

62 Ph. Touzain, *Canad. J. Chem.*, 1969, **47**, 2639.

63 A. Simon, *Z. Anorg. Chem.*, 1973, **395**, 301.

64 P. W. Kendall, *J. Nucl. Materials*, 1970, **35**, 41.

65 P. J. Gellings, G. B. Huiscamp and E. G. van den Broek, *J. Chem. Soc.* (Dalton), 1972, 151.

66 P. J. Gellings, A. Van der Scheer and W. J. Caspers, *J. Chem. Soc.* (Faraday I) 1974, **70**, 531.

67 D. A. Greenwood and V. K. Ratti, *Metal Physics*, 1972, **2**, 289.

68 J. Verhoeven, *Metallurgical Rev.*, 1963, **8**, 311.

69 S. G. Epstein, Reference 1(a), p. 527.

70 B. A. Nevsorov, Paper CN-13/40, IAEA Conf. on Corrosion of Reactor Materials, Salzburg, June 1962 (Translation AEC-tr-5412); Geneva Conference A/CONF. 28/P/343, USSR, May 1964.

71 E. Veleckis, K. E. Anderson, F. A. Cafasso and H. M. Feder, *Proc. Int. Conf. on Sodium Technology and Large Fast Reactor Design*, Report ANL-7520, Part I, 1968, p. 295.

72 R. J. Pulham, P. Hubberstey, A. E. Thunder, A. Harper and A. T. Dadd, *Proc. 2nd Int. Conf. on Liquid Metal Technology in Energy Production*, April 1980, Richland (American Nuclear Society).

73 R. M. Yonco and M. I. Homa, *Trans. Amer. Nucl. Soc.*, 1979, **32**, 270.

74 B. Longson and A. W. Thorley, *J. Appl. Chem.*, 1970, **20**, 372.

75 R. Ainsley, L. P. Hartlib, G. Long, A. Pilbeam and R. Thompson, *Proc. BNES Conf. on Liquid Alkali Metals*, Nottingham, April 1973, p. 143.

76 R. Ainsley, L. P. Hartlib, P. M. Holroyd and G. Long, *J. Nucl. Materials*, 1974, **52**, 255.

77 D. C. Gehri, Report AI-AEC-1286, 1970; ASEAEC Report ANL 7868, 1971, p. 58.

78 R. Thompson, IAEA Report I.W.G.F.R./33, Harwell, 1979, p. 6.

79 G. L. Johnson *et al.*, *J. Chem. Thermodynamics*, 1973, **5**, 57.

80 R. Thompson, *J. Inorg, Nucl. Chem.*, 1972, **34**, 2513; AERE Report R 6566, 1970.

81 A. Mainwood and A. M. Stoneham, *Phil. Mag.* (B), 1978, **37** (2), 263.

82 A. S. Dworkin, H. R. Bronstein and M. A. Bredig, *J. Phys. Chem.*, 1962, **66**, 572.

83 A. T. Dadd and P. Hubberstey, *J. Chem. Soc.* (Faraday Trans. 1), 1981, **77**, 1865.

314

84   C. C. Addison and B. M. Davies, *J. Chem. Soc.* (A), 1969, 1831.
85   C. C. Addison, M. Dowling and R. J. Pulham, unpublished results.
86   F. A. Kanda, R. M. Stevens and D. V. Keller, *J. Phys. Chem.*, 1965, **69**, 3867.
87   C. C. Addison, G. K. Creffield, P. Hubberstey and R. J. Pulham, *J. Chem. Soc.* (A), 1971, 2688.
88   P. R. Bussey, P. Hubberstey and R. J. Pulham, *J. Chem. Soc.* (Dalton), 1976, 2327.
89   R. Lorenz and R. Winzer, *Z. Anorg. Chem.*, 1929, **179**, 281.
90   E. Rinck, *Compt. Rend.*, 1931, **192**, 1378.
91   C. H. Mathewson, *Z. Anorg. Chem.*, 1906, **48**, 193.
92   W. Klemm and D. Kunze, *The Alkali Metals*, Chemical Society Special Publication 22, London, 1967.
93   F. L. Bett and A. Draycott, *Proc. 2nd UN Int. Conf. on Peaceful Uses of Atomic Energy*, Geneva, 1958, **7**, 125.
94   M. Hansen and K. Anderko, *Constitution of Binary Alloys*, 2nd Edn, McGraw-Hill, 1958.
95   R. P. Elliott, *Constitution of Binary Alloys*, First Supplement, McGraw-Hill, 1965.
96   F. A. Shunk, *Constitution of Binary Alloys*, Second Supplement, McGraw-Hill, 1969.
97   C. E. Messer, E. B. Damon, P. C. Maybury, J. Mellor and R. A. Seales, *J. Phys. Chem.*, 1958, **62**, 220.
98   J. A. J. Walker, E. D. France, J. L. Drummond and A. W. Smith, *The Alkali Metals*, Chemical Society Special Publication 22, 1967, p. 393.
99   D. R. Vissers, J. T. Holmes, L. G. Bartholme and P. A. Nelson, *Nucl. Technol.*, 1974, **21**, 235.
100   C. A. Smith, *Proc. BNES Conf. on Liquid Alkali Metals*, Nottingham, 1973, p. 101.
101   R. B. Holden and N. Fuhrman, USAEC Report ANL 7520, (Part 1), 1968, p. 262.
102   P. F. Adams, M. G. Down, P. Hubberstey and R. J. Pulham, *J Less-Common Metals*, 1975, **42**, 325.
103   R. A. Davies, J. L. Drummond and D. W. Adaway, *Proc. BNES Conf. on Liquid Alkali Metals*, Nottingham, 1973, p. 93.
104   D. R. Vissers and L. G. Bartholme, USAEC Report ANL 7868, 1971.
105   R. J. Pulham and P. A. Simm, *Proc. BNES Conf. on Liquid Alkali Metals*, Nottingham, 1973, p. 1.
106   S. A. Meacham, E. F. Hill and A. A. Gordus, USAEC Report APDA 241, 1970.
107   R. J. Newcombe and J. Thompson, *J. Polarogr. Soc.*, 1968, **14**, 104.
108   K. M. Mackay, *Hydrogen Compounds of the Metallic Elements*, E. and F. N. Spon, London, 1966.
109   P. F. Adams, P. Hubberstey and R. J. Pulham, *J. Less-Common Metals*, 1975, **42**, 1.
110   M. N. Arnol'dov. M. N. Invanovskii, V. A. Morozov, S. S. Pletenets and V. V. Sitnikov, *Proc. Acad. Sci. USSR, Metals*, 1973, **1**, 74.
111   M. N. Arnol'dov, M. N. Ivanovskii, V. A. Morosov, S. S. Pletenets and V. I. Subbotin, *At. Energy*, 1970, **28**, 18.
112   E. L. Compere and J. E. Savolainen, *Nucl. Sci. Eng.*, 1967, **28**, 325.
113   E. Veleckis, R. M. Yonco and V. A. Morani, *J. Less-Common Metals*, 1977, **55**, 85.
114   M. A. Bredig and J. W. Johnson, *J. Phys. Chem.*, 1960, **64**, 1899 and references therein.
115   P. Hubberstey, A. T. Dadd and P. G. Roberts, *Proc. Int. Conf. on Material Behaviour and Physical Chemistry in Liquid Metal Systems*, Karlsruhe, 1981.

116 M. N. Arnol'dov, Yu. V. Bogdanov, M. N. Ivanovskii and B. A. Morozov, *Izv. Akad. Nauk. SSSR Met.*, 1977, **76**, 30.
117 F. J. Smith, J. F. Land, G. M. Begun and A. M. Batistoni, *J. Nuclear Chem. Soc.*, 1979, **41**, 1001.
118 T. D. Claar, *Reactor Technology*, 1970, **13**(2), 124.
119 W. P Stanaway and R. Thompson (AERE Harwell), *Proc. 2nd int. Conf. on Liquid Metals in Energy Production*, Richland, Washington, 1980, pp. 18–54 (available from US Dept. of Commerce, Springfield, Virginia).
120 W. P Stanaway and R. Thompson, *Proc. Int. Conf. on Material Behaviour and Physical Chemistry in Liquid Metal Systems*, Karlsruhe, 1981, p. 421.
121 R. A. Baus, A. D. Bogard, J. A. Grand, L. B. Lockhart, R. R. Miller and D. D. Williams, *Proc. Int. Conf. on Peaceful Uses of Atomic Energy*, Geneva, 1955, **9**, 356 (United Nations, New York, 1956).
122 G. W. Horsley, *J. Iron and Steel Inst.*, 1956, **182**, 43.
123 M. G. Barker and D. J. Wood, *J. Less-Common Metals*, 1974, **35**, 315.
124 J. O. Cowles and A. D. Pasternak, USAEC Report UCRL-50647, 1969.
125 G. Long, *Proc. BNES Conf. on Liquid Alkali metals*, Nottingham, 1973, p. 31.
126 A. T. Dadd and P. Hubberstey, *J. Chem. Soc.* (Faraday Trans. 1), 1981, **77**, 1865.
127 A. T. Dadd, P. Hubberstey and P. G. Roberts, *J. Chem. Soc.* (Faraday Trans. 1), 1982, **78**, 2735.
128 D. Anthrop, *Solubilities of Transition Metals in Liquid Alkali and Alkaline Earth Metals, Lanthanum and Cerium* (a critical review of the literature), Lawrence Radiation Laboratory, Livermore, Report UCRL-50315, 1968.
129 R. J. Pulham and P. Hubberstey, *Proc. Int. Corrosion Forum*, National Association of Corrosion Engineers, Houston, 1982, paper T-2-1, 13.
130 P. Hubberstey and R. J. Pulham, *J. Chem. Soc.* (Dalton Trans.), 1974, 1541.
131 P. Hubberstey and R. J. Pulham, *J. Chem. Soc.* (Dalton Trans.), 1972, 819.
132 Ph. Touzain, *Canad. J. Chem.*, 1969, **47**, 2639.
133 A. Simon, *Z. Anorg. Chem.*, 1973, **395**, 310.
134 A. S. Dworkin, H R. Bronstein and M. A. Bredig, *J. Phys. Chem.*, 1962, **66**, 572.
135 M. A. Bredig, *Mixtures of metals with Molten Salts* in *Molten Salt Chemistry* (Ed. M. Blander), Interscience Publ., 1964.
136 M. A. Bredig, J. W. Johnson, *J. Phys. Chem.*, 1960, **64**, 1899.
137 M. A. Bredig, J. W. Johnson and W. T. Smith, *J. Amer. Chem. Soc.*, 1955, **77**, 307.
138 C. G. Allan, *Proc. BNES Conf. on Liquid Alkali Metals*, Nottingham, 1973, p. 159.
139 W. S. Clough, *J. Nucl. Energy*, 1971, **25**, 417.
140 T. Nakajima, R. Minami, K. Nakanishi and N. Watanabe, *Bull. Chem. Soc. Japan*, 1974, **47**, 2071.
141 E. E. Konovalov, N. I. Seliverstov and V. P. Emel'yanov, *Izv. Akad. Nauk SSSR, Met.*, 1968, **3**, 77.
142 M. G. Down, P. Hubberstey and R. J. Pulham, *J. Chem. Soc.* (Dalton), 1975, 1490.
143 F. J. Smith, *J. Less-Common Metals*, 1974, **35**, 147.
144 J. H. Hildebrand and R. L. Scott, *The Solubility of Non-Electrolytes*, Reinhold Publishing Corp., New York, 1950.
145 B. W. Mott, *Phil. Mag.*, 1957, **2**, 259.
146 R. Kumar, *J. Mater. Sci.*, 1972, **7**, 1409.
147 M. G. Down, P. Hubberstey and R. J. Pulham, *J. Chem. Soc.* (Faraday Trans.) 1975, **71**, 1387.

316

148   T. R. Cuerou and F. Tepper, American Rocket Society Conference, Santa
      Monica, Calif., Sept. 1962.
149   R. Hultgren, R. L. Orr, P. D. Anderson and K. K. Kelley, *Selected Values of
      Thermodynamic Properties of Metals and Alloys*, John Wiley, New York, 1963.
150   B. R. Orton, B. A. Shaw and G. I. Williams, *Acta Met.*, 1960, **8**, 177.
151   R. J. Pulham and G. Meredith, Department of Chemistry, University of
      Nottingham.
152   O. J. Foust (Ed.), *The Sodium-NaK Engineering Handbook*, Gordon and
      Breach, New York 1972, p. 37.
153   Reference 152, pp. 30–34.
154   G. M. B. Webber and R. W. B. Stephens, *Physical Acoustics* (Ed. W. P. Mason),
      Academic Press, 1968, p. 53.
155   M. G. Kim and S. V. Letcher, *J. Chem. Phys.*, 1971, **55**, 1164.
156   J. Jarzynski and T. A. Litovitz, *J. Chem. Phys.*, 1964, **41**, 1290.
157   F. Tepper, J. King and J. Greer, *Int. Symp. on the Alkali Metals*, University of
      Nottingham, 1966, Chemical Society Special Publication 22, p. 23.
158   H. F. Halliwell and S. C. Nyburg, *Trans.Faraday Soc.*, 1963, **59**, 1126.
159   R. Thompson, *Proc. BNES Conf. on Liquid Alkali Metals*, Nottingham, 1973,
      p. 47.
160   M. A. Bredig and J. W. Johnson, *J. Phys. Chem.*, 1958, **62**, 604.
161   M. A. Bredig and J. W. Johnson, *J. Amer. Chem. Soc.*, 1958, **77**, 1454.
162   P. Hubberstey and A. T. Dadd, *J. Less-Common Metals*, 1982, **86**, 55.
163   Reference 42, p. 367 et seq.
164   T. E. Faber, *Theory of Liquid Metals*, Cambridge Monographs on Physics,
      Cambridge University Press, 1972.
165   C. C. Addison, G. K. Creffield, P. Hubberstey and R. J. Pulham, *J. Chem. Soc.*
      (A), 1969, 1482.
166   P. Hubberstey, *Proc. 3rd Int. Conf. on Liquid Metals*, Bristol, 1976, Institute of
      Physics Conference Series 30, p. 539.
167   P. Hubberstey and A. T. Dadd, *J. de Physique* (Paris), 1980, **41**, C8–531.
168   J. F. Freedman and W. D. Robertson, *J. Chem. Phys.*, 1961, **34**, 769.
169   L. I. Aksenova, D. K. Belaschenko and A. I. Pertsin, *Teplofiz. vys. Temp.*, 1971,
      9, 1159.
170   P. Hubberstey and R. J. Pulham, *J. Chem. Soc.* (Faraday 1), 1974, **70**, 1631.
171   P. Hubberstey and P. R. Bussey, *J. Less-Common Metals*, 1978, **60**, 109.
172   C. F. Bonilla, D. I. Lee and P. J. Foley, *Advances in Thermophysical Properties
      at Extreme Temperatures and Pressures*, Amer. Soc. Mech. Eng., New York,
      1965, p. 207.
173   C. C. Addison, G. K. Creffield and R. J. Pulham, *J. Chem. Soc.* (A), 1971, 2685.
174   Kurchakov and Nikitinsky; original reference unknown.
175   P. F. Adams, M. G. Down, P. Hubberstey and R. J. Pulham, *J. Chem. Soc.*
      (Faraday 1), 1977, **73**, 230.
176   M. N. Arnol'dov, M. N. Ivanovskii, V. I. Subbotin and B. A. Shmatko, *Teplofiz.
      vys. Temp.*, 1967, **5**, 812.
177   C. C. Addison, G. K. Creffield, P. Hubberstey and R. J. Pulham, *J. Chem. Soc.*
      (Dalton), 1976, 1105.
178   H. R. Bronstein, A. S. Dworkin and M. A. Bredig, *J. Chem. Phys.*, 1962, **37**, 677.
179   G. Brauer, *Z. Anorg. Chem.*, 1947, **255**, 101.
180   A. Thorley and C. Tyzack, UKAEA Report R. Chem. R.C./P.104.
181   C. Oberlin and P. Saint Paul, *Material Behaviour and Physical Chemistry in
      Liquid Metal Systems* (Ed. H. U. Borgstedt), Plenum Press, New York, 1982, p.
      275.
182   E. M. Mitkevich and B. A. Shikov, *Russian J. Inorg. Chem.*, 1966, **11**, 1289.
183   Ph. Touzain, *Compt. Rendu*, 1973, **276C**, 1583.

184   Ph. Touzain, *Compt. Rendu,* 1974, **279C**, 41.
185   B. A. Shivov, *Russian J. Inorg. Chem.,* 1967, **12**, 545.
186   J. P. Maupre, Report CEA, R4905, 1978.
187   H. Migge; see volume described in Reference 119, pp. 18–19.
188   T. C. Waddington, *Advances in Inorganic Chemistry and Radiochemistry: Vol. 1,* Academic Press, 1959, p. 194.
189   P. F. Adams, P. Hubberstey and R. J. Pulham, unpublished results.
190   M. N. Arnol'dov, M. N. Ivanovskii, V. A. Morozov, S. S. Pletenets and V. I. Subbotin, *At. Energ.,* 1970, **28**, 18.
191   H. Ullmann, K. Teske, F. A. Kozlov and E. K. Kiznekov, *Kernenergie,* 1977, **20**, 80; 1979, **22**, 25. See also reports IAEA-SM-236/80 (1979) and ZfK-422 (1980), p. 123.
192   H. Ullmann, volume described in Reference 181, p. 375.
193   R. A. Davies, J. L. Drummond and D. W. Adaway, volume described in reference 105, p. 93.
194   C. A. Smith and A. C. Whittingham, volume described in Reference 181, p. 365.
195   A. C. Whittingham, C. A. Smith, P. A. Simm and R. J. Smith, volume described in Reference 119, p. 16–38.
196   S. Bauer, with W. B. Woollen, D. Scott and F. R. Dell, AERE Report CE/R 2518, 1956.
197   B. E. Deal and H. J. Svec, *J. Amer. Chem. Soc.,* 1953, **75**, 6173.
198   W. R. Irving and J. A. Lund, *J. Electrochem. Soc.,* 1963, **110**, 141.
199   J. Besson and A. Pelloux, *C. R. Acad. Sci. Paris* (C), 1967, **265**, 816.
200   Corrsin, Steinmetz and Marano, Nuclear Development Corporation Report NDA 84–19, 1959.
201   Furman, General Electric Company Report GEAP-3208, 1959.
202   Longton, UKAEA Report IGR-TN/C 418, 1956.
203   C. C. Addison and J. A. Manning, *J. Chem. Soc.,* 1964, 4887.
204   G. Cornec and J. Sannier, *C. R. Acad Sci. Paris* (C), 1967, **265**, 137.
205   A. Pelloux and J. Besson, *Rev. Chim. Miner.,* 1970, 7 (5), 955.
206   H. M. Saltsburg, Reports KAPL–1495 (1956) and KAPL–1763 (1957).
207   R. N. Newman, A. R. Pugh and C. A. Smith, volume described in Reference 105, p. 85.
208   L. F. Epstein, Report GEAP–3335, 1960.
209   D. D. Adams, G. J. Barenborg and W. W. Kendall, *A.C.S. Advances in Chemistry,* **19**, 92.
210   L. F. Epstein, Report GEAP–3272, 1959.
211   R. H. Jones, H. J. Williams and J. A. Murphy, Report AECU 3193, 1956.
212   A. J. Friedland, J. T. Hagstrom and H. Hori, Report APDA-226, 1968.
213   J. A. Bray, *J. Brit. Nuclear Energy Soc.,* 1971, **10**, 107.
214   J. A. Bray, volume described in Reference 103, p. 107.
215   K. Dumm, H. Mausbeck and W. Schnitker, *Atomkernenergie,* 1969, **14**, 309.
216   W. Schnitker, *Atomkernenergie,* 1974, **24**, 233.
217   N. N. Ivanovskii and F. A. Kozlov, *Atomnaya Energiya,* 1964, **17**, 406.
218   'Specialists' Meeting on Sodium–Water Reactions', Report ANL, IAEA/NPR–9, November 1968.
219   N. Isshiki and M. Hori, *Nippon Genshiryoko Gakkaishi,* 1972, **14**, 15.
220   R. J. Pulham and M. Dowling, unpublished results; M. Dowling, Ph.D. Thesis, University of Nottingham, 1976.
221   A. Herold, *Ann. Chim.,* 1951, **12**, 536.
222   M. R. Hobdell and A. C. Whittingham, volume described in Reference 103, p. 5.
223   G. Parry, Ph.D. Thesis, University of Nottingham, 1975.

318

224   G. Parry and R. J. Pulham, *J. Chem. Soc.* (Dalton), 1975, 1915.
225   R. O. Bremner and D. H. Volman, *J. Phys. Chem.*, 1973, **77**, 1844.
226   R. M. Bowie and J. M. Woodrow, *Phys. Rev.*, 1930, **35**, 1423.
227   A. B. Ashworth, M. Dowling, R. J. Pulham and C. C. Addison, Inorganic Chemistry Laboratories, University of Nottingham, UK; M. Dowling, Ph.D. Thesis, University of Nottingham, 1976; A. B. Ashworth, Ph.D. Thesis, University of Nottingham, 1979.
228   J. A. Ford, Report APDA–167, March 1965.
229   J. L. Henry, Report USBM–RC–1469, March 1970.
230   M. M. Markowitz, *J. Chem. Education*, 1963, **40**, 633.
231   A. D. Bogard and D. D. Williams, Report NRL–3865, Sept. 1951.
232   A. G. Newlands and W. D. Halstead, CEGB Report RD/L/N130/75, Leatherhead, Surrey, 1975.
233   *Fuel-Coolant Interactions—Some Basic Studies at the UKAEA Culham Laboratory*, J. A. Reynolds, T. A. Dullforce, R. S. Peckover and G. J. Vaughan, Report CLM/RR/S2/7, March 1976.
234   H. K. Fauske, *Nucl. Sci. Engineering*, 1973, **51**, 95.
235   S. J. Board, C. L. Farmer, and D. H. Poole, *Fragmentation in Thermal Explosions*, CEGB Report RD/B/N2423, 1972.
236   S. J. Board and R. W. Hall, CEGB Report RD/B/N2850, 1974.
237   C. C. Addison, *Phil. Mag.*, 1945, **36**, 73.
238   G. Stein, *Discn. Faraday Soc.*, 1952, **12**, 227.
239   M. Anbar, *Chem. Soc. Quart. Rev.*, 1968, **22**(4), 578.
240   D. C. Walker, *Chem. Soc. Quart. Rev.*, 1967, **21**(1), 79.
241   E. J. Hart and M. Anbar, *The Hydrated Electron*, Wiley Interscience, 1970.
242   E. A. Shaede and D. C. Walker, Chemical Society Special Publication 22, 1967, 277.
243   G. Hughes and R. J. Roach, *Chem. Comm.*, 1965, 600.
244   C. C. Addison, R. J. Pulham and E. A. Trevillion, *J. Chem. Soc.* (Dalton), 1975, 2082.
245   G. J. Moody and J. D. R. Thomas, *J. Chem. Educ.*, 1966, **43**, 205.
246   R. J. Pulham and S. E. Hill, unpublished results; S. E. Hill, Ph.D. Thesis, University of Nottingham, 1979.
247   R. L. Schroeder, J. C. Thompson and P. L. Oertel, *Phys. Rev.*, 1969, **178**, 298.
248   J. P. Lelieur and P. Rigny, *J. Chem. Phys.*, 1973, **59**, 1148.
249   U. Even, R. D. Swenumson and J. C. Thompson, Inst. Physics Conf. Ser. 30, 1977, p. 424.
250   G. T. Lindley, Ph.D. Thesis, University of Nottingham, 1969.
251   M. R. Hobdell and A. C. Whittingham, *J. Chem. Soc.* (Dalton), 1975, 1591.
252   D. Wallace, Ph.D. Thesis, University of Nottingham, 1966.
253   B. M. Davies, Ph.D. Thesis, University of Nottingham, 1968.
254   L. F. Epstein, *Proc. Int. Conf. on Sodium Technology and Fast Reactor Design*, A.N.L., 1968, Rept. ANL–7520, p. 33.
255   P. A. Cafasso and A. K. Fischer, Sodium Technol. Quart. Rept. ANL/ST–8 1971, p. 37.
256   A. K. Fischer, US Patent 3,745,068, 1973.
257   R. Bee, Ph.D. Thesis, University of Nottingham, 1970.
258   P. Bussey, Ph.D. Thesis, University of Nottingham, 1974.
259   W. Frankenburger, L. Andrussov and F. Dürr, *Z. Electrochem.*, 1928, **34**, 632.
260   M. G. Down, Ph.D. Thesis, University of Nottingham, 1975.
261   E. A. Trevillion, Ph.D. Thesis, University of Nottingham, 1972.
262   C. C. Addison and B. M. Davies, *J. Chem. Soc.* (A), 1969, 1827.
263   K. J. Kelley, E. W. Hobart, and R. G. Bjork, USAEC Report CNLM–6337, 1965.

264 P. Hubberstey, A. Harper, A. T. Dadd, D. J. Knight and K. Maughan, *Proc. 2nd Int. Conf. on Liquid Metal Technology in Energy Production*, Richland, Washington, 1980, pp. 18–27 (American Nuclear Society).

265 A. D. McQuillan and R. B. Caws, British Patent 135,7046, 1974.

266 C. van der Marel, J. Hennephof, G. J. B. Vinke, B. P. Alblas and W. van der Lugt, *Proc. Int. Conf. on Material Behaviour and Physical Chemistry in Liquid Metal Systems*, Karlsruhe, 1981, Plenum Press, New York, 1982, p. 401.

267 I. Schreinlechner and F. Holub, Reference 266, p. 105.

268 R. J. Pulham, P. Hubberstey, M. G. Down and A. E. Thunder, *Int. Conf. on Fusion Reactor Materials*, Miami, 1979; *J. Nuclear Materials*, 1979, **85**, 299.

269 C. C. Addison and B. M. Davies, unpublished results.

270 'Castner's Improvements in the manufacture of cyanides', British Patent 12,218, 1894.

271 C. C. Addison, B. M. Davies, R. J. Pulham and D. P. Wallace, International Symposium on 'The Alkali Metals', Nottingham, 1966; Chemical Society Special Publication 22, 1967, p. 290.

272 R. Ainsley, A. P. Hartlib, P. M. Holroyd and G. Long, *J. Nuclear Materials*, 1974, **52**, 255.

273 E. Veleckis; K. E. Anderson, F. A. Cafasso and H. M. Feder, *Proc. Int. Conf. on Sodium Technology and Fast Reactor Design*, ANL–7520 (Part 1), 1968, p. 295.

274 M. G. Down, M. J. Haley, P. Hubberstey, R. J. Pulham and A. E. Thunder, *J. Chem. Soc.* (Dalton), 1978, 1407.

275 A. M. Pavlov, N. A. Sokolov, Yu. I. Dergunov and V. G. Golov, *Trudy Khim, i. Khim. Tekhnol.*, 1973, **2**, 27.

276 N. A. Sokolov, M. K. Safonova, V. A. Shushunov, Yu. I. Dergunov and V. G. Golov, *Trudy Khim. i. Khim. Tekhnol.*, 1973, **2**, 18.

277 A. T. Dadd and P. Hubberstey, *J. Chem. Soc.* (Dalton), 1982, 2175.

278 M. G. Barker, P. Hubberstey, A. T. Dadd and S. A. Frankham, *J. Nuclear Materials*, 1983, **114**.

279 M. G. Barker and S. A. Frankham, *J. Nuclear Materials*, 1982, **107**, 218.

280 M. G. Barker and S. A. Frankham, unpublished results; S. A. Frankham, Ph.D. Thesis, University of Nottingham, 1982.

281 M. R. Hobdell and L. Newman, *J. Inorg. Nucl. Chem.*, 1970, **32**, 1443.

282 C. C. Addison, M. R. Hobdell and R. J. Pulham, International Symposium on 'The Alkali Metals', Nottingham, 1966; Chemical Society Special Publication 22, 1967, p. 270.

283 C. C. Addison, M. R. Hobdell and R. J. Pulham, *J. Chem. Soc.* (A), 1971, pp. 1700, 1704, 1708.

284 C. C. Addison, M. R. Hobdell, G. Parry and R. J. Pulham, *Proc. BNES Conf. on Liquid Alkali Metals*, Nottingham, 1973, p. 13.

285 G. Parry and R. J. Pulham, *J. Chem. Soc.* (Dalton), 1975, 2576.

286 G. C. Bond, *Catalysis by Metals*, Academic Press, New York, 1962, p. 292.

287 G. B. Barton and H. P. Maffei, *Proc. Int. Conf. on Sodium Technology and Large Fast Reactor Design*, 1968, Part 1, p. 206 (ANL 7520).

288 Reference 286, p. 296.

289 M. Sittig, *Sodium; its Manufacture, Properties and Uses*, Reinhold, New York 1956, p. 142.

290 G. Parry, Ph.D. Thesis, University of Nottingham, 1975.

291 M. Borsier and R. Setton, *C. R. Acad. Sci. Paris* (C), 1970, **13**, 732.

292 P. A. K. Clusius and H. Mollet, *Helv. Chim. Acta*, 1956, **39**, 363.

293 L. Hackspill and R. Rohmer, *Compt. Rend.*, 1943, **217**, 152.

294 D. G. Hill and G. B. Kistiakowski, *J. Am. Chem. Soc.*, 1930, **52**, 892.

295 S. E. Voltz, *J. Phys. Chem.*, 1957, **61**, 756.

320

296 Yu. S. Khodakov, V. K. Nesterov and Kh. M. Minachev, *Isz. Akad. Nauk. SSSR Ser. Khim.*, 1973, **2**, 470.
297 M. D. Zadra, US Patent 3,751,515, 1973.
298 A. W. Shaw, C. W. Bittner, W. V. Bush and G. Holzman, *J. Org. Chem.*, 1965, **30**, 3286.
299 J. B. Wilkes, *J. Org. Chem.*, 1967, **32**, 3231.
300 G. Hugel, *Bull. Soc. Chim.*, 1931, **49**, 1042.
301 Y. Ogini, *Proc. 4th Int. Conf. on Liquid and Amorphous Metals*, Grenoble, France, 1980; *J. de Physique*, 1980, **41**, p. C8–791.
302 K. M. Mackay, *Hydrogen Compounds of the Metallic Elements*, E. and F. N. Spon, 1966, p. 21.
303 C. C. Addison, M. R. Hobdell, and R. J. Pulham, *J. Chem. Soc.* (A), 1971, 1700.
304 I. L. Finar, *Organic Chemistry*, Vol. 1, 3rd Edn, Longmans, 1959, p. 353.
305 Gmelin's *Handbuch der Anorganische Chemie*, 1928, System No. 21.
306 H. N. Gilbert, *Chem. Eng. News*, 1948, **26**, 2604.
307 R. A. Miller, H. V. Knorr, H. J. Eichel, C. M. Meyer and H. A. Tanner, *J. Org. Chem.*, 1962, **27**, 2646.
308 M. R. Hobdell, Ph.D. Thesis, University of Nottingham, 1967.
309 W. Buchner, *Helv. Chim. Acta*, 1963, **46**, 2111.
310 W. Buchner, *Chem. Ber.*, 1965, **98**, 3118.
311 U. Hoffman, O. Schweitzer and K. Rinn, US Patent 2,736,752, 1956.
312 H. C. Miller, US Patent 2,858,194, 1958.
313 V. M. Sinclair, R. A. Davies and J. L. Drummond, *Int. Symposium on The Alkali Metals*, 1966; Chemical Society Special Publication 22, p. 260.
314 D. Goude, P. Hubberstey and R. J. Pulham, unpublished work.
315 M. Garrett, Ph.D. Thesis, University of Nottingham, 1972.
316 C. E. H. Bawn and C. F. H. Tipper, *Disc. Faraday Soc.*, 1947, 104.
317 A. Saffer and T. W. Davis, *J. Amer. Chem. Soc.*, 1945, **67**, 641.
318 T. L. Davis an J. O McLean, *J. Amer. Chem. Soc.*, 1938, **60**, 720.
319 H. Staudinger, *Z. Electrochem.*, 1925, **31**, 549.
320 Reference 289, p. 285.
321 P. H. Groggins, *Unit Processes in Organic Synthesis*, 3rd Edn, McGraw-Hill, New York, 1947, p. 610.
322 C. Löwig, *Chem. Zentr.*, 1852, 575.
323 D. W. F. Hardie and J. D. Pratt, *A History of the Modern British Chemical Industry*, Pergamon Press, Oxford, 1966.
324 J. L. Carey, Ph.D. Thesis, University of Nottingham, 1977.
325 Reference 286, pp. 106, 107.
326 S. Rawson, *Chem. News*, 1888, **58**, 283.
327 W. Kroll, *Z. Anorg. Chem.*, 1917, **102**, 1.
328 J. S. Spevack and A. N. Kurtz, USAEC Report A–1246, 1944.
329 J. S. Spevack, US Patent 2685501, 1954.
330 P. Pichat and D. Forest, *Bull. Soc. Chim. France*, 1967, 3825.
331 V. I. Khachishvili, T. G. Mozdokeli, B. Smolyar and Ya Asakiani, *Russ. J. Inorg. Chem.*, 1961, **6**, 767.
332 P. F. Adams, P.D. Thesis, University of Nottingham, 1973.
333 P. Hagenmuller and R. Naslain, *Compt. Rend.*, 1963, **257**, 1294.
334 R. Naslain and J. Etourneau, *Compt. Rend*, 1966, **263**, 484.
335 P. Hagenmuller, R. Naslain, M. Pouchard and C. Cros, volume described in Reference 313, p. 207.
336 R. Naslain and J. Kasper, *J. Solid State Chem.*, 1970, **1**, 150.
337 D. R. Secrist and W. J. Childs, Report TID 17149, 1962.
338 Electroschmeltzwerk Kepten, GMBH., French Patent 1461878, 1964.

339 D. R. Secrist, *J. Amer. Ceram. Soc.*, 1967, **50**, 520.
340 FLUTEC is the trade name for a range of liquid fluorocarbons produced by ISC Chemicals Ltd., Avonmouth, Bristol.
341 See R. E. Banks, *Organofluorine Chemicals and their Industrial Applications*, Ellis Norwood Ltd., 1979, p. 54.
342 R. J. Pulham, unpublished observations.
343 C. A. Smith, P. A. Simm and G. Hughes, *Nuclear Energy*, 1979, **18**, 201.
344 K. Itoh, K. Tabeda, T. Nakasuji and K. Yamemoto, *Proc. 2nd Int. Conf. on Liquid Metal Technology in Energy Production*, Richland, Washington, USA, 1980, Conf. CONF–800401, pp. 16–55.
345 R. Haus and H.-J. Weiss, *Siemens Review*, **XLII**, 1975, 225.
346 M. R. Hobdell and C. A. Smith, *J. Nucl. Materials*, 1982, **110**, 125.
347 M. R. Hobdell, P. Simm and C. A. Smith, *CEGB Research*, Nov. 1979, (10), 15.
348 A. N. Hamer, J. H. Higson, J. Mathison and R. Swinhoe, *Proc. Int. Conf. on Liquid Alkali Metals*, Nottingham, 1973, p. 59 (British Nuclear Energy Society, London).
349 R. A. Davies, J. L. Drummond and D. W. Adaway, *ibid.*, p. 93.
350 D. R. Vissers, J. T. Holmes, L. G. Bartholme and P. A. Nelson, *Nuclear Technology*, 1974, **21**, 235.
351 R. B. Holden and N. Fuhrman, *Proc. Int. Conf. on Sodium Technology and Large Fast Reactor Design*, ANL Illinois, USA, 1968; Report ANL 7520, p. 262.
352 C. A. Smith, volume described in Reference 348, p. 101.
353 Nuclear Instrumentation and Control Department, Westinghouse Electric Corporation, Baltimore.
354 J. W. Patterson, *J. Electrochem. Soc.*, 1971, **118**, 1033.
355 J. M. McKee, D. R. Vissers, P. A. Nelson, B. R. Grundy, E. Berkey and G. R. Taylor, *Nuclear Tech.*, 1974, **21**, 217.
356 H. Ullmann, K. Teske and T. Reetz, *Kernenergie*, 1973, **16**, 291.
357 H. Ullmann *et al.*, volume described in Reference 344, p. 16–1.
358 W. Haubold, J. Jung and K. L. Schillings, volume described in Reference 344, p. 15–1.
359 J. Jung, *J. Nucl. Materials*, 1975, **56**, 213.
360 R. Thompson, R. G. Taylor, R. C. Asher, C. C. H. Wheatley and R. Dawson, volume described in Reference 344, p. 16–9.
361 M. G. Adamson, E. A. Aitken and D. W. Jeter, *Proc. Int. Conf. on Liquid Metal Technology in Energy Production*, Champion, Pennsylvania, 1976, p. 866 (ANS-AIME meeting).
362 P. Roy and B. E. Bugbee, *Nucl. Techn.*, 1978, **39**, 216.
363 G. J. Licina, P. Roy, H. Nei and A. Kakuta, volume described in Reference 344, p. 16–15.
364 M. G. Barker and colleagues, University of Nottingham.
365 G. L. Hawkes and D. R. Morris, *Trans. Met. Soc. A.I.M.E.*, 1968, **242**, 1083.
366 F. J. Salzano, M. R. Hobdell and L. Newman, *Nucl. Tech.*, 1971, **10**, 335.
367 D. R. Morris, C. Aksaranan, B. S. Waldron and S. H. White, *J. Electrochem. Soc.*, 1973, **120**, 570.
368 R. Rettig and R. Irmisch, *Kernenergie*, 1977, **20**, 76.
369 J. Qafishen and J. R. Selman, *Proc. 2nd Int. Symp. on Molten Salts*, Pittsburg, USA, 1978.
370 M. R. Hobdell, J. R. Gwyther, A. J. Hooper and S. P. Tyfield, volume described in Reference 361, p. 533.
371 M. R. Hobdell, E. A. Trevillion, J. R. Gwyther and S. P. Tyfield, *J. Electrochem. Soc.*, 1982, **129**(12), 2746.

322

372 R. C. Asher, L. Bradshaw, T. B. A. Kirstein, T. H. Nixon and A. C. Tolchard, volume described in Reference 348, p. 133.
373 R. C. Asher, D. C. Harper and T. B. A. Kirstein, volume described in Reference 344, p. 15–46.
374 M. G. Barker, D. R. Moore and N. J. Moon, private communication.
375 M. G. Down and G. A. Whitlow, *J. Nuclear Mater.*, 1979, **85**, 305.
376 M. G. Down, volume described in Reference 344, p. 14–16.
377 M. G. Down and R. E. Witkowski, ISA Transactions, 1982, **21**(4), 49.
378 S. A. Frankham, Ph.D. Thesis, University of Nottingham, 1982.
379 M. G. Barker, I. C. Alexander and J. Bentham, *J. Less-Common Metals,* 1975, **42**, 241.
380 W. Charnock, C. P. Haigh, C. A. P. Horton and P. Marshall, volume described in Reference 347, p. 3.
381 M. G. Barker and A. J. Hooper, *J. Chem. Soc.* (Dalton), 1976, 1093.
382 M. G. Barker and A. J. Hooper, *J. Chem. Soc.* (Dalton), 1973, 1520.
383 M. G. Barker and S. A. Frankham, unpublished results.
384 M. R. Hobdell, E. A. Trevillion and A. C. Whittingham, volume described in Reference 344, p. 18–47.
385 Published by North-Holland Publishing Company, Amsterdam; reprinted from *Journal of Nuclear Materials* 1979 and 1981.
386 Edited by H. U. Borgstedt, and published by Plenum Press, London and New York, 1982.
387 Available from D. G. Lorentz, Pittsburg ANS Section, c/o Westinghouse Electric Corporation, Nuclear Centre, PO Box 355, Pittsburg, PA. 15230, USA.
388 Available from the National Technical Information Service, US Department of Commerce, Springfield, Virginia 22161, USA.
389 Published for the British Nuclear Energy Society by Thomas Telford Limited, at the Institution of Civil Engineers, Great George Street, London SW1P 3AA.
390 A. W. Thorley and C. Tyzack, volume described in Reference 348, p. 257.
391 E. Evans, *J. de Physique,* 1980, **41**, p. C8–775.
392 C. C. Addison, W. E. Addison and D. H. Kerridge, *J. Chem. Soc.,* 1955, 3047.
393 D. O. Jordan and J. E. Lane, volume described in Reference 313, p. 147.
394 J. H. Hildebrand, T. R. Hogness and N. W. Taylor, *J. Amer. Chem. Soc.,* 1923, **45**, 2828.
395 Knolls Atomic Power Laboratory Report KAPL–P–231, Dec. 1949.
396 L. Wilhelmy, *Annalen der Physik,* 1863, **19**, 177.
397 C. C. Addison, D. H. Kerridge and J. Lewis, *J. Chem. Soc.,* 1955, 2861.
398 C. C. Addison, E. Iberson and J. A. Manning, *J. Chem. Soc.,* 1962, 2699.
399 C. C. Addison, E. Iberson and R. J. Pulham, SCI Monograph 28: *Surface Phenomena of Metals,* 1969, p. 246.
400 B. Longson and J. Prescott, volume described in Reference 348, p. 171.
401 B. Longson and A. W. Thorley, volume described in Reference 313, p. 153.
402 J. M. Matthews, M.Sc. Thesis, University of Nottingham, 1963.
403 D. H. Kerridge and S. Ford, AERE Report M/R 2694; see also Int. Symp. on the Physical Chemistry of Metallic Solutions and Intermetallic Compounds, National Physical Laboratory, 1958.
404 J. W. Taylor, AERE Report M/R 1729, 1955.
405 C. R. Kurkjian and W. D. Kingery, *J. Phys. Chem.,* 1956, **60**, 961.
406 E. N. Hodkin, D. A. Mortimer and M. G. Nicholas, Reference 348, p. 167.
407 E. N. Hodkin and M. G. Nicholas, *The Wetting Behaviour of Sodium—Ceramic Systems*, AERE Report M2562, 1972.
408 C. C. Addison, W. E. Addison, D. H. Kerridge and J. Lewis, *J. Chem. Soc.,* 1956, 1454.
409 M. G. Barker, Institute of Physics Conference Series 30, 1977, p. 567. This

paper gives references to many informative papers by M. G. Barker and his students on this topic.

410 A. J. Hooper and E. A. Trevillion, Reference 361, p. 623.
411 M. G. Barker, G. A. Fairhall and S. A. Frankham, volume described in Reference 264, p. 18–41.
412 M. G. Barker, S. A. Frankham, P. G. Gadd and D. R. Moore, volume described in Reference 266, p. 113.
413 M. G. Barker, S. A. Frankham and P. G. Gadd, *J. Inorg. Nucl. Chem.*, 1981, **43**, 2815.
414 I. C. Alexander and M. G. Barker, *J. LessCommon Metals*, 1979, **64**, 115.
415 M. G. Barker and I. C. Alexander, *J. Chem. Soc.* (Dalton), 1975, 1464.
416 D. H. Bradhurst and A. S. Buchanan, *Austral. J. Chem.*, 1961, **14**, 397.
417 B. Longson and J. Prescott, volume described in Reference 348, p. 171.
418 C. C. Addison and E. Iberson, *J. Chem. Soc.*, 1965, 1437.
419 C. C. Addison, M. G. Barker and R. J. Pulham, *J. Chem. Soc.*, 1965, 4483.
420 E. A. Gulbransen and K. F. Andrew, *J. Electrochem. Soc.*, 1952, **99**, 402.
421 M. G. Barker and A. J. Hooper, *J. Chem. Soc.* (Dalton), 1976, 1093.
422 M. G. Barker and D. J. Wood, *J. Chem. Soc.* (Dalton), 1972, 2451.
423 M. G. Barker, A. J. Hooper and D. J. Wood, *J. Chem. Soc.* (Dalton) 1974, 55.
424 C. C. Addison, M. G. Barker and D. J. Wood, *J. Chem. Soc.* (Dalton), 1972, 13.
425 B. M. Abraham, H. E. Flotow and R. D. Carlson, *Ind. Eng. Chem.*, 1959, **51**, 189.
426 C. C. Addison, M. G. Barker, R. M. Lintonbon and R. J. Pulham, *J. Chem. Soc.* (A), 1969, 2457.
427 P. J. Macefield, Ph.D. Thesis, University of Nottingham, 1982.
428 H. E. Evans and W. R. Watson, volume described in Reference 348, p. 153.
429 M. H. Cooper and G. R. Taylor, *Nuclear Technology*, 1971, **12**, 83.
430 J. Guon, *Trans. Amer. Nucl. Soc.*, 1971, **14**, 625.
431 R. P. Colburn, *Trans. Amer. Nucl. Soc.*, 1972, **15**, 235.
432 C. G. Allan, volume described in Reference 348, p. 159.
433 N. Sagawa et al., *J. Nucl. Sci. Technol.*, 1973, **10**(9), 523; 1975, **12**(7), 413; 1975, **12**(10), 626; 1976, **13**(7), 358.
434 J. Saroul et al., *Proc. Int. Conf. on Diffusion of Fission Products*, Saclay, 1969, CEA–N–1605, 1973.
435 N. Mitsutsuka et al., *J. Nucl. Sci. Technol*, 1977, **14**(2), 135.
436 M. G. Barker, S. A. Frankham, D. R. Moore and D. J. Wood, IUPAC Conf. on Chemistry of Materials at High Temperatures, Harwell, Sept, 1981; *High Temperatures–High Pressures*, 1982, **14**, 523.
437 C. C. Addison, M. G. Barker and J. Bentham, *J. Chem. Soc.* (Dalton), 1972, 1035.
438 M. G. Barker, I. C. Alexander and J. Bentham, *J. Less-Common Metals*, 1975, **42**, 241.
439 M. G. Barker and C. W. Morris, *J. Less-Common Metals*, 1976, **44**, 169.
440 M. G. Barker and I. C. Alexander, *J. Chem. Soc.* (Dalton), 1974, 2166.
441 M. G. Barker and D. J. Wood, *J. Metals*, 1971, 365.
442 M. G. Barker, volume described in Reference 348, p. 219.
443 M. G. Barker and D. J. Wood, *J. Less-Common Metals*, 1974, **34**, 215; and references therein.
444 C. C. Addison, M. G. Barker and R. M. Lintonbon, *J. Chem. Soc.* (A), 1970, 1465.
445 C. C. Addison, M. G. Barker and D. J. Wood, *J. Chem. Soc.* (Dalton), 1972, 13.
446 J. W. Evans and A. W. Thorley, UKAEA Report 1GR–TN/C1019, 1958.
447 C. Tyzack, *Proc. Symp. N. W. Branch Chem. Eng.*, Manchester 1964, 151.

324

448   W. F. Calaway, volume described in References 264, p. 18–18.
449   M. G. Barker, *Proc. 35th Southeastern Regional Meeting of ACS.*, Charlotte, USA, 1983 (J. Wiley).
450   M. G. Barker and A. J. Hooper, *J. Chem. Soc.* (Dalton), 1975, 2487.
451   A. H. Fleitman and H. S. Isaacs, USAEC Report BNL–14230, 1970.
452   M. G. Barker and D. J. Wood, *J. Less-Common Metals*, 1974, **35**, 315.
453   P. Gross, G. L. Wilson and W. A. Gutteridge, *J. Chem. Soc.* (A), 1970, 1908.
454   M. G. Barker, *Int. Colloq. on Refractory Oxides for High Temperature Energy Sources*, CNRS France, 1977.
455   S. A. Frankham, Ph.D. Thesis, University of Nottingham, and references therein.
456   G. W. Horsley, *J. Iron and Steel Inst.*, 1956, **182**, 43.
457   P. Gross and G. L. Wilson, *J. Chem. Soc.* (A), 1970, 1913; but see also J. R. Weeks and C. J. Klamut, *Corrosion of Reactor Materials*, Int. At. Energy Authority, Vienna, 1962.
458   A. W. Thorley, volume described in Reference 266, p. 19; see also Reference 461.
459   J. H. Devan, *J. Nuclear Materials*, 1979, **85**, 249.
460   A. W. Thorley, A. Blundell and J. A. Bardsley, volume described in Reference 266, p. 5.
461   A. W. Thorley and C. Tyzack, volume described in Reference 348, p. 257.
462   United Kingdom Atomic Energy Authority, Sixth Annual Report, 1960.
463   *The Development of Fast Reactors in Britain*, and reading list therein; also, *The Prototype Fast Reactor*; UKAEA Reactor Group, Risley, Warrington, Lancs., 1971.
464   W. Charnock, C. P. Haigh, C. A. P. Horton and P. Marshall, *CEGB Research*, 1979, (10), 3; and references therein.
465   *Fusion Reactor Materials*, Proceedings of the First Topical Meeting on Fusion Reactor Materials, Florida, 1979. North-Holland Publishing Company, Amsterdam, 1979 (2 volumes); and references therein.
466   *The Chemistry of Fusion Technology* (Ed. Dieter M. Gruen), Plenum Press, New York, 1972.
467   *Review of the Chemistry of Liquid Lithium as a Blanket Material for Thermonuclear Power Reactors*, by G. Long; in *Proc. BNES Conf. on Liquid Alkali Metals*, Nottingham University, 1973.
468   A. P. Fraas and R. D. Brooks, 9th World Energy Conference, Detroit, Sept. 1974.
469   A. Hart, *Chemistry in Britain*, 1975, **11**(2), 55.
470   M. D. Hames and J. L. Sudworth, *Proc. Instn. Electrical Engineers*, 1979, **126**, 1157.
471   J. T. Kummer and N. Weber, Automotive Engineering Congress, Detroit, 1967, **670**, 179.
472   J. L. Sudworth, *High Temperature Batteries* in *Electrochemical Power Sources* (Ed. M. Barak), Peter Pereginus.
473   D. R. Vissers, Z. Tomczuk and R. K. Steunenberg, *J. Electrochemical Soc.* 1974, **121**, 665.
474   E. J. Cairns and H. Shimotake, *Science*, 1969, **164**, 1347.
475   D. A. J. Swinkels, *Advances in Molten Salt Chemistry*, 1971, **1**, 165; see also E. H. Heitbrink, J. J. Petrarts, D. A. J. Swinkels and G. M. Craig, General Motors Company Technical Report, Air Force Aero Propulsion Laboratory TR–67–89, August 1967.
476   T. G. Bradley and R. A. Sharma, Proceedings 26th Annual Power Sources Symposium, PSC Publications Committee, Red Bank, New Jersey, 1974.
477   R. A. Rightmire, J. W. Sprague, W. N. Sorenson, T. H. Hacha and J. E.

Metcalfe, *A Sealed Lithium-Chloride Fused Salt Secondary Battery*, SAE International Automotive Engineering Congress, Detroit, Jan. 1969.

478   R. S. Nyholm, *Advancement of Science* (British Association), 1967, **23**, (115), 421.

479   L. E. Humphrey, R. C. Hess and G. I. Addis, Paper delivered at a conference of the Institution of Electrical and Electronic Engineers, Pittsburg, July 1966.

480   Marshall Sittig, *Sodium; its Manufacture, Properties and Uses,* Reinhold, New York, 1956.

481   R. E. Gold and D. L. Smith, volume described in Reference 264, p. 1–14.

482   R. S. Pease, *Atom*, monthly bulletin of the UKAEA, January 1983, (315), 2.

483   E. J. Cairns, *Energy and Chemistry* (Ed. R. Thompson), Chemical Society Special Publication 41, 1982, p. 252.

484   J. A. McKnight and E. J. Burton, *Atom*, monthly bulletin of the UKAEA, December 1983, (326), 266.

Index

330